Joint Source-Channel Coding

Joint Source-Channel Coding

Andres Kwasinski
Rochester Institute of Technology
Rochester, NY, USA

Vinay Chande
Qualcomm Technologies Inc.
San Diego, CA, USA

Registered Offices
John Wiley & Sons, Inc., 111 River Street, Hoboken, NJ 07030, USA
John Wiley & Sons Ltd, The Atrium, Southern Gate, Chichester, West Sussex, PO19 8SQ, UK

Editorial Office
The Atrium, Southern Gate, Chichester, West Sussex, PO19 8SQ, UK

For details of our global editorial offices, customer services, and more information about Wiley products visit us at www.wiley.com.

Wiley also publishes its books in a variety of electronic formats and by print-on-demand. Some content that appears in standard print versions of this book may not be available in other formats.

Library of Congress Cataloging-in-Publication Data

Names: Kwasinski, Andres, author. | Chande, Vinay, 1972- author.
Title: Joint source-channel coding / Andres Kwasinski, Rochester Institute
    of Technology, Rochester, NY, USA, Vinay Chande, Qualcomm Technologies
    Inc., San Diego, CA, USA.
Description: First edition. | Hoboken, NJ, USA : Wiley-IEEE Press, 2023. |
    Series: IEEE Press | Includes bibliographical references and index.
Identifiers: LCCN 2022036505 (print) | LCCN 2022036506 (ebook) | ISBN
    9781119978527 (hardback) | ISBN 9781118693773 (adobe pdf) | ISBN
    9781118693797 (epub)
Subjects: LCSH: Combined source channel coding.
Classification: LCC TK5102.93 .K89 2023 (print) | LCC TK5102.93 (ebook) |
    DDC 621.382–dc23/eng/20220930
LC record available at https://lccn.loc.gov/2022036505
LC ebook record available at https://lccn.loc.gov/2022036506

Cover Design: Wiley
Cover Images: © Andrea Danti/Shutterstock, Tatiana Popova/Shutterstock

Set in 9.5/12.5pt STIXTwoText by Straive, Chennai, India
Printed and bound by CPI Group (UK) Ltd, Croydon, CR0 4YY
C9781119978527_021122

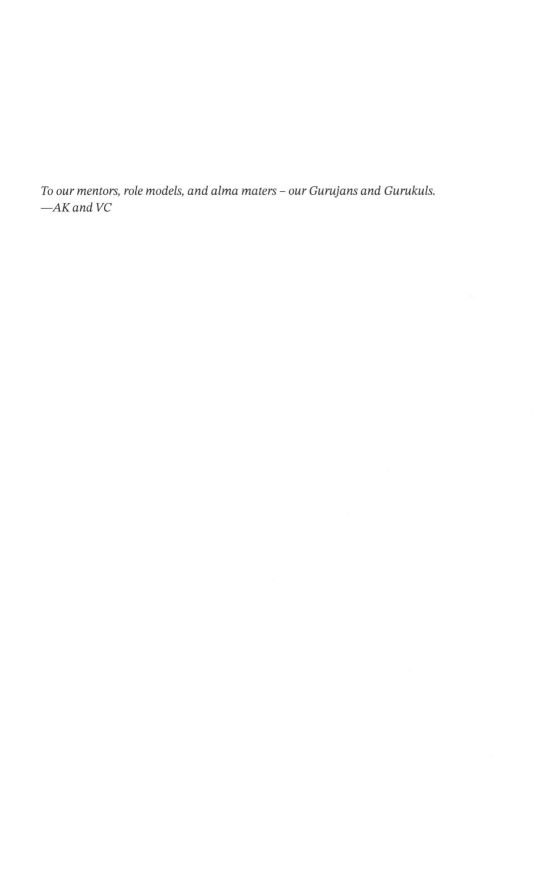

To our mentors, role models, and alma maters – our Gurujans and Gurukuls.
—AK and VC

Contents

Preface

Claude Shannon's seminal work in 1948 – "A Mathematical Theory of Communication" – gave birth to the field of Information Theory and introduced foundational deep ideas that allowed to model and understand the process of communicating information reliably between a transmitter and a receiver. The most fundamental ideas introduced by Shannon were "coding" of an information source to obtain compact representations ("source encoding") and coding for transmission in a form tailored to the communication channel characteristics ("channel encoding"). An astonishing set of results established that, for an information source, the information content can be measured as the surprise/uncertainty that is associated with the generated message. Further, communication over noisy channels with increasing reliability need not involve a diminishing rate of information transfer - that is, asymptotically error-free communication is possible over the channel using source and channel coding, provided the rate of information transfer does not exceed the "capacity" of the channel. The notion of capacity, along with the characteristics of the medium, also captures resource constraints, for example power and bandwidth.

Shannon's ideas were revolutionary in sparking the digital age that we live in (along with Shannon's Information Theory, his master's thesis, "A Symbolic Analysis of Relay and Switching Circuits," also played a large part there). Since Shannon's breakthrough work, multiple generations of communication engineers have learned these important ideas and technologies via textbooks on information theory, source coding and data compression, signal processing, and error control coding. In his paper, Shannon drew a diagram for an end-to-end communication system envisioned as a cascade of information source, source encoding, and channel encoding at the transmitter side leading to a communication channel. At the receiver side, the processing blocks followed the stages of channel decoding, source decoding, and finally source reproduction. These building blocks essentially continue to make up the modern communication systems. Modern-day end-to-end communication systems are intricate, multilayered architectures, with many devices, entities, and software and hardware components operating over diverse physical media, with heterogeneous information sources and sinks as termination points within them. For tractability, the system designers have preferred to take a modular approach where the design considerations for source coding may, physically and logically, be far removed from the aspects of channel coding.

As a consequence, it has been natural that source coding and channel coding got treated separately in dedicated textbooks or in textbooks of Information Theory. In the textbooks on communications systems, they may get treated as components of an end-to-end system, but with an understanding that each component needs to be designed and optimized independently. This approach is further rooted in Shannon's "separation theorem" which stated that in many cases of interest, coding can be done in separate source coding and channel coding stages, independently, without loss of optimality in rate. This optimality, though, is established in an asymptotic sense of increasing delays and coding/decoding complexity. However, it is recognized that blind and separate designs of various components, including source and channel coding, may have large inefficiencies in practice, especially under practical non-asymptotic limitations for delay and complexity. These inefficiencies can be reduced by paying attention to both components jointly. This recognition is not new though. Over the years, considerable work has been done by researchers that treated the source and the channel coding stages jointly at the transmitter as well as at the receiver. This field of investigation is collectively known as *joint source–channel coding* (JSCC). The works range from information theoretic studies of the validity of separation theorems to highly concrete techniques designed and applied to systems of various levels of abstractions and practicality. Yet, even as the JSCC approach keeps growing in relevance and seeing increased attention, to this day, there is not a textbook that is dedicated to presenting JSCC in a comprehensive way, leaving those willing to work on JSCC with the only option to learn by searching and reading through dozens of technical papers that have been published in the span of decades. The main motivation for this textbook is to fill this gap. Here, our aim is threefold. Firstly, we aim at presenting a fairly rich introduction to the topic, with key refreshers needed to initiate research. Secondly, we wish to provide the reader with a compendium of different classical and modern variants and approaches to JSCC. Given the vast amount of published literature over decades, this is a tall aim and we do not claim to conquer it even remotely. Yet, although the book borrows heavily from the research work of the authors, we believe that it will serve to provide broad coverage and treatment of JSCC. Thirdly, we see the book as a recipe book or procedure book, where ideas and formulations used in one context can be cross-pollinated with others or used in new contexts by the reader. As such, we intend that for newcomers in the area of JSCC, this book will serve them not only for their initiation but also to prepare them to make new contributions right away. For engineers that are already working on JSCC, we aim to provide a holistic presentation that, we hope, will spark ideas for new techniques.

To serve these objectives, the book starts with Chapters 1 through 3 that are dedicated to discussing background in communication systems, Information Theory, source coding, and channel coding, as well as to explaining the reason to use JSCC and to present the main classes of JSCC techniques to be covered later in the book. Following this introductory part, Chapter 4 delves into concatenated joint source–channel coding, as a first and basic JSCC technique. Chapter 4 is followed by six chapters where each one is dedicated to an specific approach to JSCC: Chapter 5 covers unequal error protection source–channel coding, Chapter 6 focuses on source–channel coding with feedback, Chapter 7 is dedicated to the study of the quantizer design for noisy channels, Chapter 8 covers error-resilient source coding, Chapter 9 focuses on analog and hybrid digital–analog JSCC techniques, and Chapter 10 discusses joint source–channel decoding. The book concludes in Chapter 11 with a presentation of some recent applications of JSCC, connecting the solutions with techniques seen in earlier chapters, and discussing the emerging and promising design approach for JSCC based on artificial neural networks.

This book has been written over many years of effort. We are grateful to all the staff at Wiley that has supported us in this long road. The book, as you can see it now in a finished form, reflects the investment of the most valuable of resources: time. Acknowledging that time is a finite resource that cannot be regenerated or stretched, we would like to express our deepest gratitude to colleagues, friends, and family. Without their support this book would not have been possible.

2022

Andres Kwasinski
Rochester Institute of Technology
Rochester, NY, USA

Vinay Chande
Qualcomm Technologies Inc.
San Diego, CA, USA

1

Introduction and Background

This textbook is about jointly performing source and channel coding for an information source. The idea of coding is arguably the most significant contribution from Claude Shannon in his mathematical theory of communication [1] that gave birth to information theory. Up until Shannon's work, communication engineers believed that to improve reliability in transmitting a message, all they could do was to increase the transmit power or repeat the transmission many times until it was received free of errors. Instead, Shannon postulated that a message could be transmitted free of errors when not exceeding a capacity for the communication channel. All that was required to achieve this feat was to use a suitable coding mechanism. With coding, the message from the source is mapped into a signal that is matched to the channel characteristics. The ideal coding operation is able to both represent the message from the source in the most efficient form (removing redundant information) and add redundant information in a controlled way to enable the receiver to combat errors introduced in the channel. The efficient representation of the source message is called *source coding*, and the controlled introduction of redundant information to combat communication errors is called *channel coding*.

In [1], Shannon showed that under some ideal conditions, for example, for point-to-point "memoryless" channels, asymptotically in the codeword length, there is no performance difference whether the coding operation is performed as a single block or as a concatenation of a source coding operation and a separate channel coding operation. This is typically described as a *separation theorem*. The separation theorems are a class of asymptotic results of profound importance. It may not be an exaggeration to say that those forms of results motivated the development of today's digital revolution, where every form of information – media, gaming, text, video, multidimensional graphics, and data of all forms – can be encoded without worrying about the medium or channel where it will be shared, and shared without the concern for how it was encoded. Nevertheless, there are multiple practical scenarios where the ideal conditions given by Shannon do not hold. In these cases, we pay a performance penalty for doing the source and channel coding as two separate operations instead of jointly doing source and channel coding. This book focuses on some of these scenarios and helps the reader study the theory and techniques associated with performing source and channel coding as a single operation. Before delving into these topics, in this chapter, we provide an introduction and needed background for the rest of this book.

Joint Source-Channel Coding, First Edition. Andres Kwasinski and Vinay Chande.
© 2023 John Wiley & Sons Ltd. Published 2023 by John Wiley & Sons Ltd.

1.1 Simplified Model for a Communication System

A communication system enables transmission of information so that a message that originates at one place is reproduced exactly, or approximately, at another location. To develop his mathematical theory of communication [1, 2], Claude Shannon considered a simplified model for a communication system, as shown in Figure 1.1. In this model, a communication system is formed by five basic elements:

- An *information source*, which generates a message that contains some information to be transmitted to the receiving end of the communication chain
- A *transmitter*, which converts the message into a signal that can propagate through a communication medium
- A *channel*, which constitutes the medium through which the signal propagates between the transmitter and receiver. This propagating signal is subject to different distortions and impairments, such as selective attenuation, delays, erasure, and the addition of noise.
- A *receiver*, which attempts to recover the original message (possibly affected by distortion and impairments) by reversing the sequence of operations done at the transmitter while also attempting to correct or compensate for the distortion effects introduced in the channel
- A *destination*, which receives the transmitted version of the message and makes sense of it.

The design of a communication system focuses on the transmitter and receiver. By designing the transmitter output to match the channel characteristics and requirements, the transmitter converts, or maps, the source message into a signal appropriate for transmission. This mapping operation is called *encoding*. The reverse operation, where a source message is estimated from the encoded signal, is called *decoding*. For reasons that will be seen later, the mapping operation at the encoder is frequently divided into two stages. The first stage, called *source encoding*, aims to represent the source output in a compact and efficient way that will require as few communication resources as possible while achieving some level of fidelity for the source message recovered after source decoding at the receiver. The compact and efficient representation that results from source encoding is matched

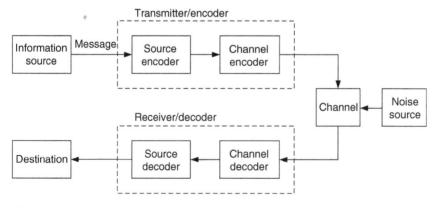

Figure 1.1 A block diagram of a general communication system.

to the source but is likely not matched to the channel characteristics and requirements. For example, the representation may be so compact that each of its parts may be critical for the recovery of the source message at the required level of fidelity, so any error introduced during transmission renders the received message completely unrecoverable. Because of this, the second stage of the transmitter, called *channel encoding*, has the function of converting the compact source representation into a signal matched to the channel. This likely results in a representation that is less compact but more resilient to the effects of channel impairments. On the receiver side, it is natural to think of the structure also separated into a sequence of decoding stages because the goal is to reverse the sequence of encoding operations that were performed at the transmitter.

Designing a communication system entails designing the transmitter and thus the mapper from the source message to the channel signal. Would it be better to design the mapping as a single operation from source directly to the channel? Or, would it be better to design the mapping as a first source encoding stage followed by the channel encoding stage? Is there any performance advantage with either approach? One answer to these questions is given in an asymptotic sense by Shannon's source–channel separation theorem, which states that under some conditions, there is no difference in performance between a transmitter designed as a single mapper and a transmitter designed as the two cascaded source and channel encoding stages. This result is appealing from a designer's perspective. It appears to be a simpler divide-and-conquer approach, involving design of the source and channel coder–decoder pairs (*codecs*) separately. Nevertheless, the tricky element of Shannon's source–channel separation theorem is that the conditions under which it holds are difficult to find in practical scenarios. To discuss this, we first need to establish some key principles from information theory. The next sections provide an overview of important information theory concepts that will be used throughout the rest of this book.

1.2 Entropy and Information

The concept of *information* as understood in Shannon's information theory originates in Hartley's paper [3] and provides a measure on how much the uncertainty associated with a random phenomenon (mathematically characterized as the outcome of a random variable) is reduced by the observation of a particular outcome. The information provided by the outcome x of a discrete random variable X is defined as:

$$I_X(x) = \log \frac{1}{P[X = x]} = -\log P[X = x], \tag{1.1}$$

where $P[X = x]$ is the probability of the outcome $X = x$ and the logarithm can be of any base in principle, but it is most frequently taken as base 2 (in which case the unit of information is the *bit*, a condensed merging of the words *binary* and *digit*). Intuitively, the rarer an event is, the more information its occurrence provides.

The notion of information as introduced in (1.1) provides a measure that is related to a specific outcome of a random variable. To measure the uncertainty associated with a random variable, Shannon introduced the concept of *entropy* (in an information theoretic context) as the expectation of the information function associated with a random variable. Then the entropy $H(X)$ of a discrete random variable X is defined as:

$$H(X) = E[I_X(x)] = -\sum_{x \in X} P[X = x] \log P[X = x], \tag{1.2}$$

where $P[X = x]$ is, in this context, the probability mass function (PMF). The entropy of a random variable can be interpreted as the mean value of the information provided by all its outcomes.

In the case of a continuous random variable with probability density function (PDF) $f_X(x)$, the concept of entropy has been extended to become the *differential entropy*:

$$H(X) = E[-\log f_X(x)] = \int_{-\infty}^{\infty} f_X(x) \log \frac{1}{f_X(x)} dx. \tag{1.3}$$

For example, it can be shown that differential entropy of a zero-mean Gaussian random variable with variance σ^2 is $(1/2) \log(2\pi e \sigma^2)$ [4].

By using joint probability distributions, one can extend the definition of entropy to calculate the joint entropy between multiple random variables. In the case of two discrete random variables, X and Y, their *joint entropy* is

$$H(X, Y) = -E\left[\log P[X = x, Y = y]\right]$$
$$= -\sum_{x \in X} \sum_{y \in Y} P[X = x, Y = y] \log P[X = x, Y = y], \tag{1.4}$$

where $P[X = x, Y = y]$ is the joint PMF and $P[X = x]$ and $P[Y = y]$ are marginal PMFs. Similarly, the *conditional entropy* of X, $H(X|Y)$ is

$$H(X|Y) = -\sum_{x \in X} \sum_{y \in Y} P[X = x, Y = y] \log P[X = x|Y = y], \tag{1.5}$$

where $P[X = x|Y = y]$ is the conditional PMF of X given that $Y = y$. Consequently, the conditional entropy can be regarded as the mean value of the information provided by all the outcomes of a random variable (X) given that the outcome of a second random variable (Y) is known.

The entropy presents a number of useful properties that in the interest of maintaining the introductory nature of this chapter we enumerate next without a detailed proof:

1. Entropy is nonnegative: $H(X) \geq 0$.
2. $H(X) = 0$ if and only if there exists a value x_0 such that $X = x_0$ almost surely (that is, the random variable X behaves as a deterministic constant).
3. Distributions with maximum entropy: $H(X) \leq \log n$ if the discrete random variable X takes with nonzero probability $n < \infty$ different values. Equality, $H(X) = \log n$, is achieved if and only if X is uniformly distributed. For a continuous random variable X with variance σ_X^2, we have the differential entropy $H(X) \leq (1/2) \log(2\pi e \sigma_X^2)$, with equality achieved when X is a Gaussian random variable.
4. Subadditivity of joint entropy: $H(X) + H(Y) \geq H(X, Y)$ with equality if and only if X and Y are statistically independent random variables.
5. $H(X, Y) \geq H(X)$.
6. Conditional entropy is the remaining uncertainty: $H(X|Y) = H(X, Y) - H(Y)$. This property states that conditional entropy measures how much uncertainty about a random variable (X) remains after knowing the outcome of a second random variable (Y).

7. Nonnegativity of conditional entropy: $H(X|Y) \geq 0$.
8. Chain rule for Shannon's entropy and conditional entropy:

$$H(X_1, X_2, \ldots, X_n) = H(X_1) + \sum_{i=2}^{n} H(X_i|X_1, \ldots, X_{i-1}).$$ (1.6)

In a simpler form: $H(X, Y) = H(X) + H(Y|X) = H(Y) + H(X|Y)$.
9. Conditioning reduces entropy: $H(X) > H(X|Y)$.

Communication is inherently a process relating two or more random variables to each other (e.g. the input and output of a channel, an uncompressed and a compressed representation of a signal). A quantity that can measure the closeness of two random variables in terms of information shared between them is the *mutual information*. For two discrete random variables X and Y, it is defined as:

$$I(X; Y) = \sum_{x \in X} \sum_{y \in Y} P[X = x, Y = y] \log \frac{P[X = x, Y = y]}{P[X = x]P[Y = y]},$$ (1.7)

Following Bayes's theorem ($P[X = x, Y = y] = P[X = x|Y = y]P[Y = y]$), the mutual information can also be written as

$$I(X; Y) = \sum_{x \in X} \sum_{y \in Y} P[X = x, Y = y] \log \frac{P[X = x|Y = y]}{P[X = x]}.$$

We can also write

$$
\begin{aligned}
I(X; Y) &= -\sum_{x \in X} \log P[X = x] \sum_{y \in Y} P[X = x, Y = y] \\
&+ \sum_{x \in X} \sum_{y \in Y} P[X = x, Y = y] \log P[X = x|Y = y] \\
&= -\sum_{x \in X} P[X = x] \log P[X = x] \\
&+ \sum_{x \in X} \sum_{y \in Y} P[X = x, Y = y] \log P[X = x|Y = y].
\end{aligned}
$$ (1.8)

Note that the first term in this result is the entropy of the random variable X and the second term can be written in terms of the conditional entropy of X. Therefore, the mutual information as in (1.8) can now be written as:

$$I(X; Y) = H(X) - H(X|Y).$$

This quantity has an intuitive interpretation as follows. If $H(X)$ was the measure of uncertainty in random variable X and $H(X|Y)$ is the remaining uncertainty in X on observing (i.e. on learning of the outcome about) Y, then $I(X; Y)$ is the amount of uncertainty reduction about X on observing Y. $I(X; Y)$, therefore, is the information about X contained in or shared by Y. By its definition, it is seen that the quantity is symmetric (i.e. $I(X; Y) = I(Y; X)$). Therefore, it is called *Mutual Information* between two random variables. The analogous quantity of mutual information can be defined for continuous random variables too – by replacing the notion of entropy and conditional entropy by differential entropy and conditional differential entropy, respectively.

Analogous to the introduction of conditional entropy, we can also think of a *conditional mutual information* between two random variables X and Y given that the outcome of a random variable Z is known. This conditional mutual information, denoted $I(X;Y|Z)$, is defined as:

$$I(X;Y|Z) = H(X|Z) - H(X|Y,Z). \tag{1.9}$$

The conditional mutual information can be understood as being the average amount of information shared by two random variables after a third random variable is known. In this case, the third random variable is often interpreted as representing side information that has been revealed through some process. The conditional mutual information allows for the definition of a chain rule to calculate the mutual information between a random variable X and two other random variables Y and Z:

$$I(X;Y,Z) = I(X;Z) + I(X;Y|Z) = I(X;Y) + I(X;Z|Y), \tag{1.10}$$

or in a more general case when having n random variables $X_1, X_2, \ldots, X_n$,

$$I(X_1, X_2, \ldots, X_n; Y) = \sum_{i=1}^{n} I(X_i; Y|X_{i-1}, \ldots, X_1). \tag{1.11}$$

1.3 Introduction to Source Coding

The communication process starts with a source generating a message intended for transmission. The message may be of very different types. For example, it may be a text (a sequence of letters and symbols), a file residing in a computer, or somebody speaking. In all cases, the message is just a container of information that needs to be transmitted. Messages of all types usually need to go through a step that efficiently represents the information contained in the message. Further, transmission of a message through a communication system as conceptualized by Shannon requires a few extra steps when the message is formed by a continuous-time signal. All these steps are collectively called "source coding." We provide next a brief overview on source coding, leaving for Chapter 2 to cover more details.

1.3.1 Sampling and Quantization of Signals

Today's "digital" age derives its name from the revolution brought about by digital representation of information – which permits unprecedented flexibility in storage, reproduction, computing, and communication. When the source of information is "analog" (that is, continuous-time and continuous-amplitude), the analog signals need to be converted to a digital form for transmission or storage. The first steps in this process are sampling and quantization.

The conversion of an analog signal into a digital form starts with a sampling operation. If the signal is a function of time, sampling consists in recording samples of the continuous-time analog signal, usually at constant time intervals. (If the signal is visual, such as an analog photograph, the sampling occurs at regular spatial intervals, but the basic principles explained here remain the same.) The process is illustrated in Figure 1.2. The left plot shows a speech signal lasting about 3 s. The plots in the middle and right

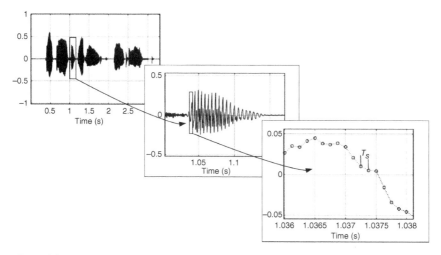

Figure 1.2 Illustration of the sampling process.

show a progressive zooming in to a portion of the signal. The signals in the first two plots can be regarded as continuous time; the signal has an infinite number of values no matter how small a time interval we examine. In the third plot, the signal is made discrete time by taking samples (the circles) of the infinite number of values in the continuous-time signal (the dotted line). For example, the first five samples in the portion of the discrete-time signal shown in Figure 1.2 are {0.027, 0.035, 0.034, 0.041, 0.045}. The samples are taken at a constant time interval, which is called the *sampling period*, denoted T_S in Figure 1.2. The inverse of the sampling period is called the *sampling frequency*, $f_S = 1/T_S$.

An idealized version of the sampling operation can be modeled as a multiplication of a continuous-time signal $x(t)$ and an impulse train of infinite duration, $p(t)$, resulting in a still continuous-time signal $x_s(t)$ that represents the sampled signal $x(t)$ after sampling:

$$x_s(t) = x(t)p(t) = x(t) \sum_{n=-\infty}^{\infty} \delta(t - nT_S).$$

The final conversion into a discrete-time sequence $x[n]$ of samples from $x(t)$ is achieved by passing the signal $x_s(t)$ through an ideal processing block that selects the values of the signal $x_s(t)$ at the instants when the impulses occur, resulting in a sequence with values $x[n] = x(nT_S)$ at discrete times $n \in \mathbb{Z}$.

The continuous-time signal $x_s(t)$ is an idealization that is useful for understanding the process of sampling and the relationship between samples and the sampled signal. An initial question of interest is how to choose the sampling period or the sampling frequency. The theory of Fourier analysis suggests that the spectrum of the signal $x_s(t)$ is an addition of an infinite number of shifted replicas of the spectrum of the original signal (see Figure 1.3). This arises from the properties that (i) the Fourier transform of an infinite train of impulses in time domain is another train of impulses and (ii) multiplication in the time domain implies a convolution in the frequency domain. Now if the original signal $x(t)$ is band-limited, i.e. its spectrum has nonzero power components at frequencies f restricted to a bounded frequency band $f \in [-B, B]$, then it is possible to recover or reconstruct it

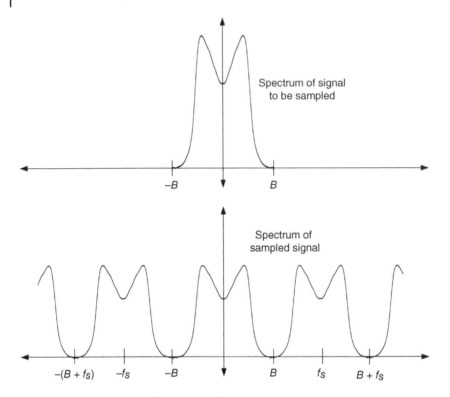

Figure 1.3 The spectrum of a sampled signal.

fully – in continuous time – without distortion from the samples $x[n] = x(nT_s)$, provided the sampling frequency is larger than twice the maximum frequency component in the spectrum of $x(t)$:

$$f_S > 2B.$$

This result, known as the Nyquist–Shannon sampling theorem, determines the conditions for the sampling frequency under ideal assumptions.

In practice, there are multiple factors of non-idealness built into the sampling process. First, it is practically impossible to generate an infinite train of impulses $\delta(t - nT_S)$ that form $x_s(t)$. Second, no real world signal is perfectly band-limited. Third, there is always noise present that prevents the samples from being perfect instantaneous representations of the input signal $x(t)$.

In a practical setting, the choice of sampling frequency needs to account for the effect of noise in the sampled signal, the performance of the filter to limit the sampled signal maximum frequency, the goals for sampled signal quality, and a number of other factors. Therefore, the sampling frequency is frequently set to a value much larger than $2B$. Also note that the Nyquist–Shannon theorem specifies a bound on the sampling frequency based on preventing the overlap of copies of the spectrum of $x(t)$. There are applications that allow the use of sampling at frequencies lower than $2B$, but this subject is out of the scope of this chapter. More information about sampling can be found in digital signal processing textbooks such as [5–7].

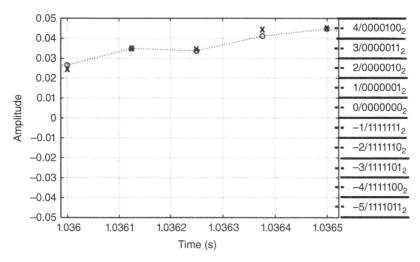

Figure 1.4 Quantization of the five left-most samples in Figure 1.2.

After sampling, the second operation in the conversion of an analog signal into a digital signal is *quantization*. While sampling converts a continuous-time signal into a discrete-time sequence of samples, quantization converts the continuous amplitudes of the samples into values from a discrete set. Figure 1.4 illustrates this operation using the five left-most samples in Figure 1.2. The values of these samples before quantization are real numbers. During quantization, these values are approximated by integer multiples of some value (in Figure 1.4, it was chosen to be 0.01). This operation divides the range of values taken by the input signal into equal-sized (0.01 in this case) intervals, known as *quantization intervals*. For the example in Figure 1.4, using prior knowledge of the signal, the quantizer was designed under the assumption that the samples to be quantized were bounded between ± 0.64 (note that Figure 1.2 shows a small portion of the full signal range). Consequently, the decision to have a quantization interval equal to 0.01 divides the complete range of sample values into $1.28/0.01 = 128$ quantization intervals of equal size. A quantizer with equal-sized intervals is called a *uniform quantizer*. During quantization, the approximation of the real-valued samples to a multiple of the quantization interval size is accomplished by a simple operation of rounding to the closest value. This operation is shown in Figure 1.4 with the original real-valued samples represented as circles and their approximation to the closest interval values as x's. The last step in quantization consists of assigning a label that identifies the quantization interval. In the case of Figure 1.4, a different seven-bit number is assigned to each of the 128 quantization intervals. All samples that are within the same quantization interval are represented by the label of the corresponding quantization interval.

1.3.2 Source Coding of Quantized Signals

If we now pause to reflect on what has been achieved through sampling and quantization, we see that these operations have converted analog signals into discrete-valued functions of discrete-valued variables. The analog source is behaving as a generator of messages that can be broken down into elementary indivisible elements drawn from a set of discrete values we call *source symbols*. We see an example of this in Figure 1.4, where the source generates messages drawn from a discrete set of 10 source symbols. In Figure 1.4, we labeled these symbols with the numbers "−5," "−4," …, "0," …, "4," but we could just as well have labeled the symbols with the letters "A," "B," …, "J," or even use more cumbersome but wholly descriptive labels such as "Amplitude in the interval [−0.04,−0.03)." Yet, whether we use the label "−4," "B," or "Amplitude in the interval [−0.04,−0.03)," the source symbol remains the same in its elementary indivisible nature (given by the quantization operation). The label itself is just an indicator of the information contained in the symbol, which in this case is that the analog signal sample amplitude is in the interval [−0.04,−0.03). At this point, after quantization, the analog source is behaving as a digital source. A textbook is also a source that generates messages composed of elementary indivisible symbols called "letters," and because of this, it is common to call the source symbols *letters* and call the set of all source symbols the *alphabet*.

As sampling and quantization convert an analog source into a digital source, at this point we will focus exclusively on the common block for all sources called the *source encoder* (and the corresponding *source decoder* on the receiver side). Figure 1.4 shows that the source symbols are represented also by their corresponding two's complement 7-bit binary number (e.g. "−4" is represented by 1111100). This operation of defining the sequence of bits that correspond to each source symbol is called *source encoding*. Formally, source encoding is the operation of mapping the blocks of source symbols (a block may be composed of a single symbol) into sequences of bits, grouped into blocks called *codewords*. The mapping from blocks of source symbols to codewords is called a *source code*.

If we pretend that the alphabet of the source for Figure 1.4 consists of only the 10 symbols shown, the source encoding operation illustrated is wasteful. It uses seven bits when only four would be needed. Moreover, in general, there is no reason why all codewords have to be of the same length; since it is reasonable to assume that not all symbols have the same probability of occurrence, we could strive for more efficiency in the source encoding by mapping more frequent symbols to shorter codewords and rare symbols to longer codewords (as in Morse code). In more formal terms, we would like that the source code will result in a source representation that is efficient in requiring the least number of bits on average (for long sequences of source messages). In the case of *lossless* source coding, the source output is represented with the least number of bits in such a way that after source decoding the reconstructed source message is identical to the original one that was encoded. However, there are cases where it is desirable to have a representation that uses even fewer bits than those needed for the ideal lossless source coding. In this case, perfect reconstruction of the source message after decoding is not possible; some of the information conveyed by the source has been irreplaceably lost. The result presents some level of distortion. This second case is called *lossy* source coding to indicate that some information is lost in the process of source encoding and decoding. Over the next few paragraphs, we examine the minimum number of bits needed to represent the source using lossless source coding, and the magnitude of source distortion when using lossy source coding.

In Figure 1.1, the source encoder compresses the representation of the source with the goal of generating a bitstream that is compatible for transmission over a channel of limited capacity. In the case of lossless coding, the compression operation involves the removal of redundant information from the source, leaving only the source's essential information content, which is quantified as the source entropy. In the case of lossy compression, the removal of information includes some of the essential information, leading to the intro-duction of source distortion. The source entropy stems from the stochastic behavior of the source.

A key metric in source coding is the *mean codeword length*, which measures the expected length of a codeword. Let a *sourceword* x^n be a sequence $x_1, x_2, \ldots, x_n$ of source symbols, and suppose that $l(x^n)$ is the length of the codeword resulting from source encoding x^n. Then the mean codeword length is as follows:

$$L_n = E[l(X^n)] = \sum_{x^n} P[X^n = x^n] l(x^n). \tag{1.12}$$

The mean codeword length is measured in units of bits per codeword. When it is multiplied by the rate at which codewords are generated per unit time, we obtain the source code rate in units of bits per unit time (usually bits per second).

The link between the source entropy (that is, the quantification of the source stochastic characteristics we saw in previous sections) and the minimum source code rate needed for its lossless coding stems from a theorem called the *asymptotic equipartition property* (AEP). For the collection of sourcewords x^n of length n, the log-likelihood of the sequence $-\log_2(P[X^n = x^n])$ itself is a random variable.

At its simplest, the information theoretic result, AEP is an application of the law of large numbers to the log-likelihood random variable $-\log_2(P[X^n = x^n])$ for increasing n.

Consider a source (e.g. one in Figure 1.1) that generates an output in the form of a sequence of i.i.d. discrete random variables $X_1, X_2, \ldots, X_n$. Such a source is called a *discrete memoryless source* (DMS). Using $P_X(x) = P[X = x]$ to denote the PMF for any of the random variables in the sequence, the joint PMF for a DMS output sequence (a sourceword) is as follows:

$$P_{X^n}(x_1, x_2, \ldots, x_n) = \prod_{i=1}^{n} P_X(x_i). \tag{1.13}$$

The AEP studies the asymptotic properties of the information contained in the sequence $X_1, X_2, \ldots, X_n$ as the sequence length grows to infinity. The AEP states that if the sequence $X_1, X_2, \ldots, X_n$ is the output from a DMS with entropy $H(X)$, then for any $\epsilon > 0$,

$$\lim_{n \to \infty} \left\{ \left| -\frac{1}{n} \log_2 P_{X^n}(X_1, X_2, \ldots, X_n) - H(X) \right| > \epsilon \right\} = 0. \tag{1.14}$$

Using $\mathcal{X}^n$ to denote the set of length-n sequences $x_1, x_2, \ldots, x_n$ generated by a DMS, we define the *typical set* $A_{\epsilon}^{(n)}$ with respect to $P_{X^n}(x^n)$ as the set that contains sequences that meet the property:

$$2^{-n(H(X)+\epsilon)} \leq P_{X^n}(x^n) \leq 2^{-n(H(X)-\epsilon)}. \tag{1.15}$$

The typical set $A_{\epsilon}^{(n)}$ has a number of properties that will be useful in characterizing the source code for lossless compression that is optimal in yielding the smallest mean codeword length:

- A first property follows directly from (1.15) and states that all elements of the typical set are nearly equiprobable. Formally, the property indicates that the probability of a sequence $x^n = (x_1, x_2, \ldots, x_n) \in A_\epsilon^{(n)}$ is narrowly squeezed by close-by upper and lower bounds $2^{-n(H(X)-\epsilon)}$ and $2^{-n(H(X)+\epsilon)}$.

- A second property indicates that $P[A_\epsilon^{(n)}] > 1 - \epsilon$ for a sequence length n large enough, which means that the probability of the typical set is almost equal to 1.
- Lastly, a third property states that the cardinality of the typical set, denoted $|A_\epsilon^{(n)}|$, is approximately equal to $2^{nH(X)}$. Mathematically, this property is expressed as $(1 - \epsilon)2^{n(H(X)-\epsilon)} \leq |A_\epsilon^{(n)}| \leq 2^{n(H(X)+\epsilon)}$ for n sufficiently large on the lower bound and for all n on the upper bound.

In summary, through the AEP and its derived properties we learn that for source sequences long enough, there is a set of probability almost one composed of nearly $2^{nH(X)}$ length-n sequences that are all approximately equiprobable.

The AEP motivates a conceptually simple source code for the DMS that generates source-words of length n. The source code is constructed by splitting all length-n sequences into those that are and those that are not in the typical set. The codewords for entries in the typical set are of length nearly $nH(X)$ bits. Those not in the typical set are sorted and assigned a code based on the order the sequence occupies in the sorted list.

Following this procedure, it can be shown, [4], that the resulting code is a lossless one-to-one (a decodable) mapping with the property that for sufficiently large n we have that $E[(1/n)l(X^n)] \leq H(X) + \epsilon$. In other words, a sufficiently long sequence of length n can be represented with no loss of information using a code with mean codeword length $nH(X)$.

The AEP and its implications are useful to understand the behavior of the source codes – lower and upper bounds on the mean length achievable – in the limit of the sourceword length going to infinity (that is, for sufficiently long sequences). A part of this conclusion that has important consequences for the subject of this book is that for any source coding involving finite and small length source sequences, the AEP has limited use in code construction.

The AEP as stated earlier applies to sequences of i.i.d. random variables. The notion and the techniques have been extended for sequences with characteristics other than i.i.d. For certain sequences of random variables, it is possible to calculate a magnitude called the *entropy rate*, which leads to results akin to the AEP. For a discrete-time random process $\{X_i\}$, the entropy rate, $H(\mathcal{X})$, or simply the *entropy* of the random process, is defined as:

$$H(\mathcal{X}) = \lim_{n \to \infty} \frac{1}{n} H(X_1, X_2, \ldots, X_n), \tag{1.16}$$

when the limit exists. The entropy rate should be thought of as a definition of entropy for general random processes. Consistent with this, when the random process is a sequence of i.i.d. random variables we have that:

$$H(\mathcal{X}) = \lim_{n \to \infty} \frac{1}{n} H(X_1, X_2, \ldots, X_n) = \lim_{n \to \infty} \frac{1}{n} n H(X_i) = H(X_i). \tag{1.17}$$

The definition of an entropy rate allows the characterization of a general AEP for a stationary ergodic random process. Specifically, for a stationary ergodic random process $\{X_i\}$, the general AEP states that:

$$-\frac{1}{n} \log_2 P_{X^n}(X_1, X_2, \ldots, X_n) \to H(\mathcal{X}) \quad \text{with probability one.} \tag{1.18}$$

Following the parallels between the general AEP and the AEP for an i.i.d. sequence, it can be shown that for a stationary ergodic random process there exists a typical set composed of approximately $2^{nH(\mathcal{X})}$ sequences of length n, which can be represented using approximately $nH(\mathcal{X})$. From here, what is known as the *source coding theorem* follows; it states that for lossless source coding, the minimum mean codeword length L_n^* is upper and lower bounded as

$$\frac{H(X_1, X_2, \ldots, X_n)}{n} \leq L_n^* \leq \frac{H(X_1, X_2, \ldots, X_n)}{n} + \frac{1}{n}, \tag{1.19}$$

and, as n grows to infinity (source sequences growing infinitely long), we have that $L_n^* \to H(X)$ for a DMS, and $L_n^* \to H(\mathcal{X})$ for a stationary stochastic process.

1.3.3 Distortion and Rate-distortion Theory

The source coding theorem describes the minimum mean codeword length for a source code that would be optimal for lossless coding. In many cases, this minimum mean codeword length (or the one that can actually be achieved under practical conditions as, for example, with small source sequence lengths n) is still too large for a particular application. In these cases, it is necessary to use a source code with a smaller mean codeword length. The use of such source codes, however, comes at the cost of losing information in the encoding. We call this operation lossy source coding. It leads to distortion of the source message recovered after source decoding. We encountered a simple example of this situation in the quantization operation illustrated in Figure 1.4. In this example, information is lost when mapping a sample's value measured with infinite resolution (the analog sample amplitude) into a label drawn from a finite set that identifies quantization intervals. Information is lost and distortion is introduced when all the analog source values belonging to the same quantization interval are mapped to the same reconstruction value during decoding. With fewer quantization intervals, fewer labels are needed to identify each of them, but also the quantization distortion is larger because the range of source analog values associated with each quantization interval has increased. Quantization as a lossy source coding operation will be discussed more in detail in the next chapter. At this point, we merely emphasize that there is a relation between quantization distortion and the alphabet size of the quantized source. Of course, quantization is not the only instance of lossy coding. To transmit a digital source through a communication channel of limited capacity, it is common to do lossy source coding on the digital source by removing information in a controlled way. The removal of some information during source coding leads to a distortion observed in the decoded source. The relation between the amount of information that is kept after source encoding and the distortion of the decoded source is characterized through the *rate-distortion function*.

The first step in the study of the rate-distortion function is the definition of distortion metrics. A distortion metric, measure, or function, is a mapping from the combined domains of source alphabet and reconstruction alphabet into nonnegative real numbers. A distortion metric is a function that takes in as arguments a source symbol x and a source decoded symbol $\hat{x}$ and outputs a measure of the cost associated with representing x by $\hat{x}$. We use $d(x, \hat{x})$ to denote a distortion metric. Two common distortion measures are the Hamming distortion and the squared-error distortion. The Hamming distortion is defined as:

$$d(x, \hat{x}) = \begin{cases} 0 & \text{if } x = \hat{x}, \\ 1 & \text{if } x \neq \hat{x}. \end{cases} \tag{1.20}$$

The Hamming distortion essentially measures the probability of error because its expected value is $P[x \neq \hat{x}]$. The squared-error distortion is as follows:

$$d(x, \hat{x}) = (x - \hat{x})^2. \tag{1.21}$$

These two distortion metrics measure the distortion between two symbols, the source symbol and its reconstructed version after decoding. However, since source codecs tend to operate on sequences of source symbols, it is more commonplace to encounter the per-symbol distortion metrics used to measure the distortion between the source symbol sequence and the reconstructed version of the source sequence after decoding. For this, the distortion between the two length-n sequences x^n (the source sequence) and $\hat{x}^n$ (its reconstructed version) is defined as:

$$d(x^n, \hat{x}^n) = \frac{1}{n} \sum_{i=1}^{n} d(x_i, \hat{x}_i). \tag{1.22}$$

The distortion between two sequences is the average distortion between their constituent symbols. When the symbol distortion metric is the squared error as in (1.21), the distortion between two sequences is the *mean squared error (MSE)* distortion measure.

In continuing toward the definition of a rate-distortion function, it turns out that the AEP and its relation to lossless representation of a sequence of source symbols can be extended to the case when source distortion is introduced during lossy coding. For pairs of length-n sequences $(x^n, \hat{x}^n)$ of source and reconstructed symbols that are drawn i.i.d. following the joint probability $P_{X^n, \hat{X}^n}(x^n, \hat{x}^n)$ from a support product set $\mathcal{X} \times \hat{\mathcal{X}}$, we can define the *distortion ϵ-typical set* as the set satisfying

$$D_{\epsilon}^{(n)} = \left\{ (x^n, \hat{x}^n) \in \mathcal{X} \times \hat{\mathcal{X}} : \tag{1.23} \right.$$

$$\left| -\frac{1}{n} \log_2 P_{X^n}(x^n) - H(X) \right| < \epsilon,$$

$$\left| -\frac{1}{n} \log_2 P_{\hat{X}^n}(\hat{x}^n) - H(\hat{X}) \right| < \epsilon,$$

$$\left| -\frac{1}{n} \log_2 P_{X^n, \hat{X}^n}(x^n, \hat{x}^n) - H(X, \hat{X}) \right| < \epsilon,$$

$$\left. \text{and } \left| \frac{1}{n} d(x^n, \hat{x}^n) - E[d(x, \hat{x})] \right| < \epsilon \right\},$$

where $E[d(x, \hat{x})]$ is the expectation with respect to the probability distribution $P_{X^n, \hat{X}^n}(x^n, \hat{x}^n)$. From this, the following convergence properties hold:

$$\lim_{n \to \infty} \left\{ \left| -\frac{1}{n} \log_2 P_{X^n}(X_1, X_2, \dots, X_n) - H(X) \right| > \epsilon \right\} = 0 \tag{1.24}$$

$$\lim_{n \to \infty} \left\{ \left| -\frac{1}{n} \log_2 P_{\hat{X}^n}(\hat{X}_1, \hat{X}_2, \dots, \hat{X}_n) - H(\hat{X}) \right| > \epsilon \right\} = 0 \tag{1.25}$$

$$\lim_{n\to\infty}\left\{\left|-\frac{1}{n}\log_2 P_{X^n,\hat{X}^n}(X_1,\hat{X}_1,\ldots,X_n,\hat{X}_n)-H(X,\hat{X})\right|>\epsilon\right\}=0 \tag{1.26}$$

$$\lim_{n\to\infty}\left\{\left|\frac{1}{n}d(x^n,\hat{x}^n)-E[d(x,\hat{x})]\right|>\epsilon\right\}=0. \tag{1.27}$$

Note that these properties expand the description of the AEP in two ways compared with the characterization of the AEP in the lossless representation case. The first difference is that now the typical set is defined in terms of the joint statistics for two sequences of i.i.d. random variables: the sequence of symbols generated by the source (X) and the sequence of reconstructed symbols after decoding ($\hat{X}$). This extra element of the definition makes this case a *joint* AEP. The second difference is a statement of consistency in the measurement of the distortion between the two sequences of random variables when using the metric defined in (1.22).

Serving as a preliminary step, the AEP for lossy representation leads to *Shannon's rate-distortion theorem* for memoryless sources. The theorem states that for a sequence $X_1,X_2,\ldots,X_n,\ldots$ generated by a DMS with alphabet $\mathcal{X}$ that is encoded and subsequently decoded into a sequence $\hat{X}_1,\hat{X}_2,\ldots,\hat{X}_n,\ldots$ with alphabet $\hat{\mathcal{X}}$, given a distortion measure as in (1.22) satisfying $\max_{(x,\hat{x})\in\mathcal{X}\times\hat{\mathcal{X}}} d(x,\hat{x})<\infty$ (bounded per-source symbol distortion), the source's rate-distortion function, $R(D)$, is given by:

$$R(D)=\min_{P_{\hat{X}|X}:E[d(X,\hat{X})]\leq D} I(X;\hat{X}), \tag{1.28}$$

where $P_{\hat{X}|X}$ denotes the probability distribution of $\hat{X}$ conditioned on X. In essence, the rate-distortion function characterizes the convex hull resulting from reducing as much as possible the average amount of information shared by a source output and its reconstruction after coding and decoding, subject to a maximum distortion. As the rate-distortion function is defined in terms of the mutual information between a source symbol and its reconstruction after decoding, the rate is measured in units of bits per source symbol (the units of bits follows our use of logarithm base 2 when measuring information and entropy).

While in many cases it is not possible to compute the rate-distortion function as a closed-form expression, there are a few notable (and fortunately useful) cases where this is possible. One such case is a binary DMS source, modeled with an alphabet $\mathcal{X}=\{0,1\}$ with X being a Bernoulli random variable with PMF $P[X=1]=p$. Under the Hamming distortion measure, the rate-distortion function is given by:

$$R(D)=\begin{cases} H(p)-H(D), & \text{for } 0\leq D\leq\min\{p,1-p\}, \\ 0, & \text{for } D>\min\{p,1-p\}. \end{cases} \tag{1.29}$$

Figure 1.5 illustrates this rate-distortion function for different values of the parameter p. Due to the symmetry in this rate-distortion function, the cases with $p=0.25$ and $p=0.75$ yield the same function. Of all the choices for p, the case with $p=0.5$ yields the largest value of rate for a given distortion. This is intuitively natural, since we expect that a source exhibiting the largest uncertainty in the symbol that it generates will require the largest rate to represent its output for the same level of source encoding distortion. When distortion is made equal to zero, this becomes the lossless coding case, and the rate is equal to $H(p)$, consistent with the previously discussed source coding theorem.

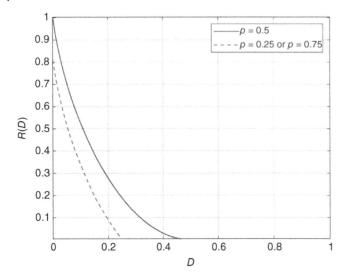

Figure 1.5 Rate-distortion function for binary sources following a Bernoulli distribution with parameter p.

While in this section we have so far focused on sources with a discrete alphabet and assumed a bounded distortion measure, the definition of the rate-distortion function in (1.28) can be extended to sources with a continuous alphabet and unbounded distortion measures. We could encounter this situation when the source output consists of samples from an analog signal. In this case, the rate-distortion function will characterize the source encoding operation as encompassing the quantization operation and the proper source encoding operation all in one operation block. A useful member of the class of sources with continuous alphabets is the Gaussian source. The Gaussian source output consists of i.i.d. samples from a random process that follows a Gaussian distribution with mean μ and variance σ^2. Using a squared-error distortion, the rate-distortion function for a zero mean, variance σ^2, Gaussian ($\mathcal{N}(0, \sigma^2)$) source is given as follows:

$$R(D) = \begin{cases} \frac{1}{2}\log_2\left(\frac{\sigma^2}{D}\right), & \text{for } 0 \le D \le \sigma^2, \\ 0, & \text{for } D > \sigma^2. \end{cases} \tag{1.30}$$

It is often useful to work with the inverse of the rate-distortion function, that is, a function which tells the distortion that will be expected when encoding at a certain rate. This inverse function, called the *distortion-rate function*, is:

$$D(R) = \sigma^2 2^{-2R} \tag{1.31}$$

for the case of the i.i.d. Gaussian source. From this expression, we learn that the distortion is reduced by a factor of 4 with each one-bit increase in the source coding rate. This improvement is independent of the variance of the source. In fact, considering the variance as the power of the source signal, we can see in (1.31) that source coding rate R results in a distortion that equals the source power divided by the rate R-power of 4. Figure 1.6 illustrates rate-distortion and distortion-rate functions for Gaussian sources with different variances.

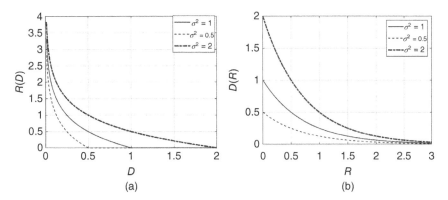

Figure 1.6 (a) Rate-distortion functions and (b) distortion-rate functions for Gaussian sources of different variances.

Now, recall from (1.28) that the functions $R(D)$ or $D(R)$ are defined based on the minimum rate to achieve a limit on distortion. There are many source codes that will not perform as well as the one associated with the rate-distortion function and will need a higher rate to achieve the same distortion. It can be shown that the limit performance characterized by the rate-distortion function is achieved when using source codes on very long blocks of source symbols. We encountered this condition for achieving a limit performance earlier: the AEP is achievable in the limit of very long sequences. This recurrent condition of achievability is of note; it leads to the main motivation for using joint source–channel coding techniques.

1.4 Channels, Channel Coding, and Capacity

One of the important elements in Shannon's model for a communication system (Figure 1.1) is the channel over which the transmission of information between transmitter and receiver takes place. As the source–channel coding operation maps source messages into signals appropriate for the channel characteristics, the channel plays a critical role in the source–channel codec design. Many different channel models exist; they differ by the physical conditions they model, their level of abstraction, or their level of simplifying assumptions. This section provides a brief introduction to a selected subset of important channel models.

1.4.1 Channel Models

A common channel model involves noise added to the transmitted signal; the channel output is written as $Y = X + Z$, where Y, X, and Z are the samples of, respectively, the channel output signal, the channel input signal, and the additive noise signal. The use of signal samples implies that this is a discrete-time model. The noise is often modeled as a random process originating from the combined effect of multiple noise sources (e.g. thermal noise at the receiver's multiple electronic components, background radiation), in which case it is assumed that each noise sample follows a Gaussian distribution. Furthermore, it is assumed that noise samples are independent from the channel input sample and that at different times they are also uncorrelated from each other or, equivalently, the noise process power spectrum density is constant across all frequencies on which the receiver operates. Consequently, this channel model is called the *additive white Gaussian noise* (AWGN) model. This model only accounts for the effects of additive noise and does not consider the effects of signal attenuation or interference, for example.

Other models take an approach that abstracts the effects of the channel on the transmitted information. One such case is the *discrete memoryless channel* (DMC). Assume that the input to the channel is a discrete random variable X drawn from a set of symbols represented as letters from the alphabet $A_I = \{x_0, x_1, x_2, \ldots, x_M\}$ and the output is a discrete random variable Y from a set given by the alphabet $A_O = \{y_0, y_1, y_2, \ldots, y_N\}$. The DMC is described by the collection of conditional probabilities $P(Y = y_O | X = x_I)$ that characterize the probability of measuring at the channel output the letter $y_O \in A_O$, given that the channel input was $x_I \in A_I$. The "memoryless" in the model's name indicates that when the input is a sequence of letters, the channel affects the transmission of each letter independently of each other. Because of this, if the input sequence is $\vec{X} = \{X_0, X_1, \ldots, X_K\}$ and the output sequence is $\vec{Y} = \{Y_0, Y_1, \ldots, Y_K\}$, the conditional probability $P(\vec{Y}|\vec{X})$ that describes the channel effect can be calculated as:

$$P(\vec{Y}|\vec{X}) = \prod_{i=0}^{K} P(Y_i | X_i). \tag{1.32}$$

When transmitting a single symbol at a time, and when the input alphabet size M is equal to the output alphabet size N, the DMC can be used to model situations where after demodulating and detecting a block of $\log_2 M$ received bits there is a probability (given by the conditional probabilities $P(Y_i = y_O | X = x_I)$) that some of these bits may be estimated with a value different from the transmitted ones. An important case of this model is when the input/output alphabets have two possible values, which are usually the binary digits "0" and "1." It is usually assumed that the probability of introducing an error during the transmission over the channel is the same regardless if a "0" or a "1" was transmitted (that is, $P(Y_i = \text{'0'}|X = \text{'1'}) = P(Y_i = \text{'1'}|X = \text{'0'})$). This model is called the *binary symmetric channel* (BSC) and is illustrated in Figure 1.7a, where ϵ is the probability of making a receiver decision error on the transmitted bit (a probability that often is considered equivalent to the *bit error rate* [BER]).

The case of the DMC when $N = M + 1$ allows us to model what is known as an *erasure channel* by introducing an extra letter in the output alphabet. Rather than being a true letter, this addition to the output alphabet can be seen as a state of the receiver operation where there has been a transmission error and the received symbol is discarded (that is, erased). The simplest case, illustrated in Figure 1.7b is the *binary erasure channel* (BEC), where the input alphabet has two letters ("0" and "1") and the output alphabet has three letters ("0," "1," and "e" for "erasure"). Figure 1.7b shows a general case for the BEC case where the erasure probability is different depending on whether the transmitted symbol was a "0" or a "1" (probabilities equal to α and β, respectively). If the two probabilities are equal, the channel becomes a binary symmetric erasure channel.

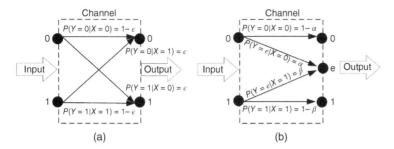

Figure 1.7 (a) The binary symmetric channel and (b) the binary erasure channel.

1.4.2 Wireless Channels

Wireless channels are affected by noise, often modeled as AWGN. In addition, wireless propagation of electromagnetic – radio – signals is characterized by complex physical phenomena of reflection, refraction, absorption, scattering, and diffraction from surrounding environmental components that lead to unknown and varying attenuation, or *fading*. The characterization of fading is divided into *large-scale propagation effects* and *small-scale propagation effects*. These refer to the scale of distance over which the effect is noticeable. This distance is relative to the radio signal wavelength (which is related to the signal frequency by $\lambda = c/f$, where λ is the wavelength, f is the frequency, and $c = 3 \times 10^8$ m/s is the velocity of light in a vacuum). Small-scale fading creates differences that are noticeable at distances in the order of half the wavelength, while large-scale effects are noticeable at distances of several times the wavelength.

Large-scale Propagation Effects
Three large-scale propagation effects are usually considered, with each of them causing a loss of power in the received radio signal. The effects are measured as the ratio between the transmitted and received signal power. The first of these large-scale propagation effects is the *path loss*. For a radio signal propagating in free space, the path loss effect arises because the energy of the emitted radio wave is spread over a surface of space that expands as the radio wave propagates away from the emitter. After measurement at a receiver, the net effect is an attenuation of the signal. A radio wave emitted or received near the Earth's surface is more difficult to model because of the presence of copies of the emitted signal that arrive to the receiver after reflecting off surfaces or for other reasons. In all these cases, the received power is attenuated as a function of the distance from the transmitter at a larger rate than that of free space propagation. In general, for a system comprising a transmitter and a receiver each with a single omnidirectional antenna, the path loss is typically modeled as a function of the following form:

$$\Gamma_{dB} = 10v \log(d/d_0) + C_p, \tag{1.33}$$

where Γ_{dB} is the path loss Γ measured in decibels (dB), d is the distance between transmitter and receiver, v is the path loss exponent, C_p is a constant, and d_0 is the distance to a power measurement reference point (sometimes embedded within the constant C_p). In many practical scenarios, this expression is not an exact characterization of the path loss but is used as a sufficiently good and simple approximation. The path loss exponent v is a statistical characterization of the rate that signal power decays with distance, taking values in the range of 2 (corresponding to free space decay) to 6. The magnitude of the path loss exponent, which captures the effects of reflection, refraction, and scattering, is also roughly a function of the relative size of the "clutter" – the elements contributing to fading – and the wavelength of the signal. In a city environment, typical values for the path loss exponent are 4 for distances of several hundred meters and 3 for shorter distances for radio frequencies of the order of a few gigahertz. The constant C_p includes parameters related to the physical setup of the transmission such as signal wavelength and antennas height.

A second path loss effect characterizes the loss of signal power when propagating through sizeable objects in the path between the transmit and receive antennas. This impairment is named *shadowing loss* or *shadow fading*. Since the nature and location of the obstructions causing shadow loss cannot be known in advance, the path loss introduced by this effect is modeled as a random variable, typically with a log-normal distribution (a random variable that follows a Gaussian distribution when measured in decibels). Together the path and shadowing losses are modeled through the power loss expression:

$$\Gamma_{dB} = 10\nu \log(d/d_0) + S + C_p, \tag{1.34}$$

where S denotes the shadowing loss (a zero-mean Gaussian random variable because the power loss expression is given in decibels).

In a particular case when at least the receiver is indoors and one wants to include in the model a more detailed characterization of the wireless signal propagating into the building, a third term is added to (1.34) yielding:

$$\Gamma_{dB} = 10\nu \log(d/d_0) + S + L_p + C_p. \tag{1.35}$$

The new term L_p is called the *penetration loss* and models, usually as a constant value, the attenuation in the received radio signal power after penetrating a building (following assumptions of the structure and materials for the walls of the building).

Small-scale Propagation Effects
Small-scale propagation effects model differences in radio signal fading that can be noticed at the scale of a wavelength. Often called small-scale fading, it arises because radio signals arrive at the receive antennas after being affected by countless random reflectors, scatterers, and attenuators encountered during propagation. Radio signals do not follow a single direct path between transmitter and receiver. Instead, multiple copies of the transmitted signal arrive at the receiver antenna after each having followed a different path. Because paths are of different lengths, the copies of the transmitted radio signal arrive at the receiver at slightly different times. Such a channel where a transmitted signal arrives at the receiver with multiple copies is known as a *multipath channel*. Several factors influence the behavior of a multipath channel. One is the random presence of reflectors, scatterers, and attenuators. The speed of the mobile terminal, the speed of surrounding objects, and the transmission bandwidth of the signal are other factors determining the behavior of the channel. If there is motion of the transmitter, the receiver, or the surrounding objects, the multipath channel changes over time. The effect of multiple copies of the radio signal arriving at the

receiver is modeled as a linear system (a time-varying tap-delay line) through the following expression:

$$y(t) = \sum_{i=1}^{L} h_i(t)x(t - \tau_i(t)),$$
(1.36)

where $y(t)$ is the received signal, $x(t)$ is the transmitted signal, $h_i(t)$ is the attenuation of the ith path at time t, $\tau_i(t)$ is the corresponding path delay, and L is the number of resolvable paths at the receiver. As can be seen in (1.36), because each of the multiple copies of the transmitted signal is received with a different amplitude, phase, and delay, when they are combined at the receiver they create patterns of constructive or destructive interference between them. As (1.36) is a linear model, the channel impulse response is given as follows:

$$h(t, \tau) = \sum_{i=1}^{L} h_i(t)\delta(t - \tau_i(t)).$$
(1.37)

In many scenarios, it is possible to assume that the channel will not change for the duration of the transmission of interest. In this case, (1.36) and (1.37) can be written in a simpler, not time-varying form, as follows:

$$y(t) = \sum_{i=1}^{L} h_i x(t - \tau_i),$$
(1.38)

$$h(t) = \sum_{i=1}^{L} h_i \delta(t - \tau_i).$$
(1.39)

When studying digital communication systems, it is convenient to convert the channel characterization into a discrete-time baseband-equivalent model, for which the input–output relation derived from (1.36) can be written as:

$$y[m] = \sum_{k=l}^{L} h_k[m]x[m - k],$$
(1.40)

where $x[m]$ is the discrete-time equivalent of the transmitted signal (a sampled version of $x(t)$), $y[m]$ is the discrete-time equivalent of the received signal, and $h_k[m]$ are the channel coefficients. This is the same type of modeling seen in Figure 1.1, although the model considered here has gained in mathematical detail. The conversion to a discrete-time model combines all the paths that have an arrival time within one sampling period into a single-channel response coefficient $h_l[m]$. Since the nature of each path, its length and the presence of reflectors, scatterers, and attenuators are all random, the channel coefficients h_k of a time-invariant channel are random variables (and note that the redundant time index need not be specified). If, in addition, the channel changes randomly over time, then the channel coefficients $h_k[m]$ are random processes. Channel coefficients could be considered non-time-dependent when the duration of a transmission is shorter than the time it takes for the channel to exhibit noticeable changes. The rate at which the channel (and its impulse response coefficients) may be changing over time depends on the radio environment settings for the transmitter, the receiver, and objects that affect the multiple signal paths. In particular, a major factor influencing the change over time of a channel characteristic is the velocity at which the transmitter, the receiver, and the objects engendering the multiple signal paths are moving with respect to each other. The duration of time over which the channel coefficients can be considered to practically remain unchanged is quantified through the *channel coherence time*. The channel coherence time is the time difference that makes the correlation between two realizations of the channel impulse response be approximately zero.

If we assume that the channel is not changing over time (that is, we consider the channel model over a time shorter than the coherence time), we can write (1.40) based on the non-time-varying channel impulse response coefficients h_k as:

$$y[m] = \sum_{k=l}^{L} h_k x[m - k].$$ (1.41)

Multipath effects arise from combining copies of the same signal that arrive at the receiver through paths of different lengths. Because of the path length differences, the copies arrive with different attenuations and phase shifts. This effect is reflected in (1.41) by modeling the channel impulse coefficients h_k as complex-valued random variables. The coefficients can be represented using Cartesian coordinates as $h = h_I + jh_Q$ (where h_I and h_Q are the in-phase and quadrature-phase components, respectively) or using polar coordinates as $h = r\, e^{j\theta}$ (where r is the coefficient magnitude and θ the phase).

Several models exist to characterize the statistical properties of the coefficients. One of the most common models for the random channel coefficients, known as the *Uniform Scattering Environment*, is derived from an assumed configuration for the transmitter, the receiver, and the objects that affects the multiple signal paths. This model is also known as Clarke's model or Jakes' model in honor of R.H. Clarke who first introduced it and W.C. Jakes who developed it. The model assumes that the radio signal arrives at a receiver after being scattered by a very large number of scatterers randomly located on a circle centered on the receiver. The radius of this circle is much smaller than the distance between transmitter and receiver. As a result, the received radio signal (1.41) is made of the superposition of many copies of the transmitted signal arriving from the scatterers at angles that are uniformly distributed between 0 and 2π. The model assumes that there is no radio signal that is received through a direct path between transmitter and receiver (known as the line-of-sight [LOS]). It can be shown [8] that under the uniform scattering model, the in-phase and quadrature-phase components of the coefficients are each a zero-mean Gaussian random variable all with the same variance σ^2, which means that the coefficients themselves are circularly symmetric complex Gaussian random variables with zero mean and variance $2\sigma^2$. By applying a change of variables to convert the coordinate system from Cartesian to polar, the phase of a channel coefficient θ is a uniform random variable taking values between 0 and 2π and the magnitude of the channel coefficient (the amplitude gain for the corresponding path), r, is a Rayleigh random variable with PDF given by:

$$f(r) = \frac{r}{\sigma^2} e^{-r^2/(2\sigma^2)}, \quad r \geq 0.$$ (1.42)

Because the magnitude of the channel coefficients follows a Rayleigh distribution, this model is frequently called a *Rayleigh fading* model.

Since the power gain for a path is a random variable $G = r^2$, its PDF can be calculated by applying the transformation of random variables theory and find that it is distributed as an exponential random variable with PDF:

$$f_G(g) = \frac{1}{\sigma^2} e^{-g/\sigma^2}, \quad g \geq 0. \tag{1.43}$$

Moreover, the sum of the squared magnitudes of the channel coefficients, $\sum_i |h_i|^2$, results in a Chi-square random variable with $2L$ degrees of freedom, where L is the number of channel coefficients in the sum. The PDF of this distribution is as follows:

$$f(x) = \frac{x^{L-1}}{(L-1)!} e^{-x}, \quad x \geq 0. \tag{1.44}$$

If we assume that the uniform scattering environment model also includes an LOS path, approximating both the in-phase and quadrature-phase components of a channel coefficient as zero-mean Gaussian random variables is no longer valid. Now, the two components are Gaussian random variables but one has mean A, which is the peak amplitude of the signal from the LOS path, and the other still has mean zero. The analysis for this case leads to the conclusion that the magnitude of a channel coefficient follows a Rician distribution with a PDF given by:

$$f_r(z) = \frac{z}{\sigma^2} e^{-\left(\frac{z^2}{2\sigma^2} + K\right)} I_0\left(\frac{2Kx}{A}\right), \quad z \geq 0, \tag{1.45}$$

where $I_0(x)$ is the modified Bessel function of the first kind and zero order, defined as:

$$I_0(x) = \frac{1}{2\pi} \int_0^{2\pi} e^{x\cos\theta} d\theta. \tag{1.46}$$

In (1.45), K is a parameter of the Rician distribution, defined as $K = A^2/(2\sigma^2)$. When $K = 0$, the Rician PDF becomes equal to the Rayleigh PDF, consistent with the fact that $K = 0$ means there is no LOS path.

1.4.3 Channel Coding and Channel Capacity

The revolutionary innovation that Shannon introduced with his mathematical theory of communication was the use of coding (the mapping from a source message into an appropriate signal for the channel characteristics) to improve the communication reliability of a message. The central result of Shannon's mathematical theory of communication is an indisputable justification for the use of coding: when the code used follows certain characteristics in relation to the channel, it is possible to achieve an arbitrarily small probability of error when transmitting a message. This central result, known as Shannon's noisy channel coding theorem, [1], is the subject of this section.

Figure 1.8, which emphasizes the encoding and decoding operations as well as their relation with the channel, is a simplified version of the general communication system diagram in Figure 1.1. We assume that the source is a DMS and that, without loss of generality, its alphabet is the set of indexes $\{1, 2, \ldots, M\}$. A message W formed by a single letter (index) from the source is encoded, resulting in a codeword $\vec{X}(W) = \{X_0, X_1, \ldots, X_n\}$. This codeword is transmitted through the channel and received as a sequence $\vec{Y} = \{Y_0, Y_1, \ldots, Y_n\}$, which, because of the stochastic impairments introduced into the transmitted signal

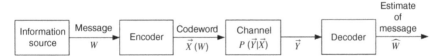

Figure 1.8 Distilled view of the message communication process.

by the channel, is a random sequence that follows the probability distribution $P(\vec{Y}|\vec{X})$. The received sequence $\vec{Y}$ is input to a decoder that outputs an estimate $\hat{W}$ of the transmitted message. We can describe this process with an emphasis on the role of the coding and its mathematical formulation. Given an index set $\{1, 2, \ldots, M\}$, the encoding operation of an (M, n) code is a function $X^n : \{1, 2, \ldots, M\} \rightarrow \mathcal{X}^n$, where $\mathcal{X}^n$ is the set of sequences $\{x^n(1), x^n(2), \ldots, x^n(M)\}$, each called a codeword and associated through the encoding operation with one of the indexes that forms the source output. The set of codewords is called a *codebook*. On the receiver side, the decoding operation is a function $g : \mathcal{Y}^n \rightarrow \{1, 2, \ldots, M\}$ that estimates an index from the source for each received sequence $\vec{Y}$. For an (M, n) code, its rate is defined as:

$$R = \frac{\log_2(M)}{n} \qquad \text{[bits per transmission]}. \tag{1.47}$$

We now introduce the concept of information capacity of a channel. Let the random variables X and Y be the channel input and output, respectively. The *information capacity* of this channel is defined as the maximum of the mutual information between the input and output:

$$C = \max_{p_X(x)} I(X; Y), \tag{1.48}$$

where the maximum is taken over all possible PMFs $p_X(x)$ for the input and subject to a power constraint on the input. For a continuous channel, the capacity definition involves the PDF $f_X(x)$:

$$C = \max_{f_X(x), E(|X|^2) \le P} I(X; Y). \tag{1.49}$$

From the definition of mutual information, the channel capacity measures the maximum average amount of information that could be shared between the input and the output of the channel. Informally, the channel capacity indicates the maximum amount of information that could be retained from the random variable at the input of the channel after being transmitted through a channel. Linking these thoughts and the use of a code for transmission, Shannon established the relation between the channel capacity and the rate of the code used when seeking a communication process with an arbitrarily small probability of error. This relation is established in Shannon's *noisy channel coding theorem* [1], which states that it is possible to communicate over a DMC with an arbitrary small probability of error by using a code that has a rate R that is less than the channel capacity C. In this way, through his noisy channel coding theorem, Shannon provided a functional meaning to the channel capacity as the maximum transmission rate (measured in bits per channel use) at which it is possible to send information over the channel with an arbitrary low probability of error. Without delving into details (which the interested reader can find in [4]), it is important to note that the proof for the noisy channel coding theorem follows the definition of a jointly typical set between the random variables X and Y at the channel input and output, respectively, and a resulting joint AEP. The joint AEP in this case is analogous to what we have seen earlier with the rate-distortion theory, but without the presence of a distortion metric, and, importantly, it is now defined in terms of the joint statistics between the random sequences at the input and output of the channel. Specifically, we can define a *jointly typical set* as the set satisfying:

$$A_\epsilon^{(n)} = \left\{ (x^n, y^n) \in \mathcal{X}^n \times \mathcal{Y}^n : \right. \tag{1.50}$$

$$\left| -\frac{1}{n} \log_2 P_{\vec{X}}(\vec{x}) - H(X) \right| < \epsilon,$$

$$\left| -\frac{1}{n} \log_2 P_{\vec{Y}}(\vec{y}) - H(Y) \right| < \epsilon,$$

$$\left. \left| -\frac{1}{n} \log_2 P_{\vec{X}, \hat{Y}}(\vec{x}, \vec{y}) - H(X, Y) \right| < \epsilon \right\}.$$

The noisy channel coding theorem uses the identification of a jointly typical set between the input and output of the channel. It does this to establish that for codewords of large length, a sequence at the output of the channel will be jointly typical with a codeword at the input of the channel with a probability very close to 1, and for the remaining codewords the probability of being jointly typical will be very small. This property allows for a decoding of the received sequence that undoes transmission errors resulting in an arbitrarily small probability of transmission error. The condition $C > R$ together with (1.48) or (1.49) simply determines the cardinality of the jointly typical set that can be allowed by the channel characteristics. For the purpose of this book, the important implication of these observations is that the condition of an arbitrarily small probability of error seen in the noisy channel coding theorem implicitly depends on the use of codes of infinite decoding complexity and operation with sequences of length n that approach infinity.

Having introduced the concept of capacity and the noisy channel coding theorem, we can now apply the definition of channel capacity and derive its expression for some of the channel models discussed earlier. For the AWGN channel, we can write (using differential entropy properties):

$$I(X; Y) = H(Y) - H(Y|X) = H(Y) - H(X + Z|X) = H(Y) - H(Z|X).$$

Since noise and channel input samples are independent of each other, we have that $I(X; Y) = H(Y) - H(Z)$. Also, for noise that is a Gaussian random variable with zero mean and variance σ_Z^2, we have $H(Z) = (1/2)\log_2(2\pi e \sigma_Z^2)$. Next, recall the property that the differential entropy is bounded as $H(Y) \leq (1/2)\log_2(2\pi e \sigma_Y^2)$. Since Y has zero mean, its variance is:

$$\sigma_Y^2 = E[Y^2] = E[(X + Z)^2] = E[X^2] + 2E[XZ] + E[Z^2] = \sigma_X^2 + \sigma_Z^2,$$

where we have used the independence between X and Z, and $E[Z] = 0$, to see that $E[XZ] = E[X]E[Z] = 0$. Using $P = \sigma_X^2$ to denote the signal power and $N = \sigma_Z^2$ to denote the noise power, we have $\sigma_Y^2 = P + N$ and $H(Y) \leq (1/2)\log_2(2\pi e(P + N))$. By collecting these results, we get:

$$I(X; Y) = H(Y) - H(Z) \leq \frac{1}{2}\log_2(2\pi e(P + N)) - \frac{1}{2}\log_2(2\pi e N) = \frac{1}{2}\log_2\left(1 + \frac{P}{N}\right),$$

where P/N is the signal-to-noise ratio (SNR) and, since $C = \max_{f_X(x)} I(X;Y)$, the channel capacity for the AWGN channel is:

$$C_{AWGN} = \frac{1}{2}\log_2\left(1 + \frac{P}{N}\right) \text{ [bits per channel use].} \tag{1.51}$$

If we now consider the case of a channel that is AWGN and also behaves as a low-pass linear filter, the output of the channel is the convolution of the channel impulse response with a signal formed from the addition of noise to the transmitted signal. Since the channel has a low-pass characteristic, the output signal is band-limited (suppose to a frequency W) and can be sampled at a sampling rate of $2W$ samples per second. The sampling operation turns the channel model into the discrete-time AWGN channel for which we have just discussed the capacity. Therefore, the capacity for the band-limited AWGN channel can be written also as in (1.51), or as the popular equivalent expression:

$$C_W = \frac{1}{2}\log_2\left(1 + \frac{P}{N_0 W}\right) \text{ [bits per channel use],} \tag{1.52}$$

where N_0 is the power spectral density of the AWGN. Also, since each channel use corresponds to the transmission of one sample and there are $2W$ samples per second, the channel capacity can also be written in units of bits per second as:

$$C_W = W\log_2\left(1 + \frac{P}{N_0 W}\right) \text{ [bits per second].} \tag{1.53}$$

The maximization that defines the channel capacity in (1.48) and (1.49) is an operation of finding the supremum of all channel code rates that achieve an arbitrarily small probability of error during the process of communication. The channel code that achieves this supremum rate is ideal in the sense of involving infinite length and coding complexity. Communication systems realized with practical codes will exhibit a performance loss with respect to this ideal, manifested by the need for a larger SNR in order to achieve the same channel throughput. This effect is often modeled by approximating the throughput of a practical communication link with the same expression as in (1.53) but introducing the concept of an *SNR gap*, denoted Υ, that models the need for higher SNR. With this approach, the throughput T of a practical communication system is approximated as:

$$T = W\log_2\left(1 + \frac{\Upsilon P}{N_0 W}\right) \text{ [bits per second].} \tag{1.54}$$

1.5 Layered Model for a Communication System

We will now pause on the abstractions of information theory and consider how practical systems are usually architected. In particular, we consider communicating a message through a network. Multiple interoperating components are needed for a network to effectively deliver a message from source to destination. Historically, network design components were organized into modular functional groups. This divide-and-conquer approach (simplifying a problem by breaking it apart into simpler subproblems) also exposed a hierarchy in the functions of the modular groups. Some groups provide a function needed by another group. For example, a component that deals with transmitted bit errors depends on another component that converts a stream of bits into electrical signals that are sent into the communication channel. Collectively, the symbiosis of the functions provided by the different groups build up a network from the most basic functions of electrical signals to complex functions of managing traffic load at network nodes. The modular grouping of functions and the relations between them define a stack of layers, where each layer groups components with similar functions and dependencies. Layers lower in the stack provide services to higher layers. The layer stack forms a bottom-up building of network functionality, where basic lower-layer functions are needed by the more complex functions at higher layers.

In the 1970s, recognizing a widespread use of layered architectures for network protocol design, the International Organization for Standardization (ISO) led the development of the reference standard for the definition of the different layers and the features they provide to the overall network functioning. This standard is known as the *Open Systems Interconnection* (OSI) reference model for the layered architecture of networks [9]. The well-thought-out design behind the model has led to its robust longevity. The OSI model influence extends to the present time; all networks are based on the same layered architecture and retain most of the layers and their specification from the OSI model. Figure 1.9 shows the five-layer protocol stack used in the Internet. From the base of the stack to the top, the five layers are as follows:

- *Physical layer*: The *PHY layer* provides the services involved in the transmission of individual bits. It is at the physical layer that logical bits are converted into an electrical signal to be sent over a communication medium.
- *Data link layer*: The data link layer, or link layer, provides functions for establishing a connection between a transmitter and a receiver, and for the reliable communication of a structured block of bits (called a *frame*). The direct point-to-point communication connection between a transmitter and receiver is called a *link* and forms the minimum unit of interconnection in a network. In some protocol layer architectures, the data link layer is divided into further layers. Some of the functions provided by this layer are checking for bit errors at the receiver, retransmitting frames with unrecoverable errors, and arbitrating between multiple potential transmitters on access to the communication medium (called *Medium access control* [MAC]). With the services provided by the physical and link layers, it is possible to set up reliable communication between a transmitter and receiver.
- *Network layer*: This layer connects multiple transmitter–receiver links into a network with multiple nodes. The chaining of links establishes a *path* in the network. The network layer implements functions, such as routing, that enable information to

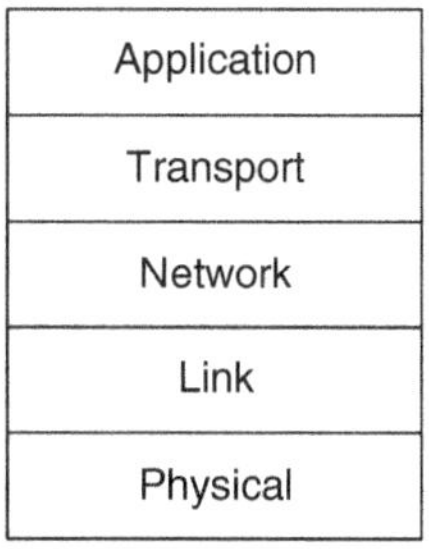

Figure 1.9 The Internet protocol stack consisting of five layers.

| Application |
| Transport |
| Network |
| Link |
| Physical |

reach the intended destination by following a multi-node path through the network. The functions implemented at the network layer expand the topology of the network from a point-to-point (transmitter–receiver) connection to a network consisting of two end nodes (an originator of information and a receiver of the information) and multiple interconnected intermediate nodes. While devices that implement the bottom two layers of the protocol stack are only able to establish a direct connection with another device, those devices that also implement a network layer protocol become able to communicate information over a true network of paths consisting of many links.

- *Transport layer*: This layer enables an abstracted view of the network between the end nodes by providing functions for the reliable end-to-end transfer of information. At the transport layer level, the network in between the two end nodes is seen as a single entity that encompasses all the networks with their devices and components over which the two end nodes are interconnected. At the link layer, the concept of reliability concerns errors that occur during signal transmission over a medium. In contrast, end-to-end reliability at the transport layer addresses communication errors associated with considering the whole network between the end nodes as a single element. Such errors include packet losses at congested intermediate nodes or packets belonging to a stream that arrive in a different order from how they were sent.

- *Application layer*: At the top of the protocol stack, the application layer is associated with protocols for the different applications being run by end users. Applications intended to run over a network are designed with the implicit configuration of running in a distributed manner. For example, a distributed web browsing application has one part running on the end-user side, where web page queries are issued, and a complementary part that runs on the web server side, responding to the user queries. The protocols at the application layer provide the network interface for applications that run at the end nodes. Because of this, these applications are often considered part of the application layer. The source encoders and decoders from Shannon's model for a communication system are one example of applications.

Layers are often simply named with a number (e.g. PHY layer is Layer 1, data link layer is Layer 2). The modular approach combined with the high complexity of the functions implemented at each layer has led to a specialization of both layer design and implementation. In a typical networked device, one finds the two lower layers implemented within a chipset or circuit board (known as the network interface card [NIC]), while layers 3 and 4 would be a group of processes running in the kernel of the device's operating system, and the application layer would be a group of processes running within the user space of the device's operating system. Other than the particular case of control messaging, a message to be communicated originates at the application layer. From the application layer, the message is passed to the transport layer and then the network layer through a series of procedure and functions calls. The transfer of the processed message from the network layer (which resides in the device's operating system kernel) to the data link layer (which resides in a peripheral chipset) is done through function calls to the device driver (software in the operating system kernel that provides interfacing functions). The modular approach in the layered protocol architecture allows for the design or adaptation of certain layers to be better suited to specific communication environments.

Modern-day communication systems extend the basic layered architecture in multiple ways: capturing additional nuances of function, information ownership, service guarantees, and protocol termination points. The layered architecture also follows the software paradigms of separation of responsibilities, encapsulation, and hiding of details, multiple layers of abstraction, and separation of control procedures and data transmissions. Considerable thought is devoted to the design of the functions and responsibilities in each layer and the services they provide to and expect from other layers. Further, communication procedures and protocols between nodes and abstract entities are designed completely within a layer. With various network components owned, designed, and developed by altogether different entities, many of these architectures go through a process of standardization. Today's wireless communication standards are a triumph of technological innovation and large-scale multiparty collaboration in a common interest, introducing spectacular new technological features at a rapid pace, while leaving sufficient room for creativity and differentiation for all the parties involved. This is in no small part due to the idea of clean layered architectures. The layered architectures of today enable multiple access, unicast and multicast, single- and multihop communications, various quality-of-service (QoS) offerings on throughput, latency and reliability, security and privacy, and mixing of highly disparate physical deployments, e.g. frequency bands and heterogeneous nodes and form factors. As an example, Figure 1.10 shows the schematic layered architecture 5G-NR stack.

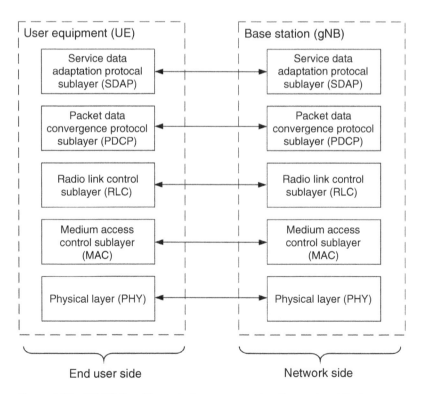

Figure 1.10 5G-NR User Plane stack as an example of layered architecture.

Yet, despite the layered architecture and the presumed independence of the layers, higher efficiency is often achieved when the layers are designed *jointly*, keeping in consideration the end-to-end service/application use case and the requirements of their QoS.

Joint source–channel coding paradigms understand the ground reality from the outset that joint designs may not only be better, but may even be unavoidable. Most designs of layered architectures today have a strong joint design element in their evolution. To appreciate this, it is useful to learn the notions of QoS and quality of experience (QoE) and their relationship with distortion metrics.

1.6 Distortion, Quality of Service, and Quality of Experience

Assessing communication system performance through a measurement of distortion or quality is of fundamental importance for the subject of this book. Measurements of distortion or quality can be divided along many different lines, according to whether they are objective or subjective, attempting to hew closely to properties of human perception or not, or involving comparison to a reference or not. Objective measurements of distortion are quantifiable calculations of the degradation of a signal or message in comparison with a reference. For example, an image compressed by JPEG is compared to the original image (the reference) as captured by the digital camera, and the distortion measure aims to quantify the degradation between them.

The term *quality* is often taken to be, roughly speaking, a kind of inverse of distortion. So an image that has low distortion typically has high quality. However, an important difference between distortion and quality is that, while distortion implies a comparison with a reference, the term quality is very often used without a reference. For example, one might consider an image to be high quality if it is sharp with good contrast, and say it is low quality if it is blurry and low contrast, irrespective of any comparison with a reference. Also, if one is given a reference image that is blurry and low contrast, and after some processing the output image is still blurry and low contrast, one would have to say that the output image is rather faithful to its reference, which means it is both low distortion (relative to its reference) and low quality (in an absolute sense).

A distortion metric, in comparing a processed signal with a reference signal, implicitly assumes that any deviation from the reference is a bad thing. In a case where the processing is, for example, image sharpening or contrast enhancement, the resulting processed image might be better looking than the original reference image. One might then have a situation where the processed output image is both high distortion (relative to the low-quality reference) and also high quality (in an absolute sense).

So these are examples where quality and distortion are not acting like inverses. A quality metric in which a processed signal is compared with a reference is more properly called a *fidelity metric*. So, a distortion metric measures in some sense how far away one is from a reference, and a fidelity metric measures how close one is to a reference, so these tend to move in opposite directions, or be roughly inverses. In this book, we will tend to use the term *quality metric* both to mean a *fidelity metric* where there is an explicit comparison with a reference, and also use it in the more general sense.

1.6.1 Objective Measurements of Distortion or Quality

For objective measurements of distortion, the calculations of signal or message degradation with respect to a reference are based on the mathematical framework of measurements. A measurement is a function from the product space formed from the supports of the two quantities to be compared, into a positive real number. Using the set $\mathcal{X}$ to denote the support of both a message and its reconstructed estimate, the distortion measure d is a function:

$$d : \mathcal{X} \times \mathcal{X} \to [0, \infty). \tag{1.55}$$

Let $x, \hat{x}, y \in \mathcal{X}$. To be a proper measure (in the mathematical sense), a distortion measure needs to satisfy the following three axioms:

- $d(x, \hat{x}) = 0$ if and only if $x = \hat{x}$.
- $d(x, \hat{x}) = d(\hat{x}, x)$.
- $d(x, \hat{x}) \leq d(x, y) + d(y, \hat{x})$.

There exist numerous measurements of distortion or quality, some tailored to specific cases. We already introduced the MSE between two sequences in (1.22). When calculating the MSE between two random variables (e.g. a source symbol X and its received estimate $\hat{X}$) the MSE is:

$$MSE = E[(X - \hat{X})^2], \tag{1.56}$$

where the expectation is over the joint probability distribution of X and $\hat{X}$. For calculating the MSE between two sequences of random variables, $\vec{X}$ and $\vec{\hat{X}}$, the expression is:

$$MSE = E[(\vec{X} - \vec{\hat{X}})^T (\vec{X} - \vec{\hat{X}})] = \sum_{k=1}^{N} E[(X_k - \hat{X}_k)^2]. \tag{1.57}$$

The MSE is the most commonly used distortion measure. Another common measure is the mean absolute error between two sequences of random variables, $\vec{X}$ and $\vec{\hat{X}}$:

$$MAE = E[|\vec{X} - \vec{\hat{X}}|]. \tag{1.58}$$

These measures of distortion are not simply converted into measures of quality (or fidelity) by calculating an inverse. Earlier, we saw that the SNR provides a sense of channel quality by relating the transmitted signal power to the added noise power; we can extend this idea to measure the quality of a signal that has undergone some distortion. This requires that the distortion could be modeled as modifying the signal through some additive process. Let us first consider the case of a random variable X modeling a source symbol. After undergoing some distortion, it becomes the random variable $\hat{X}$. Assuming for simplicity that all random variables are zero mean, the SNR is calculated as:

$$SNR = \frac{\sigma_X^2}{E[(X - \hat{X})^2]}, \tag{1.59}$$

where σ_X^2 is the variance of X and the denominator is the MSE between X and $\hat{X}$. In other circumstances, we may be interested in calculating the SNR for a signal that is presented as a sequence of N samples $\{x_1, x_2, \ldots, x_N\}$, which has been distorted into the sequence of samples $\{y_1, y_2, \ldots, y_N\}$. In this case, the signal power is $\sum_{k=1}^{N} x_k^2$, and the SNR is calculated as:

$$SNR = \frac{\sum_{k=1}^{N} x_k^2}{\sum_{k=1}^{N} (x_k - y_k)^2}. \tag{1.60}$$

For signals that are images or videos, instead of calculating the SNR, it is more common to calculate the peak signal-to-noise ratio (PSNR). The PSNR measures the square of the maximum value that the signal could take, relative to the power associated with the distortion (measured as the MSE). For the common case where the image or video frame is an eight bit per pixel grayscale image, the maximum value that a pixel can have is 255. The PSNR is typically calculated in decibels as:

$$PSNR = 10\log_{10}\left[\frac{255^2}{\frac{1}{NM}\sum_{i=1}^{N}\sum_{j=1}^{M}(x_{i,j} - y_{i,j})^2}\right], \tag{1.61}$$

where $x_{i,j}$ are the pixels of the original image or video frame and $y_{i,j}$ the pixels of the distorted image or video frame. The PSNR is also used to measure video quality either by examining the PSNR across time for each frame in a video sequence or by calculating a PSNR for the whole video. As having any single frame to be nearly (or completely) free of distortion causes the PSNR for that frame to be huge (or infinite), averaging PSNR values over video frames can be problematic. Instead, calculating a single PSNR value for an entire video usually involves averaging the MSE (the denominator) across all frames and then converting that to PSNR by taking the ratio of 255^2 to that average MSE.

1.6.2 Subjective and Perceptually Based Measurements of Distortion or Quality

Since many messages are intended for people, and people experience a signal or message through the senses (e.g. sight for images or video, or hearing for audio or speech), the measure of distortion or quality should often be related to human perception. One approach is to retain simple objective measurements and rely on individual experience to develop a correspondence between the objective measure and human perception. For example, a person with experience examining images and measuring their PSNR may know that images with a PSNR of 40 dB have very good perceived quality, while ones with a PSNR of 32 dB are good (better than mediocre). However, for some images (such as ones with smooth and uniform texture), 32 dB corresponds to very good quality, while for other images (possibly those with more texture and sharp or contrasting edges) 32 dB might correspond to mediocre perceived quality. So, while simple objective measurements such as MSE and PSNR may provide some sense of the distortion or quality perceived by a generic human subject, they are not an accurate way to assess quality or distortion as perceived by a person.

Because of this, a great deal of research and development has involved both *subjective measurements* of distortion or quality (in which people provide subjective ratings, based on their perceptions) and *perceptually based objective measurements* (in which computable objective metrics are refined, aiming for good correspondence with human subjective ratings). Perceptually based measures of quality or distortion tend to incorporate the sensitivity of human senses to different stimuli. Certain levels of distortion may be imperceptible to human senses, and humans do not have uniform sensitivity. For example, people can tolerate (or not even perceive) distortion of a sound with two nearby frequency components where one of them has a distinctly larger power than the other. This property is exploited by the MP3 audio codec to achieve compression with distortion that is very hard to perceive. For information communication, factors beyond signal representation also impact people's end-to-end perception of quality. For example, end-to-end delays of more than about 150 ms in interactive conversational communication result in mediocre or poor perceived end-to-end quality. As a result, various means to measure the subjective quality or distortion of a communication process as perceived by human end users were developed over time.

Subjective measurements of distortion have been used for quite some time in communication systems. One of the first uses was in landline (wireline) telephone systems to evaluate voice quality over different systems. The established procedure in the telephone industry was to ask a group of people to listen to voice recordings processed by the equipment under test (for example, a new speech codec at the time, such as ITU-T G.711) and rate the speech quality. The trained listeners used a scale of five possible quality scores to provide their individual opinions: Excellent (5), Good (4), Fair (3), Poor (2), and Bad (1). The scores were averaged across listeners, producing the *mean opinion score* (MOS) as the final measurement result. Over time, this evaluation procedure was standardized by the International Telecommunications Union (ITU) as ITU-T recommendation P.800 "Methods for subjective determination of transmission quality."

With the acceleration of technology, the need arose for a more automated way to measure or estimate MOS without the laborious process of convening and training listeners. Algorithms were developed to estimate the MOS of a signal that had been degraded by some processing operations compared to the signal reference in an unprocessed form. Figure 1.11 illustrates the Perceptual Evaluation of Speech Quality *(PESQ) model*, an automated MOS estimation algorithm. This model was defined in ITU-T recommendation P.862 "Perceptual evaluation of speech quality (PESQ): An objective method for end-to-end speech quality assessment of narrow-band telephone networks and speech codecs." Its block diagram in Figure 1.11 is typical of this type of algorithm; it has stages of signal conditioning, filtering, and transforms that modify the signals based on the sensitivity of human hearing to

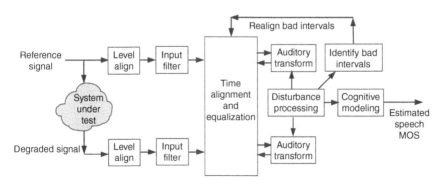

Figure 1.11 Block diagram for speech MOS estimation using the ITU-T recommendation P.862 Perceptual Evaluation of Speech Quality (PESQ).

different stimuli. The aim is to boost, or put more weight, on sounds for which human hearing is more sensitive and remove sounds that humans cannot hear. Interestingly, while the MOS scale is between 1 and 5, the PESQ algorithm generates results between 1 and 4.5 so that PESQ results better correlate with the normal range of MOS values seen in tests.

The PESQ speech MOS estimation algorithm is known as an *intrusive* or *full-reference* (FR) method of measuring signal quality. With intrusive or FR methods, the estimation of quality requires access to the original signal. With the evolution of telephony into digital technology and, specially, packet voice in the form of voice-over-IP (VoIP) services, network operators needed a nonintrusive method to estimate MOS speech quality to manage the network based on the inferred quality. *Nonintrusive* or *No-Reference* (NR) methods to estimate quality do not require access to the original, unprocessed, signal, allowing instead for subjective quality estimation to be done using parameters of the system (e.g. codec used) and objective measurements of traffic and network performance. These objective performance measurements, known as *QoS* metrics, include transmission throughput, packet (or bit) error rate, and transmission delays. One nonintrusive method is the E-model (defined in the ITU-T recommendation G.107), which estimates speech quality using the *R-factor*, which is calculated using the formula:

$$R = R_0 - I_s - I_d - I_e + A, \tag{1.62}$$

where R_0 is related to the SNR, I_s represents the combined effects of different impairments that may occur during voice communication (including quantization distortion), I_d represents the combined effects of communication delay (e.g. echo or too long end-to-end delay), I_e represents the performance of specific voice codecs against packet loss errors (e.g. error concealment), and A describes the end-user tolerance to impairments due to the convenience of the technology in use. While the R-factor is not in the same range as MOS, it is possible to convert from one to the other through the relation $MOS = 1 + 0.035R + 7 \times 10^{-6}R(R - 60)(100 - R)$ for $0 < R < 100$, $MOS = 1$ for $R \leq 0$, and $MOS = 4.5$ for $R \geq 100$, as illustrated in Fig. 1.12.

The term *QoS* refers to an objective performance measurement of network parameters (e.g. packet loss rate, bit rate, throughput, delay, jitter, outage rate) that indicates the

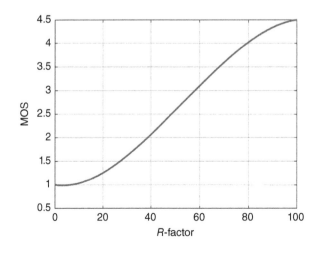

Figure 1.12 Conversion from *R*-factor into MOS.

quality of the networking service provided by the network. Driven by an increased focus on end-user network design and management, subjective and perceptually based quality measurements grew into the concept of *QoE*, which are measurements (derived from people or derived from computable metrics seeking to emulate human responses) of the quality of a network service (or traffic type) as subjectively experienced by end users. The QoE depends on the QoS the network provides. A direct method to measure QoE in voice traffic is the MOS. As with the development of VoIP leading to MOS estimation algorithms, the growth of non-voice services over networks drove the development of MOS estimation algorithms for other types of network traffic/service, not only for those related to human perception (e.g. video streaming) but also for those not experienced through senses (e.g. a file download). For MOS estimation in the case of a file download service, one formula that has been used is as follows:

$$MOS = a\log_{10}(bR(1 - \epsilon)),$$
(1.63)

where R is the bit rate and ϵ is the packet error rate. The constants a and b are parameters to adjust the regression estimate in the curve to the concept behind an MOS result. They are calculated using the conditions that MOS should equal 4.5 when R equals the rate that the end user has subscribed to and the packet error rate is zero, and that MOS equals 1 when the rate is at its minimum assumed value. For video streaming services, many models have been developed. One simple approach uses the following mapping between PSNR and MOS [10]:

$$MOS = \begin{cases} 5, & \text{for } PSNR > 37, \\ 4, & \text{for } PSNR \text{ between 31 and 37,} \\ 3, & \text{for } PSNR \text{ between 25 and 31,} \\ 2, & \text{for } PSNR \text{ between 20 and 25,} \\ 1, & \text{for } PSNR < 20. \end{cases}$$
(1.64)

Other approaches improve the MOS estimate by using other functions, e.g. logarithmic. In [11], the approach to model MOS of video streaming first characterizes the relation between PSNR (measured in decibels) and source encoding rate R as:

$$PSNR_{dB} = a + b\left(\sqrt{\frac{R}{c}} - \sqrt{\frac{c}{R}}\right),$$
(1.65)

where a, b, and c are parameters that fit the model to a video stream or sequence. Next, the PSNR is mapped to MOS through the following formula:

$$MOS = \begin{cases} 1 & \text{for } PSNR \leq PSNR_1, \\ d\log(PSNR) + e & \text{for } PSNR_1 < PSNR < PSNR_{4.5}, \\ 4.5 & \text{for } PSNR \geq PSNR_{4.5}, \end{cases}$$
(1.66)

where the model fitting parameters d and e are defined as:

$$d = \frac{3.5}{\log(PSNR_{4.5}) - \log(PSNR_1)},$$

$$e = \frac{\log(PSNR_{4.5}) - 4.5\log(PSNR_1)}{\log(PSNR_{4.5}) - \log(PSNR_1)}.$$

Combining (1.65) into (1.66) results in a model that relates source encoding rate and MOS. An alternative model between PSNR and source encoding rate of the form $\text{PSNR} = a \log(R) + b$ can be found in [12]. Using this logarithmic relation between PSNR and source encoding rate and the logistic regression model between MOS and the logarithm of the source encoding rate presented in [13], the relation between video MOS and PSNR can be expressed as [14]:

$$\text{MOS} = \frac{a}{1 + e^{b(\text{PSNR}-c)}},$$

where a, b, and c are model fitting parameters.

There exist many more proposed alternative measures to measure quality and distortion, with different approaches for capturing perceptual effects. The simple objective measurements (such as MSE, PSNR, and mean absolute error (MAE)) continue to be used because of mathematical tractability or simplicity. But the subjective measurements and complex perceptually based computable metrics perform better at describing how human end users will subjectively experience the quality of a communication process. In connection with Section 1.5, it is worth noting that the inherent end-to-end nature (or "mouth-to-ear" as it is sometimes called in telephone technology) of QoE measures lend their use naturally to cross-layer design approaches, of which joint source–channel coding (JSCC) represents one important subclass.

1.7 Shannon's Separation Principle and Joint Source–Channel Coding

The alert reader may have realized that earlier parts of this chapter ordered topics aligned with the general communication system block diagram in Figure 1.1. We started with the foundational concepts of entropy and information in Section 1.2, followed by a discussion in Section 1.3 of information content in a source, how to represent this information, and the consequences of removing some of it. Then we focused in Section 1.4 on channel models and how much information can be communicated through them. All these elements now come together in the study of a complete system for communicating a source.

We start with lossless communication. We have seen that to be lossless, the source encoding of a source of entropy H bits per source sample needs to use a code with a rate R_s bits per source sample that satisfies $R_s > H$. Consider the communication of this source over a channel where n channel uses are permitted per source sample. We have also seen that to achieve transmission with an arbitrarily low probability of error over a channel with capacity C bits per channel use, the channel code used needs to satisfy the condition on the information rate R_c (bits per channel use) as $R_c < C$. The two conditions, one for lossless representation of source information and one for a mapping matched to the capability of a channel to reliably carry information, can be combined into the condition $H < R_s = nR_c < nC$. If, without loss of generality, $n = 1$, then the above condition becomes $H < R < C$, where $R_s = R_c \overset{\text{def}}{=} R$. This condition leads to two questions. Is the condition $H < C$ necessary and sufficient for the transmission of a given source over a channel with arbitrarily small

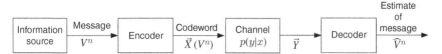

Figure 1.13 Distilled view of the message communication process with joint source and channel coding.

probability of error to be realizable? And how does one approach the design of a code with rate R, such that $H < R < C$, which succeeds in losslessly representing the source and matching the channel capacity so that the source transmission has an arbitrarily small probability of error? These questions are addressed by Shannon's *source–channel coding theorem* and its implications.

Consider the setup in Figure 1.13. A discrete source associated with a finite alphabet $\mathcal{V}$ generates a random sequence of symbols V^n. The sequence is encoded into a codeword $\vec{X}(V^n)$ that is transmitted through a DMC of capacity C and characterized with the probability distribution $P(\vec{Y}|\vec{X})$, where $\vec{Y}$ is the received codeword. This received codeword is decoded to obtain an estimate $\hat{V}^n$ of the original message. Shannon's source–channel coding theorem states that when the sequence V^n satisfies the AEP and $H(\mathcal{V}) < C$, there exists a source–channel code for which it is possible to send the source message through the channel with an arbitrarily small probability of error (probability that $\hat{V}^n \neq V^n$). Conversely, when $H(\mathcal{V}) > C$, there is no code that would make possible the transmission of the source message with an arbitrarily small probability of error.

Shannon's source–channel coding theorem states that the condition to be able to transmit a stationary ergodic source with an arbitrarily small probability of error is that its entropy rate needs to be less than the channel capacity. The proof of achievability for the theorem has important implications for the subject of this book. In the proof, we first consider that since the source sequence V^n satisfies the AEP, there will be a typical set $A_\epsilon^{(n)}$ of probability $P[A_\epsilon^{(n)}] > 1 - \epsilon$ formed by $2^{n(H(\mathcal{V})+\epsilon)}$ distinct length-n sequences. Next, it is assumed that only the sequences in the typical set are encoded. Sequences not in the typical set are discarded. This will introduce an error with probability at most ϵ (the probability of the set that is the complement of the typical set). To encode the sequences in the typical set, they are listed and indexed. The lossless representation of the index can be achieved using $n(H(\mathcal{V}) + \epsilon)$ bits because there are at most $2^{n(H(\mathcal{V})+\epsilon)}$ sequences to be indexed. We can now draw from the noisy channel coding, where it was assumed that the source alphabet was a set of indexes and assert that we can transmit the source message, an index, with a probability of error no larger than ϵ if the rate of the code used, R bits per channel use, is less than the channel capacity, or, by including the representation of a transmitted index, if $H(\mathcal{V}) + \epsilon = R < C$. As the length n of the sequence grows to infinity, the probability of error introduced when encoding the source (at most ϵ) and the probability of transmission error (also at most ϵ) go to zero. As a consequence, messages from the source can be transmitted with an arbitrarily low probability of error when $H(\mathcal{V}) < C$.

The proof of achievability for the source–channel coding theorem gravitates around the existence of a typical set with properties as described in the AEP. The proof applies the AEP in two instances. The first is the lossless encoding of the source using at most $n(H(\mathcal{V}) + \epsilon)$ bits. The second instance uses the joint AEP to ensure that if the code rate R is less than the channel capacity C, it will be possible to achieve an arbitrarily small probability of error during transmission. The two instances of applying the AEP are separate because they do not depend on any common information quantity; the first instance depends on the statistical characteristics of the source, and the second depends on the channel statistics (as given in terms of the input and output of the channel). This application of the AEP in two separate instances implicitly introduces a design for the communication system where the message undergoes two separate encoding operations before transmission. The first encoding, which we have called *source encoding*, performs a representation of the message that is as compact as possible while being lossless. The second encoding, which we called *channel encoding*, represents the source-encoded message in such a way that the transmitted codeword will have a probability of almost 1 of being jointly typical with the received sequence, and, thus, there will be a very high probability of decoding the received sequence with no errors. While there is nothing new in what we are describing here (these concepts were described earlier in the chapter, see Figure 1.1), the important new implication that follows from the source–channel coding theorem (and its proof) is that *there is no performance penalty to be paid* when doing the encoding in two separate stages: a first stage optimized for source encoding and a second stage tuned to undoing transmission errors during decoding. This principle – there is no performance penalty in designing source and channel coding separately – is known as the *source–channel separation theorem*.

Separating the design of source and channel coding subsystems has been prevalent in the development of communication systems. This prevalence has been motivated more by considerations of system modularity, flexibility, and design simplicity, rather than by an interest in taking advantage of the lack of a performance penalty promised by the theorem, because it is generally acknowledged that the separate design of source and channel coding does entail a performance penalty for practical communication systems. The lack of a performance penalty stated in the source–channel separation theorem is achieved at the hidden cost of using ideal codes of infinite complexity and delay because they operate on sequences of infinite length.

Furthermore, codes operating on very large sequences may not turn out to have such ideal high performance under practical circumstances. A practical source may change its statistical characteristics over time frames shorter than those associated with a source encoder input sequence, turning an initially optimal source coder design into a suboptimal one. For example, a speech source could change its statistical characteristics over a few tens of milliseconds, which might correspond to no more than two thousand bits. Similarly, a wireless channel may change its statistical characteristics on time frames shorter than those associated with a channel encoder input sequence, turning an initially optimal channel coder design into a suboptimal one. Depending on the transmitter–receiver velocities, the coherence times for a wireless channel may be as small as a few milliseconds, and typical values may be close to a hundred milliseconds.

Also, it might not be the best design approach to strive for a source encoder capable of extracting all redundant information in the source, achieving the most compact representation possible. The resultant source-encoded data would be extremely fragile to transmission errors because every part is indispensable for the reconstruction of the source at the receiver. Residual source redundancy that remains in the source-encoded data stream could be useful in dealing with transmission errors at the receiver. Moreover, separating the design of source and channel coding subsystems entails another non-apparent performance penalty in terms of the rate that an optimal code's probability of error decreases with its block length – also known as the *error-exponent*. It has been shown in [15] that even for discrete memoryless sources and channels, the joint source–channel codes can achieve as much as twice the rate of error-exponent decrease as a separate source and channel coding design, even in the regime of an infinite input sequence length. We will explore all of these concepts in great detail later in this book.

As seen in the foregoing discussion, Shannon's source–channel coding theorem follows from combining the implications of the AEP in the lossless source coding theorem and the joint AEP in the channel capacity theorem. We can also combine in a similar way the joint AEP in the rate-distortion theorem with the joint AEP in the channel capacity theorem and obtain a *lossy source–channel coding theorem*. Assume that a discrete source associated with a finite alphabet $\mathcal{V}$ generates a random sequence of symbols V^n. The sequence is encoded into a codeword $\vec{X}(V^n)$ that is transmitted through a DMC of capacity C and characterized with the probability distribution $P(\vec{Y}|\vec{X})$, where $\vec{Y}$ is the received codeword. This is decoded to obtain an estimate $\hat{V}^n$ of the original message. In this system, it is now assumed that the encoding and decoding operations perform following a rate-distortion function $R(D)$. Let

$$D = E[d(V^n, \hat{V}^n)] = \frac{1}{n} \sum_{i=1}^{n} E[d(V_i, \hat{V}_i)] \qquad (1.67)$$

be the average *end-to-end distortion* resulting from the combined source and channel coding as in Figure 1.13. The lossy source–channel coding theorem states that the distortion D is achievable if and only if $R(D) < C$. Of course, since higher rates lead to smaller distortions, the theorem could be regarded as establishing the smallest source distortion that could be achieved for a channel of a known capacity. Together, the lossless and lossy source–channel coding theorems say what source distortion can be achieved over a channel. If the channel capacity is larger than the source entropy, source distortion will be zero. If the channel capacity is less than the source entropy, lossy compression of the source message will be needed leading to some distortion.

In (1.67), we introduced the concept of end-to-end distortion that measures the difference between the sequence of symbols generated by the source and the estimate of those symbols after the message has been decoded by the recipient. In measuring the message difference between the two ends of the communication system, the end-to-end distortion incorporates the effects of the transmission through the channel as well as all the encoding and decoding stages in the processing chain of the communication system. Consequently, the end-to-end distortion is a metric that encapsulates the performance trade-offs between different stages in the communication of a message, including the trade-off within the source–channel coding theorem. Thus, the distortion measure is at the center of most joint source–channel design techniques and will be of frequent use in this book.

The lossy source–channel coding theorem also leads to the definition of the *optimal performance theoretically attainable* (OPTA) curve. The OPTA curve represents the smallest distortion that can be achieved over a channel as a function of the received signal power (or, equivalently, the channel SNR). It is defined as the distortion value D that satisfies

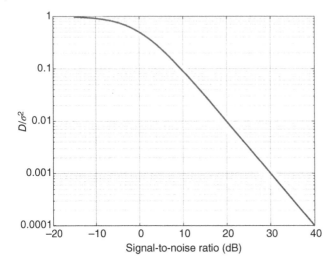

Figure 1.14 The Optimal Performance Theoretically Attainable (OPTA) curve for the Gaussian source over AWGN.

$R(D) = C(P)$. For Gaussian sources over AWGN channels, combining (1.30) and (1.52) through the definition of the OPTA curve (for nonzero rate and channel capacity) yields:

$$\frac{1}{2}\log_2\left(\frac{\sigma^2}{D}\right) = \frac{1}{2}\log_2\left(1 + \frac{P}{N_0 W}\right).$$

Thus, after denoting the channel SNR as $\gamma = P/(N_0 W)$, the OPTA curve is the function:

$$\frac{D}{\sigma^2} = \frac{1}{1+\gamma}. \tag{1.68}$$

where the distortion is expressed in the customary normalized form D/σ^2. Figure 1.14 illustrates the OPTA curve as expressed in (1.68). More discussion on the OPTA curve can be found in [16].

1.8 Major Classes of Joint Source–Channel Coding Techniques

In the previous section, we discussed a number of circumstances under which Shannon's source–channel coding theorem does not hold. These conditions are common in practical systems and include source and channel dynamic change, and coding complexity and delay constraints. When Shannon's source–channel coding theorem does not hold, there is a performance penalty to be paid when designing the source codec separately from the channel codec, and it is advantageous to follow a *JSCC* design approach. There are many approaches to joint design. We can identify the following classes of JSCC:

- *Concatenated joint source–channel coding*: This approach aims at striking a balance between retaining the advantages of separate source and channel codes design, in terms of modularity, flexibility, and design specialization and at the same time attaining the performance improvement associated with joint source–channel coding design. In this approach, the source codec and channel codec are two separate processing blocks within the communication system, and joint design is realized by jointly determining their configuration parameters. The joint design entails distributing the channel capacity (or achievable throughput) between the source and channel encoders, aiming to minimize end-to-end distortion. Since end-to-end distortion measures the difference between the source message and its reconstruction after undergoing source and channel encoding and decoding, as well as transmission through a channel, it couples the source and channel coding operation by incorporating in a single metric the effects associated with source coding and with transmission errors. This coupling effectively establishes a trade-off between the source and channel coding rates, which is balanced through this JSCC design approach. As a result, the essence of the design in this approach consists in finding the amount of source compression that reduces the bit rate from the source (by eliminating redundancy and possibly introducing some distortion) enough to make room for the controlled addition of structured redundant bits that will help combat transmission errors so that end-to-end distortion is minimized while the resulting overall source and channel coding rate equals the channel achievable throughput.

- *Unequal error protection source–channel coding*: In many scenarios, the output from the source encoder is frequently not uniform but exhibits sections of different impact on the reconstructed source distortion. Some sections of source encoder output may result in large reconstruction distortion after being affected by transmission errors, while under the same circumstances other sections may yield small reconstruction distortion. Unequal error protection source–channel coding incorporates the nonuniform characteristics of the source-encoded output into the design by applying channel codes with stronger protection against transmission errors to those sections of the source-encoded output that would result in large distortion if received with transmission errors, and applying weaker channel codes to sections that, if received with errors, have a small impact on the reconstructed source distortion. Typical examples of more sensitive parts of compressed sources are control or synchronization information or portions where errors may lead to error propagation. This type of paradigm is also useful when the channel is changing or not fully known at the instant of source/channel encoding.

- *Constrained joint source–channel coding*: This class encompasses JSCC techniques where the source codec design incorporates constraints that account for the transmission of the source-encoded codeword over a noisy channel. One JSCC technique within this class is channel-optimized quantizer design, where the distortion introduced by transmission errors is included within the performance metrics used to design a quantizer. In another approach to constrained JSCC, the source encoder is designed so that its output presents an inherent resiliency against channel errors. Constrained JSCC techniques are not limited to the source encoder design; they also include receiver-side techniques that exploit the fact that under conditions when the source–channel separation theorem does not hold, the source and channel decoders do not behave as completely independent entities and can benefit from information exchange.

- *Analog and hybrid digital–analog JSCC*: In this case, the communication system does not use digital symbols and, instead, applies the analog model of the channel across the source–channel design.

References

1 Shannon, C.E. (1948). A mathematical theory of communication. *Bell System Technical Journal* 27: 379–423.

2 Shannon, C.E. (1949). Communication in the presence of noise. *Proceedings of the IRE* 37 (1): 10–21.

3 Hartley, R.V.L. (1928). Transmission of information. *Bell System Technical Journal* 7 (3): 535–563.

4 Cover, T. and Thomas, J. (1991). *Elements of Information Theory*. Wiley.

5 Oppenheim, A.V. and Schafer, R.W. (2013). *Discrete-time Signal Processing*. Pearson.

6 Manolakis, D.G. and Ingle, V.K. (2011). *Applied Digital Signal Processing: Theory and Practice*. Cambridge University Press.

7 Vetterli, M., Kovačević, J., and Goyal, V.K. (2014). *Foundations of Signal Processing*. Cambridge University Press.

8 Goldsmith, A. (2005). *Wireless Communications*. Cambridge University Press.

9 Zimmermann, H. (1980). OSI reference model - the ISO model of architecture for open systems interconnection. *IEEE Transactions on Communications* 28 (4): 425–432.

10 Gross, J., Klaue, J., Karl, H., and Wolisz, A. (2004). Cross-layer optimization of OFDM transmission systems for MPEG-4 video streaming. *Computer Communications* 27 (11): 1044–1055.

11 Saul, A. and Auer, G. (2009). Multiuser resource allocation maximizing the perceived quality. *EURASIP Journal on Wireless Communications and Networking* 2009: 1–15.

12 Kwasinski, A. and Kwasinski, A. (2015). Integrating cross-layer LTE resources and energy management for increased powering of base stations from renewable energy. *2015 13th International Symposium on Modeling and Optimization in Mobile, Ad Hoc, and Wireless Networks (WiOpt)*, pp. 498–505.

13 Hanhart, P. and Ebrahimi, T. (2014). Calculation of average coding efficiency based on subjective quality scores. *Journal of Visual communication and image representation* 25 (3): 555–564.

14 Mohammadi, F.S. and Kwasinski, A. (2018). QoE-driven integrated heterogeneous traffic resource allocation based on cooperative learning for 5G cognitive radio networks. *2018 IEEE 5G World Forum (5GWF)*, pp. 244–249.

15 Zhong, Y., Alajaji, F., and Campbell, L.L. (2006). On the joint source-channel coding error exponent for discrete memoryless systems. *IEEE Transactions on Information Theory* 52 (4): 1450–1468.

16 Berger, T. (1971). *Rate Distortion Theory*. Prentice-Hall.

2

Source Coding and Signal Compression

In Chapter 1, we introduced source coding as a fundamental block in a communication system. In this chapter, we take a deeper dive into source coding. Some of the concepts introduced in Chapter 1 are succinctly explained again to maintain a cohesive presentation within this chapter. After discussing different types of sources, the chapter explains some important lossless and lossy source coding techniques and concludes by explaining how these techniques are used in the source coding of images, video, and speech signals.

2.1 Types of Sources

A source is the characterization of an element or object that generates information to be transmitted. Many sources, such as speech or audio, generate a continuous-time wave-form for which *sampling* is usually needed to enable subsequent encoding. In sampling, the amplitude of the waveform is sampled (measured) at regular time intervals to generate a sequence of measurements, called *samples*. Under idealized conditions (that is, when the input waveform is bandlimited and the sampling operation is performed using ideal devices with infinite bandwidth), the samples could be obtained with a frequency as small as, but not smaller than, twice the maximum frequency of the input signal, without losing any useful information. The sequence of measurements resulting from the sampling operation is called a discrete-time signal. The values of the discrete-time signal are not defined at times other than the sampling instants. The sample values may be continuous or discrete. Prior to the digital revolution, most sources had continuous values. Today, this is still possible as many physical signals (for example, speech and music) are originally analog. After sampling, such a signal results in a sequence of continuous-valued samples. This is known as a *continuous source*. Figure 2.1 illustrates these points with a speech signal. Looking at the complete signal on the left, the signal seems to be continuous time, but a closer examination of any part of the signal, shown on the right, reveals that the signal is discrete time, that is, a sequence of samples. However, while the signal is discrete time, since the samples are continuous-valued, the source in Figure 2.1 is a continuous source.

The output of a continuous source is often modeled as a sequence of random variables $X = \{X_0, X_1, X_2, \ldots\}$. If the source were to generate a sequence of deterministic values, the source would not convey any information and its communication would be unnecessary.

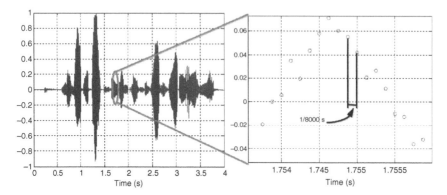

Figure 2.1 A speech signal as an example of a continuous source.

Since the different random variables that constitute the source output sequence may be correlated, the joint probability density function (PDF) of these random variables is needed to characterize the source output statistically. Knowing the joint PDF of a random continuous source for the entire sequence is the best case scenario for design and analysis because all the information about the stochastic behavior of the source is captured in the joint PDF. When this is not possible, it is still desirable to know as many statistics of the source as possible. The marginal PDF for each of the random variables (samples) at the source output is also of value to understand the behavior of a source. For example, Figure 2.2 shows the PDF of a speech source sample. At the very least, it is usually necessary to know the mean and variance of a source. The covariance between source samples is another useful statistic. Even when the PDF is known, design and analysis considerations make it convenient to approximate the PDF of a physical source, such as speech, with the PDF of well-studied random variables. For instance, the PDF of a speech source could be approximated by the PDF of a Gamma or Laplace probability distribution [1].

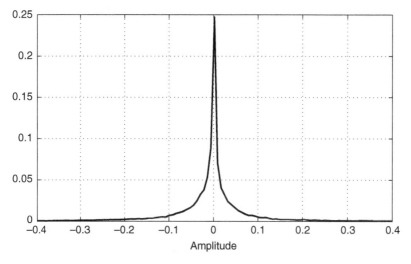

Figure 2.2 The probability density function of a speech source sample.

Synthetic sources are mathematical models of sources, which do not necessarily exist in the physical world but present statistics and characteristics that are of value for understanding a communication system. They are often used in research for analytical work or for algorithm development with readily reproducible benchmarks. An example is the *Gaussian source* which generates a sequence of random samples that follow a Gaussian distribution. Frequently, it is also assumed that the samples are uncorrelated from each other, in which case the source is called an uncorrelated source (or more precisely a time-uncorrelated source). This means that each of the source output samples, for example the *i*th sample X_i, can be described using the PDF of a Gaussian distribution with mean μ and variance σ^2:

$$f_{X_i}(x_i) = \frac{1}{\sqrt{2\pi\sigma^2}} e^{-\frac{(x_i-\mu)^2}{2\sigma^2}}.$$

The Gaussian synthetic source has been thoroughly studied and is a helpful choice for analytical studies. The uncorrelated Gaussian source conveys an upper bound on the achievable rate-distortion performance for source encoding and compression (representation). This means that in order to efficiently represent a sample and meet some maximum distortion level, the Gaussian source would require the largest number of bits of all continuous sources with same mean and variance.

A *discrete source* presents to the communication system a sequence of samples, each taking values from a finite set, called an *alphabet*. For example, a digital imaging sensor may render an image through pixels, each taking discrete values in the range from 0 to 255 (for the case of eight-bit representation). A discrete source can also be obtained from a continuous source by representing the measurement for each sample using a limited number of bits. The discrete source output sequence is modeled as a sequence of random variables that are completely characterized through the joint probability mass function (PMF). It is common to consider synthetic discrete sources that may approximate with varying degrees of accuracy the properties of actual sources. For example, the Bernoulli source generates an output with two possible values, which are represented using values $x = 0$ or $x = 1$. For this source, the PMF characterizing the probability for each output follows a Bernoulli distribution, i.e. $P_X(X = 1) = p$ and $P_X(X = 0) = 1 - p$.

Based on the correlation between successive samples, sources can be divided between *memoryless sources* and *sources with memory*. The random variables in a sequence generated by a memoryless source are independent of each other. This means that it is possible to express the joint probability distribution of the source output as the product of the marginal distributions. That is, for a continuous memoryless source, it is possible to write

$$f_{\bar{X}}(x_1, x_2, x_3, \ldots) = \prod_i f_{X_i}(x_i),$$

and for a discrete memoryless source, we can write

$$P_{\bar{X}}(x_1, x_2, x_3, \ldots) = \prod_i P_{X_i}(x_i).$$

Figure 2.3 summarizes the types of sources we have discussed. There is one more useful differentiation to make. Many sources generate an output that changes with time, and the source samples can be plotted on a graph with time as the single independent variable. We identify such a source as one-dimensional. Examples are speech and audio.

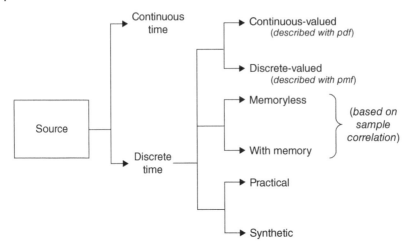

Figure 2.3 Source types.

Other sources, such as images, are functions of two independent variables (typically spatial variables) and are considered two-dimensional sources. Video is three-dimensional as it is a time sequence of two-dimensional pictures (called frames).

The rest of this chapter will focus on the different operations performed by a source codec, specifically quantization and compression. We will not consider sampling (the conversion of a continuous-time signal into a discrete-time signal) further. For our purposes, in most of this book we consider the sampling operation as taking place inside the source, before the generation of messages.

2.2 Lossless Compression

Before the output from a source can be transmitted, it is usually necessary to compress its representation and add protection against transmission errors. In this chapter, we address compression whose goal is to represent the source output using fewer bits than what would be used in an uncompressed representation. This operation is called *coding*. Formally, the coding operation is the mapping of each symbol at the output of the source to a binary sequence. Each valid binary sequence is called a *codeword*, and their collection is called a *code*. The symbols at the output of the source, which may be samples, are usually called *source symbols* or *letters*, and their collection as a set is called an *alphabet*. Coding is the assignment of a codeword from a predesigned code to a source symbol or letter (or a group of them) from a source's alphabet. The mapping operation from source symbol to codeword is performed by an *encoder* and is called *encoding*. The reverse operation of mapping a codeword to a value approximating the encoded source letter (or letters) is performed by a *decoder* and is called *decoding*. The combined operation of encoding and decoding is also called *coding*, and the combined encoder plus decoder pair is known as a *codec*.

With *reversible* mapping, by decoding any codeword it is possible to unambiguously recover the source letter that was originally mapped into the codeword. When the codec operation allows unambiguous recovery of the source letter that was originally at the input to the encoder, there is no loss of information in the codec operation and there is no distortion introduced. In this case, the coding approach is called *Lossless Compression*. We will first discuss an important family of lossless compression techniques called *Entropy Coding*. With *irreversible* mapping, the decoded codeword is a symbol that approximates a set of source letters. In this case, it is not possible to know which of the letters approximated by the decoded codeword was the one that was originally encoded into the codeword. Some information is lost in the encoding–decoding process. Generating an approximate value at the decoder introduces distortion into the recovered source at the codec output, so this coding approach is called *Lossy Compression* and will be discussed later in this chapter.

2.2.1 Entropy Coding

Entropy coding is a family of lossless coding techniques where the code mapping is designed based on the probability of occurrence of the different letters in the source's alphabet. The general idea to achieve compression is to assign shorter codewords to letters that are more probable to occur and longer codewords to letters that are less probable to occur. In this way, the average rate achieved by the code is less than what would be achieved by assigning codewords of the same length to all the letters.

One of the most popular entropy coding techniques is Huffman coding, named after its inventor in 1951. For a source for which the probability of each letter is known, the Huffman code can be generated with the aid of a tree. Consider a source that generates five possible letters, denoted "A," "B," "C," "D," and "E" whose respective probabilities of occurrence are 0.3, 0.25, 0.25, 0.1, and 0.1, respectively. Figure 2.4 illustrates the construction of the tree used to develop the Huffman code for this source. The tree is constructed from a first layer of leaves (left-most dots in Fig. 2.4), where each leaf is one of the letters organized from top to bottom in order of decreasing probability of occurrence. The tree design begins with the two letters with lowest probability of occurrence; the value 0 is assigned to one and the value 1 to the other. Then the two letters are combined into a single new letter with probability of occurrence equal to the sum of the probability of the two constituent letters. The new combined letter and the other letters form a new source alphabet. The new alphabet is organized in a new layer of the tree, with one node corresponding to each letter, again organized from top to bottom in order of decreasing probability. Branches in the tree connect the nodes from the previous layer to the new one. A branch from each of the two letters that were combined connect to the new letter.

After forming the new layer in the tree, the procedure of assigning values 0 and 1 to the letters of lowest probability and combining them and organizing a new tree layer is repeated until only two letters are left. At this point, a value of 0 is assigned to one of the letters and a value of 1 is assigned to the other. This last assignment becomes the root of the tree (on the right side of Figure 2.4). The codeword for each of the original letters is obtained from the final tree, by traversing the branches that lead to each original letter starting from the root and reading the bit values assigned to each branch. Usually, the assignment of the values 0 and 1 to the branches is done in a consistent way; if in the first assignment the value 0 is assigned to the top branch, then the top branch in every subsequent assignment will also be assigned a value of 0.

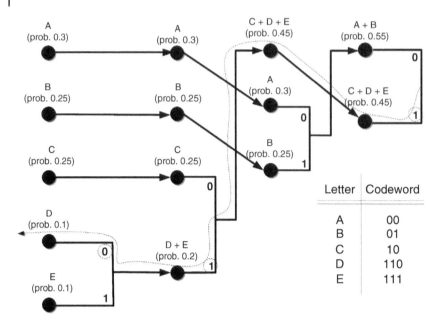

Figure 2.4 Example design of a Huffman code.

In Figure 2.4, the algorithm starts by assigning a value 0 to the letter "D" and a value 1 to the letter "E," followed by combining the two letters into a new 'D+E' letter which has a probability of occurrence equal to 0.2. Because this probability of occurrence is less than the probability of any other letter, the sorting of the new alphabet is, from top to bottom, "A," "B," "C," and "D + E." Next, the algorithm assigns a value 0 to the letter "C" and a value 1 to the letter "D + E." The letters "C" and "D + E" are combined into a new letter "C + D + E" with probability of occurrence $0.2 + 0.25 = 0.45$. The new alphabet composed by the letters "A," "B," and "C + D + E" is sorted in order of decreasing probability of occurrence. After this, the two letters with lowest probabilities are "A," with probability 0.3, and "B," with probability 0.25. Then, the letter "A" is assigned a value 0 and the letter "B" a value 1. After the assignment, "A" and "B" are combined into a new letter "A + B," with a probability of occurrence 0.55. After sorting the resulting alphabet ("A + B," with a probability of occurrence 0.55 and "C + D + E" with probability of occurrence 0.45), the final assignment results in the letter "A + B" being assigned a value 0 and the letter "C + D + E" being assigned a value 1. This last assignment concludes the construction of the tree. Figure 2.4 shows with a dashed line how the codeword for the letter "D" is built by reading the values assigned at each stage as they are encountered in the path from the tree root (on the right) to the leaf for the letter "D." For the letter "D," the result is 110.

We can now see how the Huffman code achieves compression. Consider first the case where the source is encoded into codewords of constant length. In order to differentiate five different letters we will need three bits, so we can use a mapping of codeword "000" for letter "A," "001" for letter "B," "010" for letter "C," "011" for letter "D," and "100" for letter "E." Since this code uses three-bit codewords in all cases, it has a rate of three bits/symbol.

For the case of the Huffman code in Figure 2.4, the average rate is:

$$R = 0.3 \times 2 + 0.25 \times 2 + 0.25 \times 2 + 0.1 \times 3 + 0.1 \times 3 = 2.2 \text{ bits/symbol,}$$

so the compression ratio is $3/2.2 = 1.36$ times. We can calculate the entropy of the source and compare it with the rate of the Huffman code:

$$H = -0.3\log_2(0.3) - 0.25\log_2(0.25) - 0.25\log_2(0.25) +$$
$$-0.1\log_2(0.1) - 0.1\log_2(0.1)$$
$$= 2.1855.$$

We know from information theory that 2.1855 bits/symbol is the minimum rate that could be achieved for lossless compression. The Huffman code is only 0.0145 bits/symbol away from the achievable minimum. It has been shown that any Huffman code will be within $0.086+p_M$ bits/symbol from the limit set by the entropy value, where p_M denotes the probability of the most likely source letter.

Since the codewords in a Huffman code are of variable length, they will need to have a special property to allow the decoder to know when each codeword starts and ends to be able to separate a stream of several codewords into individual ones. This *prefix property* is needed by all codes with variable length codewords, and the codes that have it are called *prefix codes*. A prefix code is one where no codeword is a prefix (appears at the start) of any other codeword. This property can be verified in the code shown in Figure 2.4. The digits "00" (codeword for "A") do not appear at the beginning of any other codeword, and "01" (codeword for "B") does not appear at the beginning of any other codeword, etc. To see why the prefix property is needed, consider a modified version of the Figure 2.4 code, where we use a mapping of codeword "00" for letter "A," "01" for letter "B," "10" for letter "C," "100" for letter "D," and "111" for letter "E." This code has the same codeword lengths and so achieves the same representation efficiency of the source (same bit rate). It is a non-prefix code because the codeword "10" is a prefix of the codeword "100." If the decoder were to receive the bitstream "1000111100," there is no way for the decoder to know whether the input to the encoder was "DBEA" or "CAED."

While Huffman codes have a rate close to the minimum for lossless compression set by the entropy of the source, there is room for improvement. Information theory tells us that the efficiency of a code (the compression ratio) can be improved by having longer and longer blocks of source symbols as input to the encoder. For Huffman coding, we can achieve further compression by considering a code designed for letters formed by a block of, say, two source symbols. This increases complexity as the design of the Huffman code would require us to calculate the probability and derive the code for every possible combination of the source symbols taken two at a time (5^2 cases in the example). This problem with Huffman codes (increasing compression efficiency at the expense of increasing input block length and complexity) does not arise with another class of codes, known as *arithmetic codes*.

Arithmetic coding is an entropy coding technique designed to encode strings of many source symbols together. It is based on the idea of dividing a line segment into sub-segments of lengths proportional to the probabilities of each individual source symbol. During encoding, numbers in the interval $[0,1)$ are expressed as fixed point binary numbers. Numbers in the range [0,1) can be expressed as binary numbers using the same rules as for natural numbers with the difference that negative (instead of positive) powers of 2 are used to represent the number. In this way, the most significant (the left-most) binary digit of a positive fractional binary number has to be multiplied by $2^{-1} = 0.5$, the next binary digit to the right has to be multiplied by $2^{-2} = 0.25$, and the following binary digit to the right has to be multiplied by $2^{-3} = 0.125$, etc. This number representation implies that the decimal point is located immediately to the left of the most significant bit. For example, the binary number "01000001" is equal to $0 \times 2^{-1} + 1 \times 2^{-2} + 0 \times 2^{-3} + \cdots + 0 \times 2^{-7} + 1 \times 2^{-8} = 0.25390625$.

Arithmetic coding is based on the easily verifiable principle that any interval of length w, such that $2^{-k} \le w < 2^{-(k-1)}$ for a positive integer k, contains a number whose binary representation is a terminating binary string of length k bits or fewer. For example an interval [0.211, 0.713) of length 0.502, which satisfies $2^{-1} \le 0.502 < 2^0$, must contain a number with binary representation of $k = 1$ bit, which, in this case is 0.5. If we partition the unit interval into subintervals, then each of the subintervals can be represented by these binary strings of potentially different lengths. The idea of arithmetic coding is, first, to map a sequence of symbols into a unique segment of unit interval whose length is equal to the probability of the sequence and then to encode that interval into a binary string using the aforementioned principle. We will explain the encoding and decoding procedure through an example using a source that can output six different possible letters, denoted as symbols "A" through "F," which have probabilities of occurrence 0.4, 0.25, 0.15, 0.1, 0.07, and 0.03.

Let us consider arithmetic encoding of a message formed by the source string "ABE". Figure 2.5 illustrates the following process: We divide the line segment [0,1) into six subsegments with lengths equal to the six source symbol probabilities. Typically, when dividing the segment, the subsegments are positioned in decreasing order of associated probability. Therefore, in our example, the unit interval is divided into the subsegments [0, 0.4) corresponding to the probability of "A", [0.4, 0.65) because the probability of "B" is 0.25, [0.65, 0.8) corresponding to "C," [0.8, 0.9) corresponding to "D", [0.9, 0.97) for "E," and [0.97, 1) for "F". As "A" is the first symbol in the string "ABE," we select the subsegment

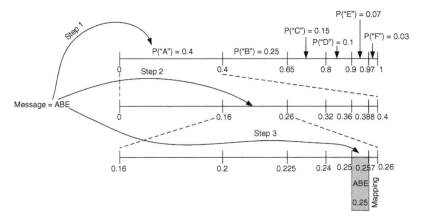

Figure 2.5 Example arithmetic encoding.

[0, 0.4) corresponding to "A." Getting ready for encoding the next letter in the message, the segment [0, 0.4) is in turn subdivided into six subsegments with lengths proportional to the probabilities of the six possible source symbols, following the same rules as for the first segmentation. The new subsegments are [0, 0.16), [0.16, 0.26), [0.26, 0.32), [0.32, 0.36), [0.36, 0.388), and [0.388, 0.4). The second encoding step will encode the second symbol in the string, "B," so we select the second subsegment: [0.16, 0.26). Next, in the third step, the subsegment [0.16, 0.26) is in turn subdivided into six subsegments with lengths proportional to the probabilities of the six possible source symbols. The new subsegments are [0.16, 0.2), [0.2, 0.225), [0.225, 0.24), [0.24, 0.25), [0.25, 0.257), and [0.257, 0.26). Since we now want to encode the symbol "E" we choose the fifth subsegment, [0.25, 0.257), because it is the one with length proportional to the probability of "E" (7% of the total segment length of 0.1 is 0.007, which is the length of the selected subsegment).

Since "E" was the last symbol in the message, we have now finished the nested segmentation process. As a result of these operations, we end up with the segment [0.25, 0.257). The complete string "ABE" has been mapped to the segment [0.25, 0.257) and the binary representation of any number within this range may be used as a codeword for the string "ABE." While any number in the range can be used, certain numbers require more significant bits to be represented than others (for example, $0.25 = $ "01" and 0.25390625="01000001"). As we stated earlier, this is an interval of length 0.007 and $2^{-8} \leq 0.007 < 2^{-7}$; hence, it must have a number with terminating binary expansion not more than eight bits long. In this case, it just happens to be 0.25, whose binary expansion is 0.01. Therefore, we can choose the codeword for string "ABE" as 0.01.

On the decoder side, if we know the number of symbols encoded, in this case 3, and the codeword, the same process is essentially traced back to discover the encoded sequence, as follows. The decoder receives the codeword corresponding to the binary representation of 0.25. The decoding operation also advances in stages, starting by segmenting the unit interval into the six subsegments previously described [0, 0.4), [0.4, 0.65), [0.65, 0.8), [0.8, 0.9), [0.9, 0.97), and [0.97, 1). Since 0.25 belongs to the subsegment [0, 0.4), which corresponds to the probability of the source letter "A," the decoder generates the first output symbol: "A." Next, the segment for the letter "A" is subdivided into its six subsegments: [0, 0.16), [0.16, 0.26), [0.26, 0.32), [0.32, 0.36), [0.36, 0.388), and [0.388, 0.4). Since 0.25 belongs to the subsegment [0.16, 0.26) corresponding to the symbol "B," the decoder outputs "B" as the second symbol of the stream. Finally, the segment [0.16, 0.26) is subdivided into its six subsegments: [0.16, 0.2), [0.2, 0.225), [0.225, 0.24), [0.24, 0.25), [0.25, 0.257), and [0.25, 0.257). Since 0.25 belongs to the subsegment [0.25, 0.257) corresponding to the symbol "E," the decoder outputs "E" as the third symbol of the stream and completes the decoding process.

An alternate decoding procedure that does not requires calculating nested segmentations uses rescaling. At the start of decoding, we consider the original six segments [0, 0.4), [0.4, 0.65), [0.65, 0.8), [0.8, 0.9), [0.9, 0.97), and [0.97, 1). The received codeword value 0.25 belongs in the first segment, so the decoder outputs "A" as before. After decoding "A," its effect is removed from the received value 0.25 by rescaling within the range 0 to 1. Specifically, 0.25 is to the segment [0, 0.4) as 0.25/0.4 = 0.625 is to the segment [0,1). We pretend next as if 0.625 were the codeword to be decoded. Of the original six subsegments [0, 0.4), [0.4, 0.65), [0.65, 0.8), [0.8, 0.9), [0.9, 0.97), and [0.97, 1), 0.625 belongs to the segment for the symbol "B," so the decoder outputs this symbol. Next, we rescale 0.625. Since 0.625

is to the segment $[0.4, 0.65)$ as $(0.625 - 0.4)/0.25 = 0.9$ is to the segment $[0,1)$, we consider 0.9 as the next value to be decoded. The decoder outputs the symbol "E" because, of the six segments $[0, 0.4)$, $[0.4, 0.65)$, $[0.65, 0.8)$, $[0.8, 0.9)$, $[0.9, 0.97)$, and $[0.97, 1)$, 0.9 belongs to the segment corresponding to "E" ($[0.9, 0.97)$).

As described earlier, the encoder reads all the input symbols in order to determine which subinterval of the unit interval the source string corresponds to. In practice, the encoder does not need to read in the entire string before beginning to output codeword bits. At any encoding stage, if the two end points of the current subinterval agree in the first m bits of their binary representation, then any value inside that subinterval will also have the same first m bits in its binary representation. So the encoder can output those m bits and then examine more source symbols and segment the interval farther. As with any entropy coder, the goal is to use short binary codewords to represent frequently occurring input symbols or strings, and long binary codewords to represent infrequent occurrences, so as to reduce the average length of the representation. The intuition behind why arithmetic coding achieves this goal lies in the fact that the subsegment lengths are taken to be proportional to the source symbol probabilities. Therefore, highly probable input symbols correspond to large subsegments, and it is less likely that a large subsegment will have both of its end points agree for many bits in their binary representation, which would cause the encoder to output bits. Highly unlikely source symbols correspond to tiny subsegments, so when they occur, the subsegment is likely to have both of its end points agree for many bits in their binary representations, so the encoder will need to output many bits. It can be shown that the average length of codewords achieved by arithmetic coding of N symbols approaches N times the entropy of the IID source.

2.2.2 Predictive Coding

Sources can be either memoryless or with memory. For a time-dependent signal, samples of a source's random process at two time instants are two random variable. If the value of one random variable depends to some extent on another random variable, then we say they are correlated. This correlation may appear as source samples that are close in time showing little change in value. Both speech and audio sources display correlation because the physical mechanisms by which they are generated create signals that change smoothly over time. Periodic components in the signal introduce another form of correlation between samples. This type of dependence can be found in speech and audio signals also since many of their sounds are based on the superposition of multiple harmonics of a fundamental sinusoidal oscillation. Similar statements could be made for sources depending on spatial variables, such as images. Video sources may exhibit sample correlation both because of time correlation between frames that occur close in time, and because of spatial correlation between neighboring samples. In general, correlation between samples can be exploited for compression levels higher than independent coding of the samples. This is also motivated by the result summarized as "conditioning reduces entropy," that is, $H(X|Y) \leq H(X)$ for any random variables X and Y with equality only if X and Y are independent.

Predictive coding is a family of techniques used to represent correlated source samples. The correlation between samples translates into predictability for sample values, which is exploited to perform lossless compression. Consider the signal in Figure 2.6, showing

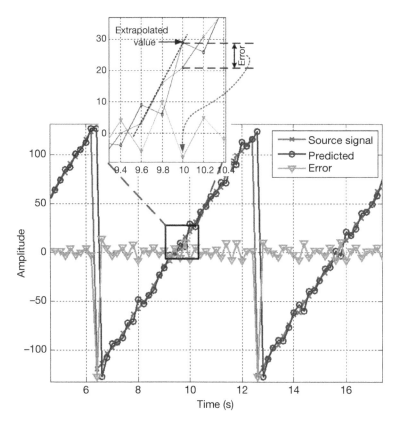

Figure 2.6 Predictive coding; the insert illustrates the prediction of one sample by using linear extrapolation of the two previous samples.

the output from a source that, while shown as continuous for ease of viewing, is in fact a discrete time and value source, with the samples represented by crosses on the curve. The source output follows a sawtooth function with an added small noise component. The smooth behavior of the source output function translates into small and, for the most part, predictable differences between successive source samples. As such, one can devise an algorithm to predict the following source sample value by using recent past samples. Assume now that the next sample is predicted through simple linear extrapolation of the two most recent samples. The result of this prediction is shown in Figure 2.6 as the curve labeled "predicted." A figure insert shows the mechanism to predict a sample and shows how the difference between the actual sample value and the predicted value results in an error. The errors for each source output sample form, over time, an error signal, as also shown in Figure 2.6. With a better prediction, the error signal will be smaller.

Why does prediction help with compression? Suppose x_n is the next source sample to be encoded losslessly, and $f(x_{n-1}, x_{n-2})$ is the prediction function which uses the past two samples to make a prediction of x_n. Then the encoder can calculate the prediction error or residual as: $e_n = x_n - f(x_{n-1}, x_{n-2})$. The encoder can then transmit e_n losslessly using Huffman coding or arithmetic coding or any other lossless coding approach. The decoder, having already losslessly decoded x_{n-1} and x_{n-2}, can compute $f(x_{n-1}, x_{n-2})$, and then, upon

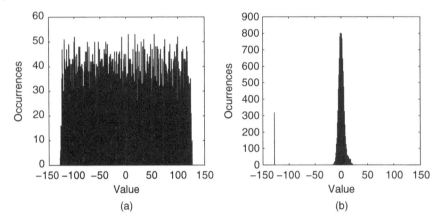

Figure 2.7 Histograms for (a) source output and (b) error signal.

receiving the lossless e_n, the decoder can reconstruct $x_n = e_n + f(x_{n-1}, x_{n-2})$. Of course the encoder could have just sent x_n losslessly in the first place, instead of e_n, so the advantage of doing this prediction approach comes down to which one of e_n or x_n takes more bits to represent.

In general, when x_n is highly correlated with the previous samples being used for prediction, an accurate prediction can be made, and this results in an error signal that can be represented using few bits. Intuitively, a good error signal would often have values equal to 0, or close to 0. Values far from 0 would be rare. Figure 2.7 illustrates this for the sawtooth example. The left plot is the histogram of the source output which is close to a uniform distribution. The right side shows the histogram of the error signal; the distribution has many values concentrated around 0 (there are also some large prediction errors corresponding to the sawtooth function discontinuities). The entropies of the source output and the error signal give an indication of how many bits will be needed for lossless representation. For the source output, the entropy is 7.96 and for the error signal it is 4.44. This reduction in entropy is due to the use of prediction to change the statistics of the source signal from roughly uniform to highly peaked.

2.3 Lossy Compression

2.3.1 Quantization

Digital systems, whether computing devices or communications systems, cannot represent internally a continuous-valued (analog) sample. Thus quantization is the first step in a digital communication system required to transmit a continuous source. With quantization, each continuous-valued sample or group of continuous-valued samples is represented using a discrete value. In general, *quantization* is the representation, or mapping, of a large (and possibly infinite, for the case of continuous-valued samples), set of sample values into a smaller set of values. Because it involves the representation of sample values with more precision into values of less precision, quantization involves a loss of precision. This loss is usually considered as the introduction of some error, called *quantization error*. Since this error works as an additive distortion to the original signal, it is common to think of it as an added noise and call it *quantization noise*.

Quantization can be done one sample at a time, known as *Scalar Quantization*, or it can be done on groups of samples, called *Vector Quantization*. We will discuss next these two options.

Scalar Quantization

In scalar quantization, the input to the quantizer is a *single* sample taking values from a large, and possibly infinite, set of values. A scalar quantizer maps an input value into an output value, called a *codeword* or *codebook index*. It is generally a many-to-one mapping; a range of source sample values map to a single codeword. The codewords generated as a result of this mapping belong to a set of smaller size than the set of possible source values. On the receiver side of a communication chain, the reverse operation to quantizing takes place. This operation, called codeword decoding or sample reconstruction, also performs a mapping. It takes as input the codeword and outputs a value representative of the range of values to which the codeword corresponds.

A simple example of scalar quantization is a pair of analog-to-digital and digital-to-analog converters (ADCs and DACs). On the quantization side, the ADC maps a range of sample values from an analog source into a digital word (the codeword) that uses a number of bits that depends on the resolution. For typical ADCs, the ranges of values that map to different codewords are generated by dividing the full range of the analog source into 2^N intervals of equal size, where N is the number of bits used in the codeword. To perform reconstruction, the DAC maps the codeword into a value that belongs to the original interval that the codeword represented. Figure 2.8 illustrates this. On the left side, the figure shows an analog source. The full range of values for the source is from $-A$ to $+A$. This range is divided into eight intervals of equal size $2A/8$. Each interval is mapped to a three-bit codeword. On the right side, the figure shows how each interval, and the associated codeword, is mapped during reconstruction to a representative value. In the example, each interval is mapped during reconstruction to the interval midpoint.

Figure 2.8 shows the example of a sample with a value close to $-0.5A$, perhaps equal to $-0.45A$, that is first quantized into the codeword associated with the interval $(-0.25A, -0.5A]$: 110. On the reconstruction side, the codeword 110 is mapped to a value of $-0.375A$. The end-to-end result is that the original value of $-0.45A$ is reconstructed as $-0.375A$. In fact, all source sample values in the interval $(-0.25A, -0.5A]$ will map to the codeword 110 and be reconstructed as $-0.375A$. This illustrates that quantization and reconstruction introduce quantization error. In this example, only if the source sample were equal to an interval midpoint such as $-0.375A$, would the quantization error equal 0.

Quantization performance and the impact of quantization error are usually measured by calculating the mean distortion. Let x be the continuous random variable representing the source sample values and let $Q(x)$ be the reconstructed value of the quantized sample. If distortion is measured using the mean squared distortion metric, the mean quantization distortion can be calculated as:

$$D_q = \int_{-\infty}^{\infty} (x - Q(x))^2 f_X(x) dx, \tag{2.1}$$

where $f_X(x)$ is the PDF of x. This expression shows that D_q is the quantization error variance when using the mean squared distortion metric. Because $Q(x)$ is discontinuous, the mean quantization distortion can be written as:

$$D_q = \sum_{i=1}^{M} \int_{y_{i-1}}^{y_i} (x - q_i)^2 f_X(x) dx, \tag{2.2}$$

where M is the total number of quantization intervals, y_{i-1} and y_i are the limits of the ith quantization interval, and q_i is the reconstruction value corresponding to that quantization interval. Equation (2.2) provides the quantization distortion. If the codewords all have the same length, the quantization rate is equal to $R = \lceil \log_2 M \rceil$. This rate can be reduced by resorting to some of the coding techniques discussed in this chapter.

The example in Figure 2.8 shows the case where all quantization intervals have the same size, a *uniform quantizer*. Uniform quantization has best performance when the source samples follow a uniform distribution, that is, when $f_X(x)$ is a constant over all the support for x. When x is not a uniform random variable, the average quantization distortion can be reduced by configuring the quantizer with intervals of different size, called *nonuniform quantization*. The design intuition for a nonuniform quantizer is to assign smaller interval size (finer resolution) for the more probable range of source values, and within each interval to assign reconstruction values so that the distortion metric is minimized. In this pdf-optimized quantization design [2], the reconstruction values are calculated by setting the derivative of (2.2) with respect to q_i to 0 and solving to find the average quantization distortion minimizer,

$$q_i = \frac{\int_{y_{i-1}}^{y_i} x f_X(x) dx}{\int_{y_{i-1}}^{y_i} f_X(x) dx}. \tag{2.3}$$

This calculation results in each reconstruction value being the "center of mass" of each interval, for the "mass" distributed according to the PDF density. Following a similar approach, and setting the derivative of (2.2) with respect to y_i to 0, it can be found that

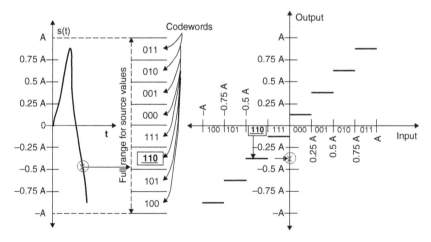

Figure 2.8 A uniform scalar quantizer.

the interval limits are given by the middle point between two neighboring reconstruction levels [2]. This is,

$$y_i = \frac{q_{i+1} + q_i}{2}. \tag{2.4}$$

The condition for this relation to hold is that the distortion measure is monotone and non-decreasing. The calculation of the reconstruction values and the interval limits can be achieved through Lloyd–Max's iterative algorithm. Algorithm 2.1 summarizes Lloyd's algorithm. Figure 2.9 illustrates the design of three-bit scalar quantizers and the output of Lloyd–Max's algorithm for two different source PDFs. On the top, the figure shows the PDF and eight reconstruction values (shown as stars) for a standard normal distribution. The bottom presents results for a uniform distribution. The optimal design for a uniform PDF has quantization intervals of equal size (the reconstruction values are separated by a fixed distance). When the input samples follow a nonuniform probability distribution, the quantization intervals are of unequal size, with smaller intervals for the range of input values that are more likely.

When studying the distortion-rate performance of scalar quantizers, expressions that are both analytically tractable and insightful are often derived by assuming that the number of quantization intervals M is very large. This is called a *high rate approximation*. Under this assumption, the source sample x can be assumed to be uniformly distributed over each of the quantization intervals because the quantization intervals are very small relative to the range of values for x. This is, within the interval y_{i-1} to y_i, $f_X(x) \approx p(q_i)/\Delta_i$, where $p(q_i)$ is the probability that the sample value will be in the interval y_{i-1} to y_i, and $\Delta_i = y_i - y_{i-1}$ is the width of the interval (which may be different depending on the value of x in a nonuniform quantizer).

When the range of values for x is infinite, but there is a finite number of quantization intervals, the source samples in the first and last quantization intervals are forced to saturate to the maximum and minimum source sample quantization values (e.g. $+A$ and $-A$ in Figure 2.8). In this case, the approximation for a uniformly distributed source sample value over a quantization interval is not truly applicable for the first and last quantization intervals, because these intervals are not small. Therefore, the high rate approximation is applied with the complementary assumption that the probability of the source sample values being within the first or last quantization interval is very small. For example, it could be assumed that A is chosen so that $A > 4\sigma_X^2$ for a symmetric PDF $f_X(x)$ where σ_X^2 is the variance of x.

Algorithm 2.1 Lloyd's algorithm

1: Choose an initial set of intervals limits $\{y_1, y_2, \dots, y_{M-1}\}$ with $y_0 = x_{min}$ and $y_M = y_{max}$.
2: **while** Current quantization distortion D_q is not stable (changes non-negligibly from previous iteration) **do**
3:     Calculate reconstruction values q_i using (2.3).
4:     Calculate intervals limits y_i using (2.4).
5:     Calculate quantization distortion D_q using (2.2).
6: **end while**

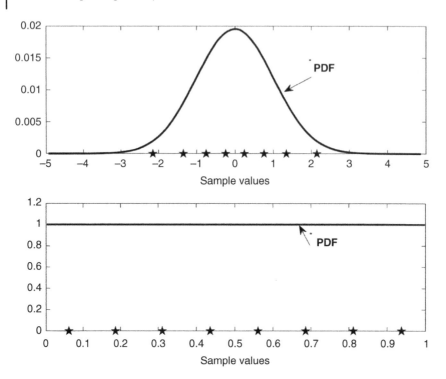

Figure 2.9 Example of the reconstruction levels of a scalar quantizer for two different PDFs.

Applying all these considerations, the distortion given by (2.2) can be approximated as:

$$D_q \approx \sum_{i=2}^{M-1} \frac{p(q_i)}{\Delta_i} \int_{y_{i-1}}^{y_i} (x - q_i)^2 dx$$

$$= \sum_{i=2}^{M-1} \frac{p(q_i)}{3\Delta_i} [(y_i - q_i)^3 - (y_{i-1} - q_i)^3].$$

From (2.3), the optimal value for q_i is the center of mass for the corresponding quantization interval. Using the assumption of uniformly distributed source samples within each interval, q_i is just the midpoint of the interval: $q_i \approx (y_{i-1} + y_i)/2$. Substituting this expression for q_i into the latest approximation for D_q yields

$$D_q \approx \sum_{i=2}^{M-1} \frac{p(q_i)}{12\Delta_i} (y_{i-1} - y_i)^3.$$

With the further assumption that the quantization intervals (except for possibly the first and last) have the same width: $\Delta_i \approx \Delta$, the approximation for the quantization distortion now becomes

$$D_q \approx \frac{\Delta^2}{12} \sum_{i=2}^{M-1} p(q_i) \approx \frac{\Delta^2}{12}, \tag{2.5}$$

where the last approximation is because $p(q_1)$ and $p(q_M)$ were assumed very small. The result in (2.5) is the quantization distortion for a uniform quantizer under the high rate approximation assumptions. This result is expressed as a function of the quantization interval width, Δ. To obtain an expression for D_q as a function of the coding rate, we need to express Δ as a function of the coding rate, which becomes a calculation specific to each quantizer design. Assuming a uniform quantizer designed following the previous example $A = 4\sigma_X^2$ and a symmetric PDF, there are $M - 2$ quantization intervals between $-4\sigma_X^2$ and $+4\sigma_X^2$. Then, $\Delta = 8\sigma_X^2/(M - 2)$, or, since M is large, $\Delta \approx 8\sigma_X^2/M$. As the source coding rate is $R = \lceil \log_2 M \rceil$ bits per sample, we can write $M = 2^R$. Applying these results in (2.5) yields

$$D_q \approx \frac{(8\sigma_X^2/M)^2}{12} = \frac{16}{3}\sigma_X^2 2^{-2R}.$$ (2.6)

Comparing this result with the distortion-rate expression (1.31) presented in Chapter 1, we see that (2.6) shows that the quantization distortion for a uniform scalar quantizer is 16/3 times larger than the distortion achieved with the optimal quantizer.

The approximate result for D_q in the high rate regime shown in (2.6) can be used to derive a popular approximate expression that provides a sense of the relation between the quantized signal quality and the number of quantization bits used. In this expression, the quantized signal quality is seen through the quantization signal-to-noise ratio, which is the ratio of the signal power to the noise power measured in decibels, $SNR_q = 10\log(\sigma_X^2/\sigma_e^2)$ [dB], where σ_e^2 is the quantization error variance. Recalling that $D_q = \sigma_e^2$, and using (2.6), the quantization signal-to-noise ratio can be written as:

$$SNR_q \approx 10\log(\sigma_X^2) - 10\log\left(\frac{16}{3}\sigma_X^2 2^{-2R}\right)$$
$$= 20R\log(2) - 10\log(16/3)$$
$$= 6.02R - 7.27 \quad (dB),$$

which tells us that the quantization signal-to-noise ratio can be improved by 6 dB for each quantization bit that is added.

Vector Quantization

In contrast to scalar quantization, a vector quantizer maps a block of samples to an output codeword. While the scalar quantizer operates by mapping a range of values from the input set into a codeword, a vector quantizer maps a portion of a surface in an N-dimensional space, where N is the number of input samples grouped together into a block. Figure 2.10 illustrates this operation when $N = 2$.

The end-to-end operation of the vector quantizer is as follows. First, the source samples are grouped into blocks of N samples. Each block is considered as a vector in an N-dimensional space, so each sample is considered as a coordinate of a vector. To perform vector quantization, the vector is compared against code-vectors that are stored in a table called the *codebook*. The code-vectors are vectors that represent all those vectors in the N-dimensional subspace formed from all possible combinations of source sample blocks. The input source vector $\vec{X}$ is compared against the code-vectors to find the closest. An index that uniquely identifies the closest code-vector is transmitted to the receiver. If $\vec{Y}_i$ denotes the code-vector with index i, the vector quantization operation can be summarized as finding the index i such that

$$\|\vec{X} - \vec{Y}_i\|^2 \leq \|\vec{X} - \vec{Y}_j\|^2, \qquad \forall \vec{Y}_j \in \text{codebook},$$ (2.7)

where the distance between vectors is typically calculated using the 2-norm, defined as:

$$\|\vec{X}\|^2 = \sum_{n=1}^{N} X_n^2,\tag{2.8}$$

for a vector with component X_n on dimension n.

On the receiver side, reconstruction is performed by using the transmitted codebook index as input to find the corresponding code-vector. The code-vector approximates the quantized block of source samples. As with scalar quantization, the difference between the code-vector and the input vector implies an error, or distortion, in the source reconstruction. The quantization operation of searching for the code-vector closest to the input vector implicitly divides the N-dimensional space into M subsets, called Voronoi regions, formed by those points that are closest to each of the M code-vectors. If we use $\mathcal{V}_i$ to denote the subset of points closest to code-vector $\vec{Y}_i$, the quantization distortion (assuming mean-squared metric) can be calculated as:

$$D_{vq} = \sum_{i=1}^{M} \int \cdots \int_{\mathcal{V}_i} \|\vec{X} - \vec{Y}_i\|^2 f_{\vec{X}}(x_1, x_2, \ldots, x_N) dx_1 dx_2 \ldots dx_N.\tag{2.9}$$

As with a scalar quantizer design, the vector quantizer design tries to find the set of code-vectors that minimize the quantization distortion. One approach is the Linde–Buzo–Gray (LBG) algorithm, shown as Algorithm 2.2, which iteratively updates code-vectors and Voronoi regions. The code-vectors are found based on the prior knowledge of the joint probability distribution of input vector components, or if this is not known, using a set of training input vectors. Once the code-vectors have been found, the algorithm finds the Voronoi regions for each code-vector (the subset of points closest to the code-vector). Figure 2.11 shows an example of Voronoi regions in a two-dimensional space. For this figure, the code vectors were generated at random with coordinates chosen from a standard normal distribution.

In terms of quantization efficiency, a vector quantizer will outperform an equivalent scalar quantizer. For the same level of quantization distortion, a vector quantizer will use on average a smaller quantization rate than the scalar quantizer. Equivalently, at the same quantization rate, the distortion from the vector quantizer will be on average less than that of the scalar quantizer. The operation of the vector quantizer uses as input longer sequences

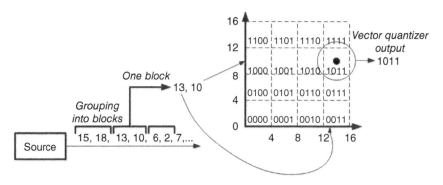

Figure 2.10 A two-dimensional vector quantizer.

Algorithm 2.2 Linde–Buzo–Gray (LBG) algorithm

1: Choose an initial set **Y** of code-vectors $\{Y_1, Y_2, \ldots, Y_{M-1}\}$.
2: **while** Current quantization distortion D_{vq} is not stable (changes non-negligibly from previous iteration) **do**
3:     Given **Y**, calculate the boundaries for the subsets of points closest to each code-vector, $\mathcal{V}_1, \mathcal{V}_2, \ldots, \mathcal{V}_M$,

$$\mathcal{V}_i = \{X : \|X - Y_i\|^2 \le \|X - Y_k\|^2, \forall k \ne i\}.$$

4:     Given the subsets $\mathcal{V}_i$, calculate the optimal reproduction points Y_i equal to (for squared error distortion),

$$Y_i = E[X|X \in \mathcal{V}_i].$$

5:     Calculate quantization distortion D_{vq} using (2.9).
6: **end while**

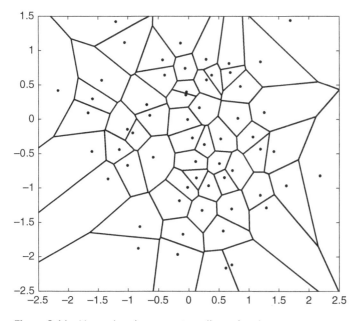

Figure 2.11 Voronoi regions on a two-dimensional space.

of values, and we know from information theory that longer input sequences yield more efficient representations. This is due to a number of factors. The first is related to nature of space and distortion metric. The relative gain, namely reduction in distortion, of using an n-dimensional vector quantizer to a scalar quantizer for uniform distribution is called *packing gain*, which is a property of the space and the distortion metric. The second form of gain is available to vector quantizers more than to scalar quantizer but is valid even for I.I.D. sources. It arises from the shape of the probability distribution function for nonuniform or non-Gaussian sources and is called "shaping gain" and is a result of nonuniform quantization. The third form of gain of vector quantization vs. scalar quantization arises from efficient use of correlation or memory in the source. If the different components of the input vector are correlated, the vector quantizer will be able to represent them using code-vectors more closely located where input vectors occur more frequently. However, vector quantizers are more complex than scalar quantizers, especially during the search for the code-vector closest to the input (Eq. (2.7)). To address this issue, a number of codebook designs have been proposed to reduce the search complexity by introducing structure into the codebook design. An example of this approach is tree-structured vector quantization [3].

2.3.2 Differential Coding

Closely related to predictive coding is differential coding, which can be seen as using a lossy quantizer to encode the error or difference signal resulting between a source sample and its predicted value. Figure 2.12 shows, on top, a differential encoder. It has two main components: a predictor and a quantizer. If the input source sample is $s(n)$ and its predicted value is $\hat{s}(n)$, then the error signal is $e(n) = s(n) - \hat{s}(n)$ and the transmitted signal is the quantized version of this error, $\tilde{e}(n) = Q(e(n))$. This approach, while simple, has an important flaw. To see this, consider that quantization introduces an error. We can write $\tilde{e}(n) = e(n) + \Delta_q$, where Δ_q is the error introduced during quantization. At the receiver, the decoder uses its own prediction of the source sample to recover the original sample from the received quantized error value. Assume the decoder can use the previously decoded source samples to obtain a prediction that is the same as that at the encoder: $\hat{s}(n)$. With this, the source sample reconstruction could be calculated by $\tilde{s}(n) = \hat{s}(n) + \tilde{e}(n) = \hat{s}(n) + e(n) + \Delta_q = s(n) + \Delta_q$. So the quantization error is transferred to the reconstruction of the source sample. Initially, this might not appear to be a problem because quantization error appears in all lossy compression methods. The problem here is that the reconstructed sample is usually used to predict future samples at the receiver. So the assumption that the decoder is able to obtain the same prediction $\hat{s}(n)$ of the sample that the encoder has is not realistic. For prediction, the encoder uses samples unaffected by quantization, whereas the decoder uses samples affected by quantization. The solution is to use during encoding the same reconstructed

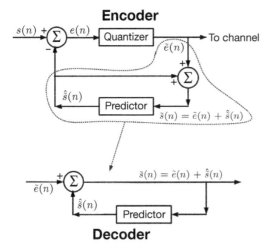

Encoder

Figure 2.12 Differential encoder and decoder.

Decoder

sample values that will be available at the decoder. This is achieved by running at the encoder a copy of the decoder, as shown in Figure 2.12. With the new configuration, the predictor at both the encoder and decoder is run using as input the source sample reconstructions that are derived from the quantized error signals.

2.3.3 Transform Coding

Differential coding is based on a powerful idea to achieve efficient signal compression: before quantizing, modify the sequence of samples so their new representation is better suited for effective quantization and compression. This idea is further developed and formalized through the concept of *Transform Coding*. In transform coding, the block of source samples is first transformed to obtain a new representation. When designed appropriately, the transformed source samples exhibit properties that make them more apt for efficient quantization.

To illustrate transform coding, Figure 2.13 shows 200 samples from a source. Each star symbol is an outcome from the underlying random process driving the source output; each sample is a two-dimensional vector $\vec{x}$ with coordinates x_1 and x_2. The coordinates of each vector (sample) follow a bivariate Gaussian distribution for which the joint PDF is:

$$f_{\vec{X}}(x_1, x_2) = \frac{1}{2\pi\Sigma} e^{-(1/2)(\vec{x}-\vec{\mu})^T \Sigma^{-1}(\vec{x}-\vec{\mu})},$$

where $\vec{\mu}$ is the two-dimensional vector with the mean of each vector coordinate μ_1 and μ_2,

$$\Sigma = \begin{bmatrix} \sigma_1^2 & \rho\sigma_1\sigma_2 \\ \rho\sigma_1\sigma_2 & \sigma_2^2 \end{bmatrix}$$

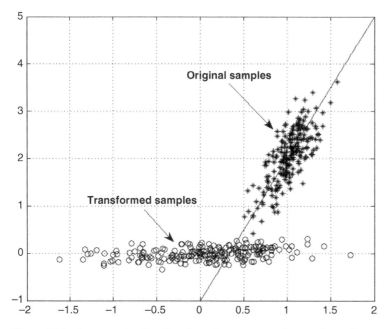

Figure 2.13 An example of coding through source sample transformation

is the covariance matrix, ρ is the correlation coefficient, and σ_1^2 and σ_2^2 are the variances for the two coordinate variables. In Figure 2.13, the values are $\mu_1 = 1$, $\mu_2 = 2$, $\rho = 0.7$, $\sigma_1 = 0.2$, and $\sigma_2 = 0.6$. The source statistics are such that the samples are clustered along the line shown in the figure. In transform coding, the samples are transformed before quantization. We choose to transform the samples, so they will cluster around the abscissa with zero mean on both orthogonal directions. Assuming that the first- and second-order statistics for the source samples are known, our transformation can be achieved by first subtracting the two-dimensional mean vector $\vec{\mu}$ from each sample and then performing a clockwise rotation with an angle equal to that of the line shown in Figure 2.13. This angle is $\theta = \arctan(\rho\sigma_2/\sigma_1)$. The transformed sample $\vec{\check{x}}_i$ is calculated by:

$$\vec{\check{x}}_i = A(\vec{x}_i - \vec{\mu}),$$

(2.10)

where

$$A = \begin{bmatrix} \cos\theta & \sin\theta \\ -\sin\theta & \cos\theta \end{bmatrix}$$

is the rotation matrix. The result of transforming each of the samples following this operation is shown in Figure 2.13 with the transformed samples represented as circles.

Since the transform operation is followed by quantization, the goal of the transformation is to obtain a new representation of the source samples that allow for a more efficient use of coded bits. For example, in Figure 2.13, while the coordinates before transformation span a range of values of two units for x_1 and and four units for x_2, the coordinates after transformation span a range of values of four units for $\check{x}_1$ and one unit for $\check{x}_2$. In this example, if we were to use four bits to represent each coordinate in all cases, the values for the coordinate $\check{x}_1$ would be coded with the same quality as the values for x_2, and the values for $\check{x}_2$ would be coded with twice the resolution of those for x_1. The coded transformed source samples would show less distortion than the samples coded without the transformation. For this example, the mean squared error between the original samples and those encoded without transformation using four bits, equals 0.16. When transforming the source samples, the distortion drops to 0.13, an improvement of approximately 18%. Use of transform coding provides either a distortion reduction for a given number of bits, or a bit reduction for a given distortion level.

To reconstruct the source samples at the decoder, it is necessary to perform the reverse operations to those done during encoding. After mapping the quantized values to their reconstructed values, it is necessary to apply the inverse transform to that done during encoding. For the example in Figure 2.13, the inverse transform is done as:

$$\vec{x}_{r_i} = A^{-1}\vec{\check{x}}_{r_i} + \vec{\mu},$$

where

$$A^{-1} = \begin{bmatrix} \cos\theta & -\sin\theta \\ \sin\theta & \cos\theta \end{bmatrix}.$$

In general, if the transform operation is represented as a matrix, as in (2.10), where A is the transform matrix, the condition for transform reversibility can be written as $A^{-1}A = c$, where c is a scalar constant. It is usually preferable to have $A^{-1}A = 1$, as this avoids scaling during reconstruction.

To faithfully recover the source samples, it is necessary to be able to reverse the applied transformations. In the example in Figure 2.13, the operations in the transformation are a translation and a rotation, both reversible. In principle, any reversible transform may be useful for transform coding. In practice, transforms into frequency domains, such as the discrete cosine transform (DCT), are very popular, as will be seen later. The transformation of source samples does not need to be driven by the source statistics. While this is an important consideration, in many cases, especially for multimedia sources, the transformation may aim at a representation related to properties of human perception. For example, the human visual system is more sensitive to distortion in the lower spatial frequencies of an image (gradual variations in intensity across an image, such as gradual shading across a sky) than to the higher spatial frequencies (many abrupt changes in intensity close together, such as grass texture or hair). Because of this, the transforms used in image coding aim to separate the lower spatial frequency components of the image from the higher ones.

2.3.4 Subband and Wavelet Coding

Subband coding separates the input signal into complementing frequency *bands*. As an idealized example, consider a source signal with bandwidth B split into two parallel paths. One path passes the signal through an ideal low-pass filter with bandwidth $B/2$. The other path passes the signal through an ideal high-pass filter with bandwidth also $B/2$. The filter output in each path is a signal with half the bandwidth of the input signal and therefore, under the ideal conditions in this example (and assuming a translation of the high-pass spectrum to baseband frequencies), they can be represented using half the number of samples (by sampling at half the original frequency). At this point, the signal is represented by two groups of samples. Each group has half the original number of samples. One group contains the low-pass component of the original signal and the other the high-pass component.

This can be generalized to the multirate analysis/synthesis subband codec shown in Figure 2.14. Here, the input signal is passed through a bank of M complementary ideal filters. Called *analysis filters*, they decompose the input signal with bandwidth B into M signals, each of which is a component of the original signal over a frequency band with width B/M. The signals at the output of each analysis filter are subsequently resampled at a sampling frequency M times slower than the original one, an operation called *downsampling* or *decimation*. The downsampling operation is implemented simply by keeping only the samples with index kM from the output of the filters, where $k = 0, 1, 2, 3, \dots$. In this way, the combined number of samples at the output of all the analysis filters remains the same as the number of samples in the input source signal.

After filtering, each of the M signals is encoded using any of the methods explained in this chapter. The structure of the subband encoder, which separates the signal spectrum into complementary ideally nonoverlapping passband signals, allows for considerable flexibility in encoder design. The encoders could be designed to exploit the properties of each of the passband signals. For example, each encoder may use a different number of bits to represent its source, with the number of bits being determined by the energy level of the corresponding passband signal, its statistical properties, or how humans perceive distortion in each subband. The flexibility provided by the subband coding structure, combined

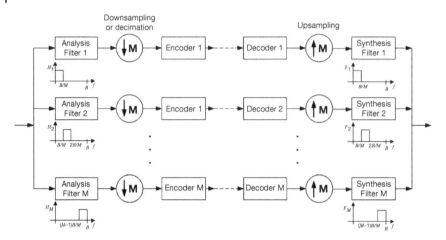

Figure 2.14 Ideal subband codec.

with the coding techniques described in this chapter, results in efficient signal compression algorithms.

After encoding, the output from all the encoders is usually transmitted or stored in a file. The decoder first decodes each of the M signals. Each signal then has $M - 1$ samples interpolated between each pair of consecutive samples, an operation called *upsampling*. The interpolation involves inserting $M - 1$ zero-valued samples in between consecutive existing samples and then filtering the resulting signal with *synthesis* filters which are, under the ideal filtering conditions assumed so far, identical to the analysis filter in the same processing branch. Following the synthesis filtering, the M resulting passband signals, with complementary spectra, are added together to obtain a reconstruction of the original source.

Up to this point, all the filtering operations were considered ideal: they completely eliminate any frequency components of the input signal not in the passband and leave unchanged those frequency components in the passband. In practice, the signal at the output of any analysis filter will have attenuated frequency components outside the passband, components that will overlap with the passband of signals at the output of other analysis filters (especially those with neighboring bands). The main problem with this is that the extra spurious bandwidth of the analysis filters output will lead to aliasing distortion during subsampling. The solution to this is to design matching synthesis filters with a frequency response that compensates for the aliasing effects. More information on filter design can be found in [4].

It is worthwhile to pause to reflect on frequency domain transforms and analysis. For transforms into the frequency domain, the most apparent choice is a Fourier transform, such as the discrete Fourier transform (DFT) or the DCT (to be revisited in the context of image coding). Both of these transforms are usually applied on blocks of input samples, and so the implementation of the DFT can be thought of as being the application of the short-term Fourier transform (STFT). When using the STFT, the segmentation of the input signal into blocks implies the use of a rectangular windowing operation. A rectangular window has the drawback of having significant energy leakage besides the main analysis frequency band. If this effect is too detrimental for the intended application, then other slightly more sophisticated windows could be used to segment the source into blocks. For any windowing operation, the uncertainty principle holds: the shorter a window is made (which is good to detect fine signal details in the time domain), the wider the analysis frequency band becomes (which is not good to detect fine details in the frequency domain). In other words, if we would like to apply a window to faithfully see details of the signal in time, with good resolution, the details of the signal representation in frequency will appear with increasingly less resolution. Analogous effects can be observed in terms of subband decomposition, especially when considering nonideal filters. Because of this, it would be useful to find a transform that allows control of the window size and thereby control of the trade-off between time and frequency resolution. If possible, it would be desirable to implement this transform as a filter bank that could be configured to have different resolutions for different parts of the spectrum. As a multiresolution analysis filter bank, it would be possible to devise signal coding mechanisms that can efficiently represent the main features of a signal while introducing less distortion due to the uncertainty principle effects. As it turns out, this technique exists and is known as the *discrete wavelet transform* (DWT) and *Wavelet Packet Decomposition*.

In a Fourier transform, the basis function used for transformation is a sinusoid with infinite duration in time. In a wavelet transform, the basis function used for transformation is a windowed sinusoid. With wavelets, the size of the window (time length of the basis function) is changed without modifying the fundamental frequency. The decomposition is implemented as a bank of cascaded analysis filters. The output from the analysis filter is followed by decimation by a factor of 2, resulting in the wavelet coefficients for the wavelet-transformed input signal. The idea is shown in Figure 2.15 where only the analysis part of the codec is shown (as with subband coding, each output from the analysis section is followed by an encoder). In the figure, the first stage applies a pair of complementary low-pass and high-pass filters. The filter transfer functions are shown as a function of the normalized frequency, where a normalized frequency of one equals half the sampling rate. This stage is called a first-level discrete wavelet decomposition. The analysis into only two subbands is usually not sufficient; a second level of wavelet decomposition is implemented by applying the same low-pass and high-pass filters (in effect, wavelet transforms) to the low-pass and high-pass subbands from the previous stage. In between the filters at each stage, there is a decimation operation by a factor of 2. The halving of the sampling rate also changes the frequency scale, halving it. This is why the second-stage filters process only in the frequency band between zero and one-half, following the first-stage low-pass filter. Similarly, the second-stage filters after the first-stage high-pass filter process only in the frequency band between one-half and one. As can be seen, the analysis structure can be expanded to further stages with each stage resulting in narrower and narrower subband analysis filters. This is frequently needed in order to achieve higher compression ratios. When the filters are applied to both the low-pass and high-pass outputs at each stage as in Figure 2.15, the structure is known as a wavelet packet decomposition.

It is possible to design the analysis section, so some subbands are more finely analyzed in frequency and others more coarsely, by adding more analysis stages only to some bands. For example, if the lower ranges of frequencies for a signal are wanted with finer frequency analysis, then the analysis branch corresponding to the low-pass filters can selectively have

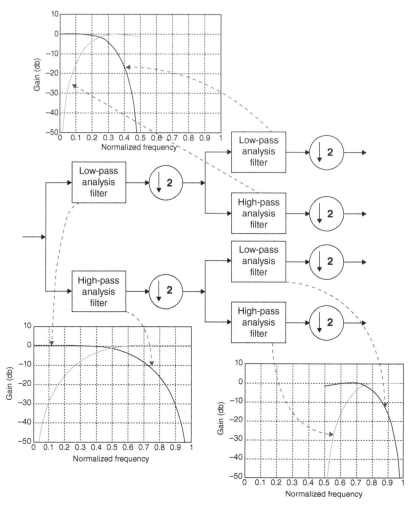

Figure 2.15 Two-level wavelet analysis and the resulting transfer functions for the cascaded analysis filters.

more stages of analysis filters. This structure is shown in Figure 2.16 and is known as a DWT.

2.4 Embedded and Layered Coding

Embedded or progressive coding refers to a procedure for encoding a source into a bit stream that can be truncated at any point and still be decoded. The technique is illustrated in Figure 2.17. A block of source samples is input into the source encoder, which puts out a stream of source-coded bits. These bits have an implicit order of importance where earlier bits in the stream are more important. The transmitter can stop transmitting the bit stream at any point, in which case the bit stream is truncated, and the bits coming after the

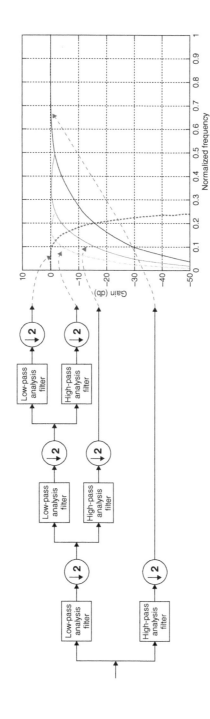

Figure 2.16 Transfer functions resulting from the cascaded filters (cascaded wavelet transforms) at different subbands. Lower-frequency subbands are analyzed with finer frequency resolution.

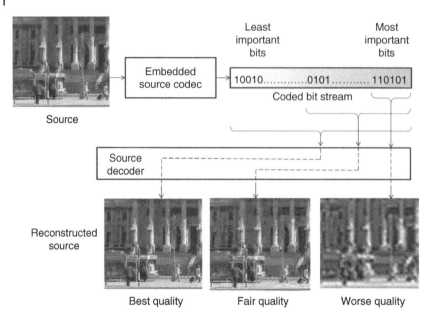

Figure 2.17 Embedded coding.

truncation point are unavailable at the decoder. The decoder can still obtain some level of reconstruction of the source, but with lower reconstruction quality if the stream was truncated earlier. A bit stream that results from embedded coding is said to have the embedded property. The embedded property does not come for free; embedded codecs usually show some loss in coding efficiency relative to non-embedded codecs because of the constraints involved in putting more important information first and making the stream decodable from any truncated portion. Unless the source has some property that naturally lends itself to embedded coding, it is usually necessary to accept some loss in coding efficiency in order to obtain an embedded coded bit stream.

In an ideal embedded codec, the bit stream can be truncated at any point. Some encoding techniques, called *Layered Coding*, result in bit streams that can be truncated only at a discrete number of points, each resulting in a different level of quality. In practice, truncation at any point is rarely needed, and it is usually possible to have the discrete number of possible truncation points large enough. Figure 2.18 illustrates layered coding. In layered coding, the representation of a source is implemented in a number of layers. If we number the layers from 0 to $L - 1$, the first layer (layer 0) is called the *Base Layer* and is different from the other layers in that it encodes the source with a very coarse resolution and low quality (large reconstruction distortion). Remaining layers, often called *enhancement layers*, encode the difference (error) between the original source and the source decoded from all the previous layers. In this way, each layer beyond the base layer adds a finer description of the errors that had been introduced through encoding up to that point. For decoding, the error signals from the enhancement layers (however, many are transmitted to achieve a desired target distortion or rate) are decoded and added to the reconstructed source from the base layer.

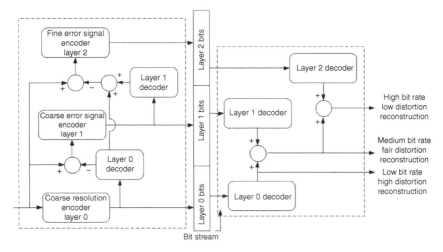

Figure 2.18   A three-layer encoder and decoder.

2.5   Coding of Practical Sources

Having described many of the techniques used to code information sources, we will next present an overview of how these techniques are used to compress practical sources. Coding of practical sources often means combining more than one coding technique.

2.5.1   Image Coding - JPEG

The most popular image compression codec to date is the JPEG image compression standard. JPEG stands for the Joint Photographic Experts Group, a joint committee between the standards organizations ISO/IEC and ITU-T (formerly CCITT) which created the compression standard. Figure 2.19 shows a block diagram of the main operations performed during encoding. JPEG image compression combines several of the techniques described in this chapter: transform coding, quantization, and entropy coding. JPEG starts by segmenting the image into square blocks of size 8×8 pixels. The blocks are processed one at a time. A block is first transformed by the two-dimensional DCT. Let B be an 8×8 matrix corresponding to one block from the input image where element b_{ij} is the pixel in row i and column j. Let C be the matrix corresponding to the DCT; its elements c_{ij} are defined as:

$$c_{ij} = \begin{cases} \sqrt{\dfrac{1}{N}} \cos \dfrac{\pi(2j+1)i}{2N} & \text{for } i = 0,\ 0 \leq j \leq N-1, \\[4mm] \sqrt{\dfrac{2}{N}} \cos \dfrac{\pi(2j+1)i}{2N} & \text{for } 1 \leq i \leq N-1,\ 0 \leq j \leq N-1, \end{cases} \tag{2.11}$$

where $N = 8$ for JPEG coding. It can be verified that $C^{-1}C = 1$, that is, the DCT is an invertible transform. Using the DCT matrix C, the transform operation on the 8×8 block can be written as $B_T = CB$, which results in another 8×8 block (matrix) B_T of transform coefficients $b_{T_{ij}}$. The definition of the transform means that the coefficient $b_{T_{00}}$ (called the DC

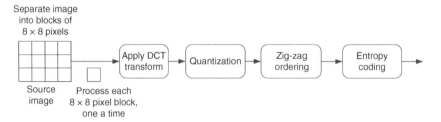

Figure 2.19 Simplified block diagram showing the main operations in JPEG image compression

coefficient) corresponds to the lowest spatial frequency (horizontal and vertical frequencies), and the coefficient $b_{T_{NN}}$ corresponds to the highest frequencies. This organization is exploited during quantization, by matching the quantization resolution for each coefficient to the sensitivity of human visual perception.

Figures 2.20 and 2.21 illustrate the results of applying the DCT to a block primarily with smooth transitions (Figure 2.20) and to a block with higher spatial frequency components in the form of sharper changes in intensity (Figure 2.21). In the case of Figure 2.20, the actual values for the pixels in the input block are:

$$
\begin{vmatrix}
87 & 80 & 72 & 74 & 68 & 66 & 66 & 61 \\
97 & 90 & 80 & 77 & 72 & 70 & 66 & 54 \\
122 & 110 & 96 & 89 & 83 & 73 & 67 & 60 \\
151 & 136 & 118 & 108 & 97 & 85 & 79 & 74 \\
173 & 167 & 151 & 139 & 126 & 115 & 111 & 116 \\
160 & 171 & 171 & 161 & 154 & 142 & 137 & 141 \\
131 & 147 & 155 & 158 & 160 & 149 & 144 & 146 \\
115 & 127 & 140 & 147 & 139 & 135 & 133 & 121
\end{vmatrix}
$$

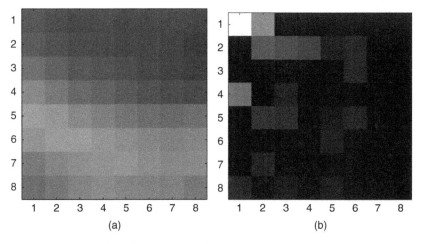

Figure 2.20 (a) An 8-by-8 pixel image with smooth variations and (b) the corresponding two-dimensional discrete cosine transformed block.

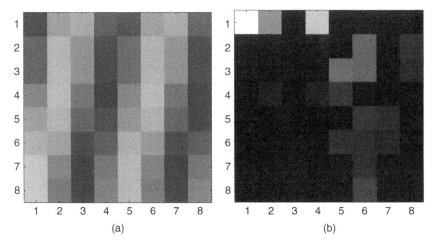

Figure 2.21 (a) An 8-by-8 pixel image with sharp variations and (b) the corresponding two-dimensional discrete cosine transformed block.

After applying the DCT to this block, the result is a block of DCT coefficients:

$$
\begin{array}{|cccccccc|}
913.75 & 95.06 & -6.14 & -0.06 & 1.50 & -0.13 & -5.03 & -0.63 \\
-230.94 & 35.31 & 28.20 & 19.76 & 3.75 & 5.17 & -0.65 & -1.26 \\
-60.98 & -64.93 & -24.97 & -2.11 & -5.34 & 5.03 & -1.56 & -0.79 \\
66.80 & -7.65 & 4.14 & -9.78 & 0.54 & -5.87 & 0.99 & -0.48 \\
-19.50 & 12.23 & 8.44 & 1.78 & 2.25 & 5.11 & -0.52 & -2.16 \\
-1.44 & -6.34 & -2.55 & 0.49 & 3.64 & -2.65 & -0.16 & 1.22 \\
1.71 & 4.35 & -0.56 & -2.64 & -0.49 & 0.93 & 0.47 & 0.39 \\
5.11 & -3.39 & 2.97 & -0.54 & 2.80 & -1.91 & -0.05 & -0.37
\end{array}
\tag{2.12}
$$

For the block with sharper changes (Figure 2.21), the pixel values in the input block are:

$$
\begin{array}{|cccccccc|}
59 & 153 & 168 & 83 & 69 & 154 & 164 & 86 \\
91 & 178 & 145 & 66 & 98 & 173 & 143 & 64 \\
90 & 179 & 146 & 65 & 100 & 174 & 142 & 68 \\
128 & 182 & 116 & 59 & 130 & 180 & 108 & 61 \\
156 & 176 & 98 & 74 & 147 & 172 & 95 & 64 \\
172 & 160 & 71 & 91 & 172 & 149 & 70 & 89 \\
189 & 128 & 61 & 118 & 180 & 124 & 59 & 114 \\
190 & 116 & 63 & 126 & 178 & 115 & 63 & 126
\end{array}
$$

After applying the DCT to this block, the result is a block of DCT coefficients:

$$
\begin{array}{|cccccccc|}
966.00 & 82.62 & -0.99 & 147.64 & -90.25 & -105.10 & -1.75 & -14.99 \\
-10.91 & -43.70 & -8.68 & -75.32 & -227.81 & 36.61 & -0.50 & 6.06 \\
-5.86 & -32.63 & 1.59 & -56.97 & 47.75 & 38.83 & 1.03 & 4.49 \\
-2.00 & 2.97 & -1.29 & 2.30 & 8.57 & -0.98 & -2.57 & -2.97 \\
-1.25 & -3.87 & 1.08 & -6.83 & -6.50 & 9.65 & 5.23 & 1.19 \\
-4.41 & -8.97 & -0.43 & -14.10 & 6.37 & 4.19 & 3.03 & 0.64 \\
-1.08 & -10.96 & 0.03 & -16.85 & -9.49 & 4.69 & -1.59 & 1.02 \\
1.71 & -5.55 & -0.47 & -12.78 & -5.83 & 13.13 & 0.43 & 0.70
\end{array}
\tag{2.13}
$$

By comparing the DCT coefficients from the smoother and the sharper blocks, we see that the block with smooth variations has the coefficients with larger values concentrated around the upper left corner of the array (2.12), that is, around the DC coefficient. In contrast, the block with sharper transitions shows relatively large values for coefficients corresponding to higher spatial frequencies (positions away from the upper left corner, including near the high frequency coefficients in the lower right corner). By transforming the input block with the DCT, the JPEG encoder represents the block in terms of its spatial frequency components.

After the DCT, the next step in JPEG coding is quantization. This step is designed to exploit the relative insensitivity of the human visual system to distortion in the higher frequencies of an image. Each coefficient in the 8×8 matrix of DCT coefficients is quantized by a uniform scalar quantizer. The step sizes of the quantizers are different. This quantization is implemented by dividing element by element the matrix of DCT coefficient with a *quantization matrix*. The result of the division is rounded to the nearest integer. Denoting by q_{ij} the elements of the quantization matrix, the quantization operation can be expressed as:

$$ b_{q_{ij}} = \text{round}\left(\frac{b_{T_{ij}}}{q_{ij}}\right). \tag{2.14}$$

Compression distortion is introduced by this quantization. During this operation, the resolution used to represent each DCT coefficient is different. Coefficients associated with lower spatial frequencies are quantized with higher resolution than those associated with higher frequencies. Since quantization is a lossy coding operation, the intent in implementing quantization with nonuniform resolution is to match the quantization error for each coefficient to the sensitivity of the human visual system. The JPEG quantization matrix is:

$$
\begin{matrix}
16 & 11 & 10 & 16 & 24 & 40 & 51 & 61 \\
12 & 12 & 14 & 19 & 26 & 58 & 60 & 55 \\
14 & 13 & 16 & 24 & 40 & 57 & 69 & 56 \\
14 & 17 & 22 & 29 & 51 & 87 & 80 & 62 \\
18 & 22 & 37 & 56 & 68 & 109 & 103 & 77 \\
24 & 35 & 55 & 64 & 81 & 104 & 113 & 92 \\
49 & 64 & 78 & 87 & 103 & 121 & 120 & 101 \\
72 & 92 & 95 & 98 & 112 & 100 & 103 & 99
\end{matrix}
\tag{2.15}
$$

In general, lower-frequency coefficients get divided by smaller numbers. Consequently, when rounding the result of the division, the lower-frequency coefficients introduce less quantization noise. Using as example the block (2.13), after quantization, the DCT coefficients become

$$
\begin{matrix}
60 & 8 & 0 & 9 & -4 & -3 & 0 & 0 \\
-1 & -4 & -1 & -4 & -9 & 1 & 0 & 0 \\
0 & -3 & 0 & -2 & 1 & 1 & 0 & 0 \\
0 & 0 & 0 & 0 & 0 & 0 & 0 & 0 \\
0 & 0 & 0 & 0 & 0 & 0 & 0 & 0 \\
0 & 0 & 0 & 0 & 0 & 0 & 0 & 0 \\
0 & 0 & 0 & 0 & 0 & 0 & 0 & 0 \\
0 & 0 & 0 & 0 & 0 & 0 & 0 & 0
\end{matrix}
.
$$

The quantized DCT coefficients show an expected pattern: when scanning the matrix of quantized DCT coefficients from the top left to the bottom right, the higher frequencies

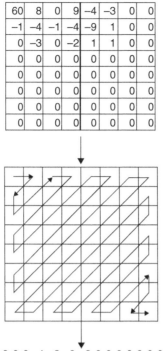

60	8	0	9	−4	−3	0	0
−1	−4	−1	−4	−9	1	0	0
0	−3	0	−2	1	1	0	0
0	0	0	0	0	0	0	0
0	0	0	0	0	0	0	0
0	0	0	0	0	0	0	0
0	0	0	0	0	0	0	0
0	0	0	0	0	0	0	0

60, 8, −1, 0, −4, 0, 9, −1, −3, 0, 0, 0, 0, −4, −3, −9, −2, 0, 0, 0, 0, 0, 0, 0, 1, 1, 0, 0, 0, 1, 0, 0, 0, ...0, 0

Figure 2.22 Converting the array of quantized DCT coefficients into a sequence by following the zigzag order.

(toward the bottom right) tend to be quantized to 0. In the next step of JPEG compression, the two-dimensional array of quantized DCT coefficients is converted to a sequence by reading the array in the zigzag order as shown in Figure 2.22. The zigzag order exploits the expected pattern for the array of quantized coefficients. It leads to sequences that tend to have long runs of consecutive zeros; the sequence can then be efficiently represented using few bits through entropy coding.

The quantization block easily allows for a controlled trade-off between image compression ratio (resulting bit rate) and distortion. More or less compression for the image, and the resulting change in distortion, can be achieved by multiplying the quantization matrix (2.15) by a constant, or by following a more sophisticated approach that changes the quantization matrix.

Many other image compression schemes are similar to JPEG in many operations. Some schemes use a different transform, and others may resort to more sophisticated coding mechanisms for the sequence of quantized bits, but most are well represented in their main elements by the JPEG image compression standard.

2.5.2 Embedded Image Coding – SPIHT

A number of JSCC techniques benefit from the use of embedded codecs. Because of their useful properties and flexibility, many different embedded image codecs have been proposed. A popular one is "Set Partitioning in Hierarchical Trees", known as "SPIHT" [5]. SPIHT is based on dividing an image into subbands of increasing frequency resolution for lower frequencies, using a wavelet transform on the image. In practice, this means applying the wavelet transform first on the sequence of samples formed by each column of image pixels, followed by the application of the wavelet transform on the sequence of samples formed by each row. Because the wavelet transform is applied on two dimensions, the resulting coefficients correspond to subbands in both the horizontal and vertical spatial frequency domains. As most wavelet-based image compression algorithms do, SPIHT applies several levels of wavelet transform analysis. Each new level of wavelet transform is applied to the horizontal and vertical low-frequency coefficients from the previous level wavelet transform. This decomposition is illustrated in Figure 2.23 with a two-level decomposition.

To achieve embedded compression, SPIHT encodes the image by implicitly sorting the wavelet coefficients in order of significance. Considering that the coefficients are represented using a fixed number N of bits, if the least significant bit is number 0 and the most significant bit is number $N - 1$, a coefficient c_{ij} is said to be significant at level n ($0 \leq n \leq N - 1$) if $2^n \leq |c_{ij}| < 2^{n+1}$. The SPIHT algorithm encodes an image by first indicating when a coefficient becomes significant. This means that coefficients are sequentially tested for significance as the level of significance is decreased by one at each iteration of the algorithm. Once a coefficient is found significant, its coordinates are added to a list where it is not tested any more for significance at later iterations; rather the bit value at the significance level is encoded (that is, significant coefficients see their less significant bits encoded one at a time with each iteration).

The result of a significance test is just one bit, with a value of 0 for no significance and 1 for significance. If the test were to be performed on each and every coefficient all the time, SPIHT would not achieve much compression. Instead, SPIHT performs the test on several coefficients together in a group, encoding a single bit answer for the whole group. Each groups is formed from a tree structure, called a spatial orientation tree, defined on the wavelet coefficients. A group has a parent node, corresponding to a coefficient at a higher level of wavelet analysis decomposition, and the descendants of the parent coefficient. The direct descendants of a coefficient are those coefficients at the immediately lower level of wavelet analysis decomposition that are in the same spatial orientation created by the filtering/decimation process. The parent–descendants relationship is illustrated in Figure 2.24. The reason for organizing the coefficients into the spatial orientation tree is that there is a spatial self-similarity between subbands. This means that if a parent coefficient tests as nonsignificant, then the descendants are likely to also test as nonsignificant. The end result is that the significance test can be encoded using very few bits to represent the results for the many coefficients in the tree.

The output stream from SPIHT encoding is embedded because the resolution of the coefficients' quantized representation improves as more bits are included. The significance test and subsequent encoding encode whether the bits representing the coefficients are 0 or 1, sequentially analyzing the most significant bit first, then the second most significant bit and so on, one level of bit significance at a time. The bit stream may be truncated at any point, and the coefficients would be represented using the most significant bits kept in the stream up to the point of truncation. If more bits are kept in the stream, then more bit positions would be used to represent the quantized coefficients.

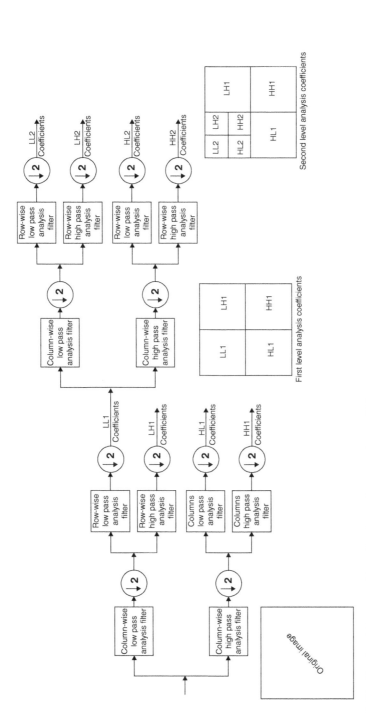

Figure 2.23 Two-level decomposition of an image into wavelet coefficients.

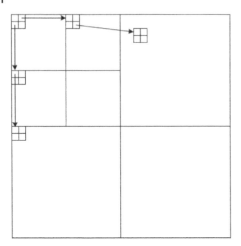

Figure 2.24 Parent–children tree relationships between wavelet coefficients.

2.5.3 Video Coding

Video coding can be thought of as an extension of image coding, where additional processing is added to exploit the temporal correlation between successive or temporally nearby video frames. A video signal needs to have frames played sufficiently close in time that their difference is small enough so that the human visual system perceives a smooth and near-continuous-time video signal. This is illustrated in Figure 2.25, which shows three consecutive video frames for the *Foreman* video sequence and their difference. The difference from frame to frame is small. Large parts of the frame do not change at all (shown as black pixels in the difference frames). There are small sections that change a little, for

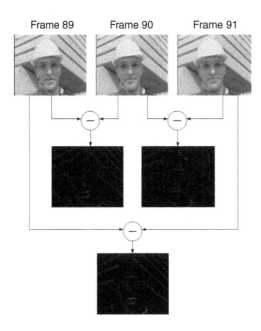

Figure 2.25 Three consecutive video frames for the "Foreman" video sequence and their difference.

example, the man's lips change because he is talking. Naturally, the difference between frames becomes larger as the time difference between frames increases.

We will focus on the H.264/AVC video codec, which shares the main operating principles with most other video coding standards, including Picture Experts Group (MPEG). The H.264/AVC standard is an evolution of the MPEG-1 and MPEG-2 standards, developed by the Joint Video Team (JVT), a partnership of the ITU-T Video Coding Experts Group (VCEG), and ISO/IEC Moving MPEG. H.264 incorporates the MPEG-4 video coding standard. Modern video coding standards operate on color video sequences, but for simplicity we will focus on the processing of luminance information (color processing follows nearly the same procedure). The H.264 standard has been followed by the H.265 *high-efficiency video coding* standard, also developed by MPEG. Although H.265 is roughly twice as efficient in compressing video as H.264, it merely refines the basic operating principles common to H.264 and also to MPEG-1 and MPEG-2. The successor to H.265 is the H.266 *versatile video coding* (VVC), which has recently been standardized and is in very early stages of deployment. This standard also shares the same basic operating principles as its predecessors, while improving compression efficiency and incorporating compression for higher resolution video formats.

The main processing blocks of the H.264 encoder are shown in Figure 2.26. The process proceeds on a frame-by-frame basis. Frames may be processed in one of two main ways. The first option is to process the frame essentially as a stand-alone picture, ignoring the time correlation that may exist between this frame and other frames. These frames are called "intra-predicted" frames or "I-Frames". The second option is to exploit the temporal correlation between the frame being processed and other frames, called "reference frames", kept in a memory buffer. The correlation is exploited by encoding the frame using a predictive coding technique. These frames are called "inter-predicted" frames. A frame being processed can choose to use as reference only frames that have appeared earlier than itself, in which case the frame being processed is called a "P-Frame" (the only Inter-predicted frames shown in Figure 2.26 are P-frames, for simplicity), or it can choose to use as reference both frames which have appeared earlier and later than itself, in which case the frame being processed is called a "B-Frame". To use frames appearing later, it is necessary to process those frames out of order, introducing extra coding delay.

Since I-frames do not exploit temporal correlation, they require higher coding bit rates than the P- or B-frames, for the same level of reconstructed quality. This is illustrated in Figure 2.27, which shows the number of bits used to encode each frame of the QCIF video sequence *Foreman* with the same quality of 36 dB peak signal-to-noise ratio (PSNR). In the figure, *Foreman* is encoded using only I- and P-frames, with one I-frame inserted for every 29 P-frames. The I-frames can easily be recognized from the P-frames, because they appear in the plot as spikes in the number of encoding bits used. Because of their need for more encoding bits, I-frames are used only from time to time; most of the encoded frames are P- or B-frames. For the case illustrated in Figure 2.27, for a video sequence with playback rate of 30 frames per second (fps) encoded using only I- and P-frames, having one I-frame followed by 29 P-frames results in one I-frame per second.

Despite the larger number of bits, it is necessary to transmit I-frames on occasion (usually at regular intervals), for several reasons. As will be discussed in later chapters, an error that may occur during transmission of an inter-predicted frame will propagate to all

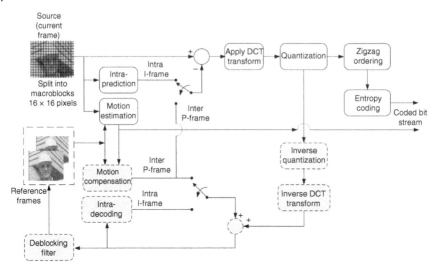

Figure 2.26 Simplified block diagram for the H.264 video encoder. The blocks with a dashed line frame and gray background are also used in the decoder.

subsequent inter-predicted frames that depend on it as a reference. This behavior leads to one of the reasons for the use of I-frames because they stop the error propagation and allow for the recovery from errors during transmission. Another reason for having I-frames is that they allow for random access into the sequence. That is, if one were to encode a video sequence with only a single I-frame at the beginning and all remaining frames as P-frames, then it would be impossible for a decoder to jump to decoding at some point 10 or 20 min into the video. With all later frames being predictively coded using various reference frames, the decoder would be unable to form the prediction, not having the reference frames.

The first major stage of encoding is prediction. Typically prediction is done for macroblocks, which are groupings of blocks. For I-frames, the pixel values of the macroblocks are predicted from pixels in neighboring macroblocks within the same frame. In the case of inter-predicted frames, macroblocks are predicted by finding similar macroblocks in the reference frames. In essence, reference frames are used to estimate the motion (change) of macroblocks as the video sequence unfolds. This operation of *motion estimation* consists of finding the macroblock within the reference frame that best matches the current macroblock being encoded. This best match will be used as the prediction of current macroblock, an operation called *Motion Compensation*. The *motion compensated frame* is created by taking, for each macroblock in the current frame being encoded, the best match macroblock (prediction) from the reference frame, and pasting it in the corresponding place. The *motion compensated difference frame* is the difference between the current frame being encoded and the motion compensated frame. The term *residual frame* can be used for either intra-predicted or inter-predicted frames as being the difference between the actual macroblocks and their predictions. The residual frame contains the prediction errors.

After prediction, the residual frames from inter-predicted and intra-predicted frames are essentially processed as still images, undergoing the same main steps as for JPEG image

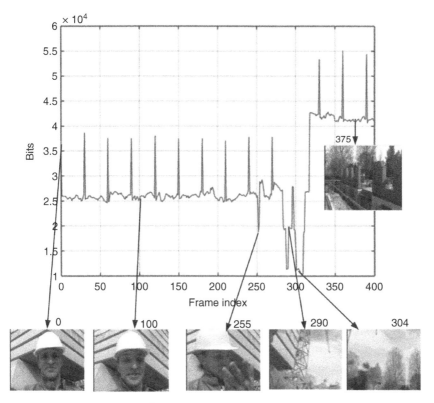

Figure 2.27 Number of bits used to encode each frame in the quarter common intermediate format (QCIF) sequence *Foreman* at a constant PSNR quality of 36 dB.

compression: DCT-transform, quantization, and entropy coding. The specific parameters of these operations are tuned to the statistics of the intra-predicted or the inter-predicted residual frames. The bit stream resulting from the encoding process includes both the encoded residual frame and the set of *Motion Vectors*, where a motion vector is the offset vector between the current macroblock location and its best match block in the reference frame. Motion vectors tend to represent the estimated movement (change in position) of the macroblocks. As in predictive coding, much of the decoding process is replicated in the encoder. This is done to match the reference signal used during the prediction at the encoder to the one that is reconstructed in the decoder and avoid introducing a mismatch error that would propagate over successive predictive coding/decoding steps. This is incorporated in the design of the video encoder as seen in Figure 2.26, where the blocks that are used both in the encoder and decoder are shown with a dashed line frame.

It is also possible to modify the H.264 video codec so that the encoded bit stream is embedded. This has been done in, for example, the MPEG-4 fine-grained scalable (FGS) video encoder (H.264 includes the MPEG-4 standard in its specification). Figure 2.28 shows a simplified block diagram for the MPEG-4 FGS encoder. Here, the video source is encoded into two layers: a base layer and an enhancement layer. The base layer is generated by processing the video source through a regular H.264 encoder using quantization parameters

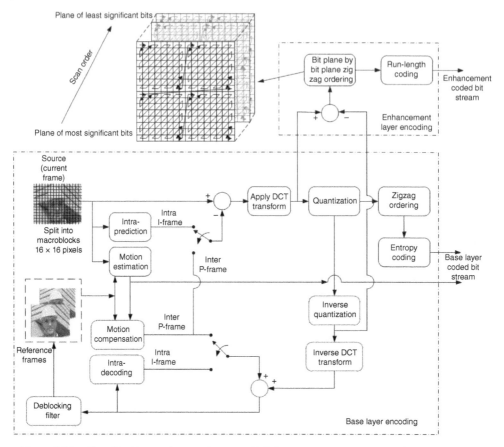

Figure 2.28 Simplified block diagram for the MPEG-4 fine-grained scalable (FGS) video encoder.

that result in a very coarse, low bit rate encoded bit stream for the video source. The base layer bit stream is not embedded and is indispensable in its totality to recover the original video source. If the encoded video source were to be reconstructed from the base layer only, the distortion would be large. As shown in Figure 2.28, a bit stream from the encoding of an enhancement layer is appended to the bit stream from the base layer. The enhancement layer stream is embedded and can be truncated at any point to achieve reconstruction of the video source at different quality levels.

The enhancement layer is the difference between the original DCT coefficients and the corresponding DCT coefficients distorted through the coarse quantization at the base layer (see Figure 2.28). To encode the enhancement layer, the coefficient error is divided into 16x16 pixel macroblocks and read following the same zigzag order used in regular H.264 coding. The difference in this case is that the zigzag reading of coefficient error is done one bit plane at a time. This is, as the coefficient error is represented using a fixed number of bits, the zigzag ordering is first done for the most significant bit of each coefficient error; after finishing with the zigzag ordering of the most significant bit plane, the zigzag ordering continues by looking only at the value of the second most significant bit plane, and so

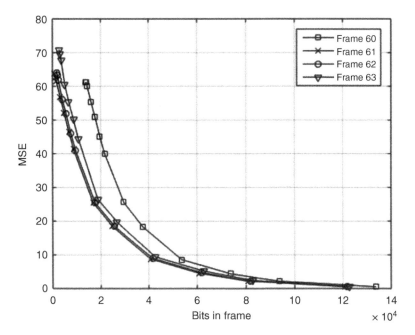

Figure 2.29 Mean-squared error as a function of encoding bits for four frames 60 to 63 in the "Foreman" QCIF video sequence. Frame 60 is an I-frame and the others are P-frames.

on until the least significant bit of the coefficient errors. This results in a bit stream that carries the finer description of the video source, adding the details missing from the coarse description in the base layer. As more bits are kept in the bit stream, finer resolution of the coefficient errors is achieved.

Figure 2.29 shows the distortion-rate curves for frames 60 to 63 in the *Foreman* QCIF video sequence. In this case, frame 60 is an I-frame and the rest are P-frames. The figure shows how the distortion-rate performance of video codecs changes from frame to frame. This is expected when comparing the curve for the I-frame to the curves for the P-frames as these two types of frames use different coding techniques. But also, P-frames, even when consecutive, may show different distortion-rate coding performance because of the changes in the frame characteristics (motion, texture, etc.) over time.

2.5.4 Speech Coding

There are two main approaches to encoding speech. One family of codecs, known as *Waveform Codecs*, treat the speech signal as just another audio signal, aiming to encode the signal waveform regardless of whether it is speech, music, or any other signal within the audio bandwidth. Typical waveform codecs use one or more of the techniques described earlier in this chapter, such as differential coding or subband coding.

The second approach, known as *Vocoders*, is based on matching a block of speech samples to a known speech production model. Once an appropriate match to a model has been found, only the parameters describing the model, and the residual difference between the model and the actual signal, are needed to represent the speech block. To decode the speech block, the parameters are used to configure a synthesizer based upon the same encoding vocal model to output a synthetic speech segment that sounds similar to the original speech block.

To understand how vocoders work, we first need to understand how people produce speech. To speak, a person pushes air from the lungs through the vocal tract and mouth. The amount of air flow determines the loudness. To make *voiced sounds*, the air flowing through the vocal tract is used to make the vocal cords vibrate. For *unvoiced sounds* or *fricative and plosive sounds*, the vocal cords let the air flow without vibrating. The sounds, either voiced or unvoiced, are further modulated (adjusted) by changing the shape of the vocal tract, the mouth, and the position of the tongue and by controlling if a portion of the air flow goes through the nose. The pitch of the voice is controlled by the frequency with which the vocal cords vibrate. During the generation of a spoken word, all these changes and adjustments vary to produce different sounds. These adjustments occur relatively slowly (on the scale of 10 msec to 50 msec), and consequently the statistics of a sequence of speech changes on the scale of 10 msec to 50 msec.

In vocoders, the physical production of speech is translated into a mathematical model. The model shown in Figure 2.30 starts by generating one of two possible discrete-time signals. If the sound to be modeled is unvoiced, the generated signal is a random sequence of white noise. When the sound to be generated is voiced, the signal generated is typically a sequence of periodic impulses (because the vocal cords vibrate by opening and closing, generating *glottal pulses*) with the period of the desired pitch. Both types of generated signals subsequently follow the same steps. They are multiplied by a gain G that models the speech volume and then they are passed through a digital filter $H(z)$ which models the shape of the vocal tract, tongue position, etc. This filter has an all-pole configuration, shown to be the most appropriate by spectral analysis. The modulation in the vocal tract emphasizes some

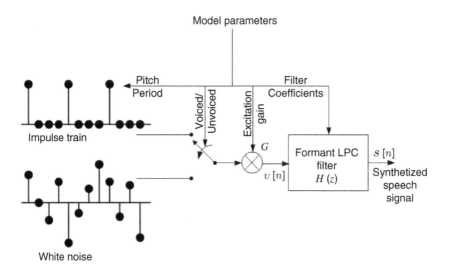

Figure 2.30 Linear predictive coding (LPC) model of speech synthesis.

frequencies more than others but does not tend to null frequencies components. A typical transfer function for the filter is:

$$H(z) = \frac{1}{1 + a_1 z^{-1} + a_2 z^{-2} + \cdots + a_N z^{-N}}, \qquad (2.16)$$

where N is typically an even number between 6 and 12, also following analysis of voice spectra, and a_i are the filter coefficients. From (2.16), the resulting synthetic speech $s[n]$ can be written as the filtered signal $v[n]$:

$$s[n] = v[n] - a_1 s[n-1] - a_2 s[n-2] - \cdots a_N s[n-N]. \qquad (2.17)$$

With the speech production model defined, the operation of a speech encoder consists of processing a block of speech samples of between 10 msec and 50 msec so that the statistics of all the samples are roughly the same (a common choice is 20 msec). The encoder processing of the block first identifies whether the speech sound is voiced or unvoiced. If the sound is voiced, the encoder also estimates the pitch period. Then, the encoder estimates the gain G that best matches the volume and the filter $H(z)$ coefficients that most closely represent the modulation of the speech sound. The coefficients can be estimated using the Levinson–Durbin linear prediction recursive algorithm that aims to minimize the mean squared error between the desired speech output and the linear combination of past outputs shown on the right-hand side of (2.17) [1]. Under ideal conditions, the encoded speech signal would be fully represented by the set of parameters estimated to best match the speech production model: voice/unvoiced sound decision, pitch period when applicable, gain G, and filter coefficients a_i. Because this procedure for estimating the speech model parameters is based on linear prediction, it is called *linear predictive coding* (LPC), and the filter $H(z)$ is called the LPC filter.

In practice, LPC cannot perfectly match the model to the input block of speech samples. Consequently, most vocoders implement an improvement to LPC called *analysis-by-synthesis* (AbS) (Figure 2.31). In this approach, the speech is synthesized at the encoder, and the error between the synthesized speech and the original speech is used to improve the estimation of model parameters. Nevertheless, there is always some residual

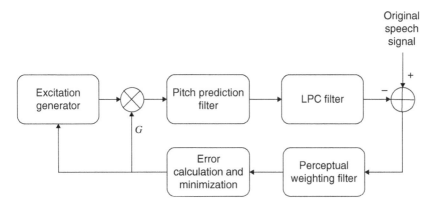

Figure 2.31 Analysis-by-synthesis encoder structure.

difference between the synthesized speech and the original speech, so this residual error needs to be compressed and transmitted to the decoder along with the model parameters. Nevertheless, because AbS schemes aim to reduce the residual error, they can achieve larger compression ratios by saving bits from the representation of the error. Finally, we note that in the AbS model in Figure 2.31, the voiced/unvoiced sequence of speech generator has been replaced by an excitation generator. In many popular vocoders, such as code-excited linear prediction (CELP), this sequence is generated from a fixed number of predefined variations of the periodic impulse signal. These possible signals are stored at the encoder in a table that is encoded and addressed using vector quantization.

References

1 Rabiner, L.R. and Schafer, R.W. (1978). *Digital Processing of Speech Signals, Prentice-Hall Signal Processing Series*. Pearson Education.
2 Sayood, K. (2000). *Introduction to Data Compression, Morgan Kaufmann Series in Multimedia Information and Systems*. Morgan Kaufmann Publishers.
3 Gersho, A. and Gray, R.M. (1992). *Vector Quantization and Signal Compression, Kluwer International Series in Engineering and Computer Science*. Kluwer Academic Publishers.
4 Vaidyanathan, P.P. (1990). Multirate digital filters, filter banks, polyphase networks, and applications: a tutorial. *Proceedings of the IEEE* 78 (1): 56–93.
5 Said, A. and Pearlman, W.A. (1996). A new, fast, and efficient image codec based on set partitioning in hierarchical trees. *IEEE Transactions on Circuits and Systems for Video Technology* 6 (3): 243–250.

3

Channel Coding

Applying the source–channel separation principle in a communication system design means that the information source is first quantized and source encoded, followed by a separate and independent processing block to implement channel coding. These coding operations seem to have conflicting functions. While source coding strives to reduce the number of bits used in representing an information source, channel coding adds bits that are redundant in terms of source representation. The reason for this is that the redundancy bits added during channel encoding have embedded a structure that can be exploited to detect or correct bit errors introduced during transmission. In principle, this type of structure is not present in the bits removed during source coding.

This book concentrates on the joint design of the source and channel coding operations. But as many of these techniques are based on source or channel codecs designed separately and then modified to work jointly, this chapter provides an overview of the main channel coding techniques that are relevant to the topics of later chapters. This chapter will concentrate on reviewing forward error correction (FEC) techniques, i.e. error-correcting codes that add redundant bits to the transmitted bit stream and related decoding techniques to detect and correct errors at the receiver. Error-correcting codes can also operate in conjunction with feedback received at the transmitter. The use of feedback in the coding process is discussed in Chapter 6. For further details, there are excellent textbooks available [1–3]. To organize the presentation, we separate channel codes into linear block codes and convolutional codes.

3.1 Linear Block Codes

Linear block codes operate on blocks of input bits or symbols, for which they generate a block of bits or symbols at the output. Figure 3.1 illustrates the idea with an (n, k) linear binary block channel encoder that takes as input a block of k bits and outputs a *codeword* with n bits.

Some block codes work with binary input and binary processing, meaning that the input is a block of binary digits, 0 or 1, and the processing is based on binary operations using modulo-2 arithmetic. Nonbinary linear block codes accept input blocks composed of nonbinary symbols. This sometimes causes confusion because symbols can be represented as

Joint Source-Channel Coding, First Edition. Andres Kwasinski and Vinay Chande.
© 2023 John Wiley & Sons Ltd. Published 2023 by John Wiley & Sons Ltd.

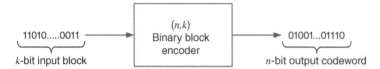

Figure 3.1 An (n, k) linear binary block channel encoder.

groups of bits. For example, consider a linear block code that accepts four possible symbols as input: α_0, α_1, α_2, and α_3. An example input block may be $\alpha_0\alpha_2\alpha_1$. At the same time, the symbol may represent a group of bits. For example, $\alpha_0 = 00$, $\alpha_1 = 01$, $\alpha_2 = 10$, and $\alpha_3 = 11$, in which case the input block becomes 001001. The main differentiating factor between binary and nonbinary codes is that all operations related to a binary code are done modulo-2, while operations related to a nonbinary code are done with the modulo of the number of different symbols. When the input uses the four symbols α_0, α_1, α_2, and α_3, the mathematical operations related to this code are done modulo-4. In more formal terms, both binary and nonbinary block codes can be studied using the mathematical framework of Galois fields (GFs). A field is an algebraic structure organizing numbers within a framework defined by axioms, properties, and the four basic mathematical operations of addition, subtraction, multiplication, and division, and for GFs the elements form a finite set, i.e. the number of elements in the field is finite. For channel coding, computational complexity considerations lead toward often using GFs where the number of elements equals a power of 2, denoted as GF(2^n). For example, the GF(2) has two elements, 0 and 1, and is used for binary block codes. For the nonbinary block code with symbols α_0, α_1, α_2, and α_3, the GF(4) is used.

Block codes operate by breaking down the input data stream into blocks of k symbols of the form $\vec{u} = \{u_0, u_1, \ldots, u_{k-1}\}$. The encoder maps each input block into a distinct codeword with n symbols, $\vec{c} = \{c_0, c_1, \ldots, c_{n-1}\}$. Each component c_i of the codeword is called a coordinate. For a binary code with $k = 2$ and $n = 3$, the encoding mapping could be $00 \rightarrow 000$, $10 \rightarrow 101$, $01 \rightarrow 011$, and $11 \rightarrow 110$ (this mapping is called an even parity-check code). When the block code operates on a GF(q), the input block symbols and the codeword coordinates can take q possible values. Then, the encoder maps q^k different possible blocks into q^k distinct codewords selected from the set of q^n possible codewords. For the binary block code with $k = 2$ and $n = 3$, Figure 3.2 illustrates the encoding mapping. In the figure, each component of the input block is considered a coordinate in a k-dimensional space, and each component of the output codeword is considered a coordinate in an n-dimensional space.

Figure 3.2 illustrates the basic concept for error detection and correction shared by all block codes. The encoder maps q^k different possible blocks into q^k distinct codewords selected from the set of q^n possible codewords. In Figure 3.2, this means that the encoder maps the four vertices of the square with coordinates (0,0), (0,1), (1,0), and (1,1), each being an input block, into only four distinct vertices of the cube defined by the eight vertices (0,0,0), (0,0,1), (0,1,0), (0,1,1), (1,0,0), (1,0,1), (1,1,0), and (1,1,1). The four vertices mapped from an input block are codewords, and the other four vertices are not codewords. A codeword gets transmitted through a channel that may add errors, so some of the transmitted codeword bits may be flipped from 0 to 1 or from 1 to 0 by the time they arrive at the decoder. When only one bit in the word is flipped, the received word at

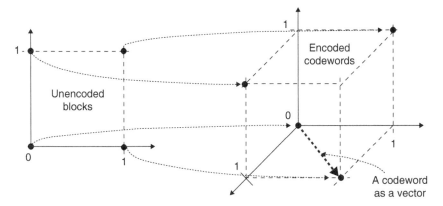

Figure 3.2 Visual representation of the mapping operation at the encoder of a binary block code with $k = 2$ and $n = 3$ (even parity-check code).

the input of the decoder will be invalid. This can be seen in Figure 3.2. Codewords are vertices that always differ from any other codeword by two of their coordinates. Changing one coordinate value will result in a vertex that is not mapped from any input block. The decoder can detect the occurrence of errors any time that the received word is invalid (not a codeword). The error detection operation is intimately related to the concept of mapping vectors from a lower dimensionality space into a higher dimensionality space. Upon detection of errors, the receiver may request the transmitter to resend the erroneous word, or it may notify the source decoder of the error so the source decoder will deal with it through error concealment, or it may attempt to correct the error using the structure in the channel code.

The same underlying vector space ideas are also used to understand the principles used to correct errors in the received word. The example in Figure 3.2, while good for visualization, has insufficient dimensionality in the codeword vector space to exemplify error correction. Suppose $n = 4$; the codewords are vectors in a four-dimensional space. If we keep $k = 2$, there are four codewords, which could have coordinates $(0,0,0,0)$, $(0,1,0,1)$, $(1,0,1,0)$, and $(1,1,1,1)$, and there are $2^4 - 4 = 12$ invalid words, such as $(0,0,0,1)$ or $(1,1,0,1)$. When attempting to decode an invalid word, error correction can be attempted by assuming that errors of smaller magnitude are more likely than errors of larger magnitude. This means that the vector associated with an invalid word should be the outcome of adding errors to the closest codeword. In other words, the decoder corrects errors by matching an invalid received word to the codeword that is closest in distance to it. For example, the codeword $(1,1,1,1)$ is closest to the invalid received word $(1,1,0,1)$.

The expansion of the dimensionality in the vector space associated with the codewords means there are more words (q^n) in the space than there are codewords. For this property, it is said that a codeword has **redundancy**. The amount of redundancy in a code or, equivalently, how much difference there is between the number of codewords and the number of all q^n words is measured through the *rate* of the code, which is defined as:

$$r = \frac{\log_q M}{n},$$ (3.1)

where M is the number of codewords. When $M = q^k$, the channel code rate becomes

$$r = \frac{k}{n}. \tag{3.2}$$

Note that $r < 1$.

3.1.1 Binary Linear Block Codes

We now narrow the review of block codes to binary codes. Binary block codes operate on the GF(2); all arithmetic uses two possible numbers, 0 and 1, and mathematical operations are modulo-2. An important issue in analyzing code performance is the number of bit errors introduced by the channel. If the errors are few, the decoder may be able to not only detect the errors but also to correct them. If there are more errors, the received invalid word may be closest to some codeword that is not the transmitted codeword. In this case, the decoder will decode to a wrong codeword, constituting a *decoder error*. Another case is where the received codeword is valid, but it is different from the transmitted one. This is called an *undetectable error*.

To calculate the likelihood of these events, it is important to measure how different one codeword is from another. Drawing from the notion of an underlying vector space, the difference between two codewords is quantified through a *distance* measure, such as the Euclidean distance between the vectors associated with the codewords. For block codes, the measure used most is the *Hamming distance*. The Hamming distance between two words of the same size is calculated as the number of coordinates in which the two words differ. For example, the Hamming distance between 00101 and 10111 is 2, because they differ in the first and fourth coordinates.

Because the error detection and correction performance of a code depends on the separation between the codewords as vectors in an n-dimensional space, the distance of codeword vectors from a reference point in the codeword space and the minimum separation between all codewords in a code are important parameters to analyze code performance. The distance of a codeword from the point with all-zero coordinates in the codeword space is called the codeword *weight*, and it equals the number of coordinates with nonzero values. The minimum separation between codewords in a code is measured through the minimum of all Hamming distances between all codeword pairs, a parameter called the *minimum distance* of a block code, denoted d_{min}. For example, for the $k = 2$ and $n = 3$ code in the previous section, the minimum distance is 2, as any codeword differs from all the others in two coordinates. The minimum distance can be used to calculate a number of parameters related to code performance. The number of errors a block code can detect is $d_{min} - 1$, and the number of errors a code can correct is $\lfloor (d_{min} - 1)/2 \rfloor$.

There exist a number of theorems that bound the minimum distance of a code based on its parameters. One of the most important is the Singleton bound that states that for an (n, k) code, the minimum distance is upper-bounded by:

$$d_{min} \leq n - k + 1. \tag{3.3}$$

Combining this bound with the limits on error detection and correction, we see that an (n, k) block code can detect a number d of errors upper-bounded by $d \leq n - k$ and can correct a number t of errors upper-bounded by $t \leq \lfloor (n - k)/2 \rfloor$.

The concept of distance is also used to define the type of decoder we have been considering so far. Considering the received word as a vector $\vec{r}$ in an n-dimensional space and a codeword $\vec{c}$ also as a vector in the same n-dimensional space, the distance between the two vectors is a function $d(\vec{c}, \vec{r})$. With this formulation, a *t-error correcting bounded-distance decoder* is one that outputs the block for codeword $\vec{c}$ that minimizes $d(\vec{c}, \vec{r}) \le t$. If there is no codeword $\vec{c}$ that satisfies this condition, the decoder outputs a decoder failure indication.

3.1.2 Generator Matrix, Parity-Check Matrix, and Syndrome Testing

So far, we have discussed how linear block codes perform encoding and decoding through operating in a GF(q). The encoding operation maps a k-symbol word into an n-symbol codeword that corresponds to a vector in an n-dimensional space. Each component of the codeword takes q possible values and is the coordinate of the vector in one of the n independent directions. This leads toward considering the encoding and decoding operations of a linear block code as linear algebra operations in the n-dimensional space. Both encoding and decoding could be achieved through mathematical operations between vectors in an n-dimensional space, with discrete coordinates, using matrices to achieve some needed transformations.

To work in a vector space, we define a basis set so as to be able to represent any vector in the vector space using a linear combination of the basis set. Each basis element determines one dimension in the vector space. Let $\{\vec{g}_0, \vec{g}_1, \ldots, \vec{g}_{k-1}\}$ be the basis of codewords for an (n, k) code operating in GF(q). Then, the codeword $\vec{c}$ can be written as a linear combination of the basis vectors, i.e. $\vec{c} = u_0\vec{g}_0 + u_1\vec{g}_1 + \cdots + u_{k-1}\vec{g}_{k-1}$. Note that the definitions here define a k-dimensional space and not a higher dimensionality n-dimensional space. This is a more formal presentation of the basic conceptual operation of a block code previously explained with Figure 3.2, where the encoder maps input blocks into a subspace with dimension k within the space n of higher dimensions. The encoding mapping is done through the use of a **generator matrix G** that has in each row one of the k basis vectors $\vec{g}_i$:

$$G = \begin{bmatrix} \vec{g}_0 \\ \vec{g}_1 \\ \vdots \\ \vec{g}_{k-1} \end{bmatrix}. \tag{3.4}$$

The generator matrix generates the codeword $\vec{c}$ for an input block $\vec{u} = \{u_0, u_1, \ldots, u_{k-1}\}$ by performing vector-matrix multiplication:

$$\vec{c} = \vec{u}\, G$$

$$= [u_0, u_1, \ldots, u_{k-1}] \begin{bmatrix} \vec{g}_0 \\ \vec{g}_1 \\ \vdots \\ \vec{g}_{k-1} \end{bmatrix}$$

$$= u_0\vec{g}_0 + u_1\vec{g}_1 + \cdots + u_{k-1}\vec{g}_{k-1}. \tag{3.5}$$

The basis $\{\vec{g}_0, \vec{g}_1, \ldots, \vec{g}_{k-1}\}$ spans the k-dimensional subspace of vectors corresponding to codewords within the space of all possible vectors with discrete coordinates in n

dimensions. Consequently, it is possible to define another subspace spanned by the vectors orthogonal to those corresponding to codewords. This subspace is called a *dual* space of the codeword space and has dimensionality $n - k$. As for all spaces, it is possible to define for the dual space a basis with $n - k$ orthogonal vectors with discrete coordinates $\{\vec{h}_0, \vec{h}_1, \ldots, \vec{h}_{n-k-1}\}$. This basis forms the **parity-check matrix H**,

$$\mathbf{H} = \begin{bmatrix} \vec{h}_0 \\ \vec{h}_1 \\ \vdots \\ \vec{h}_{n-k-1} \end{bmatrix}. \tag{3.6}$$

Because the dual space is defined as the space of vectors orthogonal to the subspace of code-words, it is possible to check whether a received word is a codeword or not by evaluating if the word vector is orthogonal or not to the dual space. If the word vector is orthogonal to the space spanned by the basis **H**, the word is a codeword. This condition means that a codeword needs to satisfy

$$\vec{c} \, \mathbf{H}^T = 0.$$

When receiving a word $\vec{r}$, which is the transmitted codeword $\vec{c}$ with possibly some errors added, by projecting it onto the dual space, the valid codeword component would be eliminated (due to orthogonality) and all that would be left is a component from errors. The received codeword can be written as $\vec{r} = \vec{c} + \vec{e}$ where $\vec{e}$ denotes the error vector. The projection of the received codeword onto the dual space is called the **syndrome vector** $\vec{s}$, and it equals

$$\vec{s} = \vec{r} \, \mathbf{H}^T$$
$$= (\vec{c} + \vec{e}) \mathbf{H}^T$$
$$= \vec{c} \, \mathbf{H}^T + \vec{e} \, \mathbf{H}^T$$
$$= \vec{e} \, \mathbf{H}^T.$$

While the syndrome can be computed from the received word ($\vec{s} = \vec{r} \, \mathbf{H}^T$), it does not depend on the transmitted codeword. It only depends on the error vector or pattern (of course, it also depends on the parity-check matrix, but this is predetermined for a given code). This is an important observation because it shows a method for decoding the block code that allows for correcting the errors. At the decoder, a previously computed table will store for each syndrome, computed as $\vec{s} = \vec{e} \, \mathbf{H}^T$, the corresponding error pattern. Then, with each received codeword, the decoder computes the syndrome as $\vec{s} = \vec{r} \, \mathbf{H}^T$ and finds in the table the corresponding error pattern. Once the error pattern is known, the errors are corrected by simply doing $\vec{c} = \vec{r} - \vec{e}$.

3.1.3 Common Linear Block Codes

One of the earliest binary linear block codes was the Hamming code, named after its inventor Richard Hamming. Hamming codes involve a parameter m that defines the number of coordinates in the input block ($k = 2^m - m - 1$) and in the codeword ($n = 2^m - 1$).

The code has $n - k = m$ of what are called *parity symbols* and is able to correct one error. For a Hamming code with codeword size $n = 2^m - 1$, the parity-check matrix is generated by having as columns all the possible nonzero binary words of size equal to m. For example, for a $(n, k) = (7,4)$ Hamming code, the parity-check matrix could be

$$\mathbf{H} = \begin{bmatrix} 1 & 1 & 0 & 1 & 1 & 0 & 0 \\ 1 & 1 & 1 & 0 & 0 & 1 & 0 \\ 1 & 0 & 1 & 1 & 0 & 0 & 1 \end{bmatrix}. \tag{3.7}$$

The generator matrix for the Hamming code can be derived from the parity-check matrix as the orthogonality of the related subspaces implies $\mathbf{GH}^T = 0$. This condition means that when the parity-check matrix is of the form $\mathbf{H} = \left[\mathbf{A}|\mathbf{I}_{n-k}\right]$, where $\mathbf{I}_{n-k}$ is the identity matrix with dimension $n - k$, the generator matrix is $\mathbf{G} = \left[\mathbf{I}_k| - \mathbf{A}^T\right]$. For the aforementioned example of the $(7,4)$ Hamming code, the generator matrix is:

$$\mathbf{G} = \begin{bmatrix} 1 & 0 & 0 & 0 & 1 & 1 & 1 \\ 0 & 1 & 0 & 0 & 1 & 1 & 0 \\ 0 & 0 & 1 & 0 & 0 & 1 & 1 \\ 0 & 0 & 0 & 1 & 1 & 0 & 1 \end{bmatrix}. \tag{3.8}$$

The need to have the parity-check matrix in the form $\mathbf{H} = \left[\mathbf{A}|\mathbf{I}_{n-k}\right]$ so as to easily derive the generator matrix is not very demanding. Because the parity-check matrix is constructed by entering as columns all nonzero binary vectors with m coordinates, the vectors building the identity matrix part of $\mathbf{H}$ can be saved for the last columns to form $\mathbf{I}_{n-k}$.

A generator matrix of the form $\mathbf{G} = \left[\mathbf{I}_k| - \mathbf{A}^T\right]$ creates an important configuration for the codewords. The presence of the identity sub-matrix $\mathbf{I}_k$ makes the first k coordinates of the codeword equal the input block (Figure 3.3). A code with this property is called a *systematic code*. Many codes besides Hamming codes can have the property of being systematic.

Cyclic codes, an important family of block codes, have the property that for any codeword $\vec{c}_0 = \{c_0, c_1, \dots, c_{n-2}, c_{n-1}\}$, any cyclic shift of the codeword components, such as $\vec{c}_{n-1} = \{c_{n-1}, c_0, c_1, \dots, c_{n-2}\}$ and $\vec{c}_2 = \{c_2, \dots, c_{n-1}, c_0, c_1\}$, are also codewords. Cyclic codes are studied by treating the codeword coordinates as the coefficients of a polynomial of degree $n - 1$. For example, the codeword from a binary code 1001011 is represented by the polynomial $c(X) = 1 + X^3 + X^5 + X^6$. When working with cyclic codes, all the structures of symbols are converted to polynomials. Polynomials of a single variable X form a vector space with basis $\{1, X, X^2, \dots, X^{n-1}\}$. Therefore, our earlier observations and characterization of codewords from block codes as vectors in an n-dimensional space are still applicable. Words, vectors, and polynomials will all be interchangeable concepts.

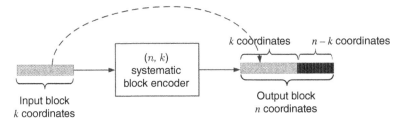

Figure 3.3 Systematic block encoder.

The construction of a specific cyclic code is based on a generator polynomial. Given an input block with k symbols $u(X) = u_0 + u_1 X + \cdots + u_{k-1} X^{k-1}$, and having designed a generator polynomial $g(X) = g_0 + g_1 X + \cdots + g_{n-k-1} X^{n-k-1} + g_{n-k} X^{n-k}$, the corresponding codeword $c(X)$ is computed as:

$$c(X) = u(X)g(X)$$
$$= (u_0 + u_1 X + \cdots + u_{k-1} X^{k-1})(g_0 + g_1 X + \cdots + g_{n-k} X^{n-k})$$
$$= c_0 + c_1 X + \cdots + c_{n-1} X^{n-1}.$$

When working with polynomials, the operations between coefficients are still done using the GF(q) chosen for the code. Consistent with polynomial coefficient operations using the corresponding GF(q) arithmetic, the multiplication of the polynomial is done in such a way that the exponents of the polynomial variable (X) are added modulo-n. This is formally expressed as performing the polynomial multiplications modulo ($X^n - 1$).

Next, it is possible to write the aforementioned cyclic code encoder operation using matrix multiplication and, in doing so, expose the generator matrix for the code:

$$c(X) = u(X)g(X)$$
$$= u_0 g(X) + u_1 X g(X) + \cdots + u_{k-1} X^{k-1} g(X)$$
$$= [u_0, u_1, \ldots u_{k-1}] \begin{bmatrix} g(X) \\ Xg(X) \\ \vdots \\ X^{k-1} g(X) \end{bmatrix},$$

and, thus, the generator matrix is:

$$G = \begin{bmatrix} g(X) \\ Xg(X) \\ \vdots \\ X^{k-1} g(X) \end{bmatrix}$$

$$= \begin{bmatrix} g_0 & g_1 & g_2 & \cdots & g_{n-k-1} & g_{n-k} & 0 & 0 & \cdots & 0 \\ 0 & g_0 & g_1 & g_2 & \cdots & g_{n-k-1} & g_{n-k} & 0 & \cdots & 0 \\ 0 & 0 & g_0 & g_1 & g_2 & \cdots & g_{n-k-1} & g_{n-k} & \cdots & 0 \\ \vdots & \vdots & \ddots & \ddots & \ddots & \ddots & \ddots & \ddots & \ddots & \vdots \\ 0 & 0 & \cdots & 0 & g_0 & g_1 & g_2 & \cdots & g_{n-k-1} & g_{n-k} \end{bmatrix}.$$

Note in this result how the multiplication of the generator polynomial with X, $Xg(X)$, results in a right rotation of the polynomial coefficients. Thus, in the generator matrix, each row is the cyclic shift of the immediately preceding row (as it should be for a cyclic code since the rows of the generator matrix are codewords).

Following the same principles discussed so far, it is possible to define for any cyclic code a corresponding parity-check matrix for which it holds true that $\vec{c}\, H^T = 0$ and $\vec{s} = \vec{r}\, H^T = \vec{e}\, H^T$. Understanding encoding as the product of two polynomials also provides an interpretation for a decoding process and the concept of syndrome. The decoder can undo the encoding operation by dividing the received word by the generator polynomial. If the received word had no errors, the division would result in a polynomial representing the original encoder block with a remainder equal to 0. If the received word had errors, the result of the division would have a nonzero remainder. This remainder, which would only depend on the error pattern, is the syndrome polynomial.

Among cyclic codes, Bose–Chaudhuri–Hocquenghem (BCH) and Reed–Solomon codes are especially popular. For BCH codes, the coordinates in the codeword take values in GF(2) (they are binary values) and the generator polynomial is the one with the lowest possible degree such that it has as roots $2t$ *consecutive* elements of the GF(2^m). Equivalently, the generator polynomial of the BCH code is constructed from factors of the polynomial $X^{2^m-1} + 1$. Technically, this definition corresponds to a subset of BCH codes, called primitive BCH codes, that are most often used. For these BCH codes, the codeword length is $n = 2^m - 1$, and the number of parity coordinates is $n - k = mt$. Reed–Solomon codes can be thought of as BCH codes where the coordinates of the codeword (equivalently, the polynomial coefficients) take values in GF(q), which means that the coordinates could take as possible values $1, 2, \ldots, q - 1$ and arithmetic operations are modulo-q.

Earlier we saw the Singleton bound which states that for block codes, the minimum distance is upper-bounded: $d_{min} \leq n - k + 1$. Reed–Solomon codes are said to be maximum-distance separable (MDS) because they achieve the Singleton bound with equality, i.e. $d_{min} = n - k + 1$.

The last type of block codes we will mention are the low-density parity check (LDPC) codes. These have seen significant research attention and success because their performance approaches the channel capacity. As with Hamming codes, LDPCs are more directly defined using the parity-check matrix. The parity-check matrix for a binary LDPC is defined by having a fixed number of coordinates equal to 1 for columns and rows. The number of 1's in the parity-check matrix is small relative to the codeword size, which makes the parity-check matrix sparse or *low density* in the number of nonzero entries. LDPC codes have an advantage in that decoding implementation makes use of a low complexity iterative belief propagation algorithm. A detailed presentation of LDPC codes can be found in [4].

3.1.4 Error and Erasure Correction with Block Codes

Earlier, we reviewed the concept of minimum distance d_{min}, which allows us to quantify the number t of errors that a code is able to correct

$$t = \left\lfloor \frac{d_{min} - 1}{2} \right\rfloor.$$

Therefore, for a binary code with bit errors that are independent of each other, the probability of a decoder failure is upper-bounded by employing the union bound which is the sum of the probabilities of the events of having more than t errors out of n bits in the codeword:

$$P_e \leq \sum_{j=t+1}^{n} \binom{n}{j} p^j (1 - p)^{n-j}, \tag{3.9}$$

where p is the probability of receiving a bit in error. For example, for binary phase shift keying (BPSK) modulation over an additive white Gaussian (AWGN) channel,

$p = Q(\sqrt{2E_c/N_0})$, where N_0 is the background noise power spectral density, E_c is the energy per codeword bit, resulting from distributing the energy per encoded bit E_b over those in the codeword, $E_c = kE_b/n$, and $Q(y)$ is the probability of a Gaussian random variable with zero mean and unit variance having a value larger than y,

$$Q(y) = \int_y^\infty \frac{1}{\sqrt{2\pi}} e^{-z^2/2} \, dz. \tag{3.10}$$

For a non-binary code, such as a Reed–Solomon code, the probability of decoder failure is derived similarly to (3.9), although instead of depending on the probability of receiving a bit in error, it depends on the probability of a symbol (coordinate) error, p_s:

$$P_e \le \sum_{j=t+1}^{n} \binom{n}{j} p_s^j (1 - p_s)^{n-j}. \tag{3.11}$$

Here, the probability of a symbol error, p_s, depends on how the symbols are transmitted. Sometimes symbols are mapped directly into an electrical signal during modulation, and p_s is the probability of the modulated symbol error. In other cases, the symbol may be converted into a sequence of bits or into a hybrid of bit groups that are next mapped into a modulated electrical signal. Each of these cases requires its own calculation for p_s. Also, for Reed–Solomon codes, their MDS property implies that $d_{min} = n - k + 1$, so the number of errors t that can be corrected is

$$t = \left\lfloor \frac{d_{min} - 1}{2} \right\rfloor = \left\lfloor \frac{n - k}{2} \right\rfloor.$$

This relation can be applied to (3.11) to obtain the probability of decoder error for Reed–Solomon codes:

$$P_e \le \sum_{j=\lfloor \frac{n-k}{2} \rfloor + 1}^{n} \binom{n}{j} p_s^j (1 - p_s)^{n-j}. \tag{3.12}$$

The MDS property of Reed–Solomon codes, along with the fact that they operate on non-binary symbols, has made them useful for correcting the loss of some packets in a group belonging to a common block of data that was fragmented for transmission. This problem gained importance years ago as communications evolved from a circuit-switched model (use of a reserved connection to transmit a possibly continuous stream of information) toward a packet-switched model where the stream of data is broken down into packets for transmission. In today's networks, packets may be lost during transmission for many reasons, which include errors introduced in the channel and packets being dropped from queues in congested network switching devices. A packet which does not arrive at the decoder is called *erased*. In contrast to an error, where a bit or codeword arrives and has the wrong value, with an erasure, the value is missing.

A packet could be erased because of a routing error, in which the packet does not arrive at its intended destination. Packet erasures can also arise from the use of very high-rate error detection codes, such as cyclic redundancy check (CRC) codes. At the receiver, the error detection code embedded in each packet is checked before the packet is passed to the main erasure-correcting code. If the error detection code detects an error, the packet is deleted and is marked as being lost (erased) for the main channel decoder. The main

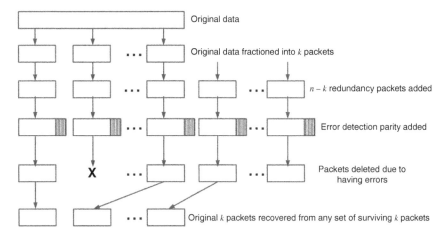

Figure 3.4 Illustration of the use of Reed–Solomon codes to recover from packet erasures.

channel decoder then operates in erasure correction mode. In this way, packets with bit errors introduced during transmission may appear to the main decoder as erased packets rather than as packets with bit errors.

Consider a block of data (e.g. a video frame) that has been partitioned into k packets of fixed size. Assume that a systematic Reed–Solomon code is used for protection. As the code is systematic, the encoder outputs the k input packets and adds $n - k$ extra parity packets. During transmission, some of the packets may be lost. Decoding aims to recover the lost packets using the redundancy in the received packets. Correcting lost packets, or erasure correction, is somewhat simpler than correcting errors because, with erasure correction, the coordinates where data was lost are known. For an (n, k) Reed–Solomon code protecting k packets, the MDS property implies that the code will be able to recover the k original packet from any k of the received packets. In other words, the entire transmitted message of length n packets can sustain the loss of any $n - k$ of the transmitted packets. Figure 3.4 illustrates the scheme for packet loss protection using Reed–Solomon codes. A similar analysis and conclusions can be reached when considering the transmission of a single codeword and the erasure of specific coordinates in the codeword.

3.2 Convolutional Codes

While block codes operate by processing blocks of input symbols, convolutional codes conceptually work on streams of input symbols. These streams may not need to terminate, but there exist techniques that allow them to be terminated when needed. The encoding process for convolutional codes operates sequentially on the input stream, processing the bits representing each symbol in sequence. The encoder in a convolutional code is implemented by a binary shift register. As shown in Figure 3.5, the encoder is formed by a cascade of D-type flip-flops, each of which can be thought of as a single-bit memory cell or unit delay block. The bit at the input of a flip-flop will appear at its output after a delay equal to one processing period (or single-bit duration). For the example illustrated

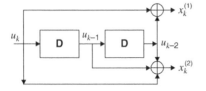

Figure 3.5 An example memory two convolutional encoder.

in Figure 3.5, the encoder has two unit delay elements. At each processing period, each encoded bit is formed by adding modulo-2 a combination of the input and flip–flop outputs. The modulo-2 additions can be implemented using an exclusive-OR gate. Each encoder output is the output from a modulo-2 adder. Therefore, if the encoder has k adders (and consequently k outputs), the rate of the code is $1/k$. For example, the encoder in Figure 3.5 generates a rate 1/2 code. In this example, the encoded bits depend on the current input and on the previous two inputs because the outputs are made from combining the input and some of the outputs from the two unit delay elements in the encoder. The convolutional encoder described so far has a simple structure with only a feed-forward data configuration. While we will focus on this configuration, other encoder configurations exist, notably those using feedback from the unit delay output, which still use the same techniques described next.

3.2.1 Code Characterization: State and Trellis Diagrams

The presence of internal memory in the encoder determines the core properties of a convolutional encoder. Importantly, the states of the memories at any time instant determine an encoding state. With each new input bit, the encoder will change from a previous state to a new state. Consider the time instant when the kth bit, u_k, is presented to the input of an encoder with M unit delay elements. The state of the encoder is designated as the output values of all memory elements: $S_k = \{u_{k-1}, u_{k-2}, \ldots, u_{k-M}\}$. Depending on whether u_k equals 0 or 1, the encoder will evolve from the state $S_k = \{u_{k-1}, u_{k-2}, \ldots, u_{k-M}\}$ into the state $S_{k+1} = \{0, u_{k-1}, \ldots, u_{k-M+1}\}$ or $S_{k+1} = \{1, u_{k-1}, \ldots, u_{k-M+1}\}$. The encoder evolves through a sequence of internal states that follow the sequence of bits presented at the input. This evolution can be represented in a state transition diagram like the one shown in Figure 3.6, which corresponds to the encoder in Figure 3.5. Each state transition diagram will depend on the number of unit delay memory elements and the encoder output structure. In the diagram, states are represented with circles and the branches represent the transition from one state to the next. The notation next to each branch indicates the value of the input that causes the transition and the values of the encoder outputs using the notation: $u_k/x_k^{(1)}x_k^{(2)}$. With a sequence of input bits, the encoding process can be visualized as a corresponding sequence of states. For example, suppose that the input to the encoder in Figure 3.5 is the sequence of bits 1, 1, 1, 0, 0. Assuming the initial state is {0,0}, with all the memory content initialized to zero, the sequence of states will be {0,0}, {1,0}, {1,1}, {1,1}, {0,1}, {0,0}. For this sequence of state transitions, the output from the encoder will be 11, 10, 01, 10, 11.

The state transition diagram is a useful tool to visualize the convolutional encoder operation and its properties. An alternate representation uses a trellis diagram that unfolds over time the state diagram by showing all possible evolutions through the different states of

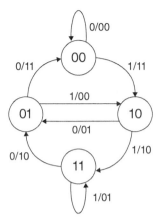

Figure 3.6 The state transition diagram corresponding to the encoder in Figure 3.5.

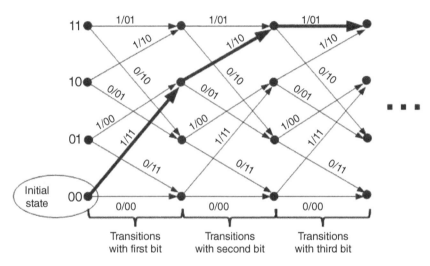

Figure 3.7 The trellis diagram corresponding to the possible state transitions for the encoder in Figure 3.5. In bold, the transitions occurring when the input bit sequence is 1, 1, 1.

the convolutional encoder. In the trellis diagram of Figure 3.7, based on the encoder in Figure 3.5, a repeated pattern of the same stage is shown as time advances from left to right. The diagram represents states with nodes and transitions from one state to the next with arrows. Each arrow includes the notation of the input bit causing the transition and the resulting output bits. Each of the repeated stages shows all the possible transitions at a given processing period from all the states at the start of that period (appearing ordered in a column at the left of the stage). The figure also shows with bold arrows the transitions that follow from the initial state (assumed to be {0,0} and appearing at the lower left corner) when the input bit sequence is 1, 1, 1. The diagram shows the sequence of states that follows from a sequence of input bits as a path through the trellis. This idea is exploited in the design of the decoder, since this operation can be thought of as estimating the path through the trellis given a sequence of received bits. Once the path is known (or estimated), the original sequence of encoded bits is known as the one that created the path. This decoding idea was formalized in the form of maximum likelihood (ML) decoding and was efficiently implemented using the Viterbi algorithm.

3.2.2 Maximum Likelihood (ML) Decoding

To study ML decoding, we start by expanding the notation. As introduced earlier, $\vec{u} = \{u_1, u_2, \ldots, u_N\}$ is the N-bit information sequence that is input to the convolutional encoder. It results in the output $\vec{x} = \{x_1^{(1)}, x_1^{(2)}, \ldots, x_1^{(K)}, x_2^{(1)}, x_2^{(2)}, \ldots, x_N^{(1)}, x_N^{(2)}, \ldots, x_N^{(K)}\}$ sequence with rN output bits, where r is the channel code rate, calculated as the ratio of the number of encoder input bits to the number of encoder output bits. This sequence of bits is modulated into an electrical signal to be transmitted. We will assume the modulation is BPSK. In terms of the digital section of the transmit–receive chain, this means that the bits in the sequence $\vec{x}$ are mapped from values 0 and 1 to values ±1. We denote the modulated sequence of bits as $\vec{v} = \{v_1^{(1)}, v_1^{(2)}, \ldots, v_1^{(K)}, v_2^{(1)}, v_2^{(2)}, \ldots, v_N^{(1)}, v_N^{(2)}, \ldots, v_N^{(K)}\}$. After going through a communication channel, where the signal is attenuated and noise is added, the received sequence is denoted as $\vec{r} = \{r_1^{(1)}, r_1^{(2)}, \ldots, r_1^{(K)}, r_2^{(1)}, r_2^{(2)}, \ldots, r_N^{(1)}, r_N^{(2)}, \ldots, r_N^{(K)}\}$. After decoding, the original sequence of bits $\vec{u}$ is estimated as the sequence $\vec{y} = \{y_1, y_2, \ldots, y_N\}$.

The function of all decoders is to use the sequence $\vec{r}$ to calculate $\vec{y}$ as the estimate for $\vec{u}$. For an ML decoder, this is done by calculating the conditional probability distributions $p(\vec{r}|\vec{y})$, for all the possible $\vec{y}$, and then choosing the particular $\vec{y}$ that results in the maximum of all the probability distributions. An alternate approach, called the *maximum a posteriori* (MAP) decoder, selects the estimate $\vec{y}$ that maximizes the probability distribution $p(\vec{y}|\vec{r})$. The probability distributions for the ML and MAP decoders are related by Bayes' rule: $p(\vec{r}|\vec{y})p(\vec{y}) = p(\vec{y}|\vec{r})p(\vec{r})$. When all input sequences of bits have the same probability of occurrence (sources often follow this model, and this is a common assumption), then it is straightforward to show that ML and MAP are equivalent.

To develop the ML decoder, then, it is necessary to compute the probability distributions $p(\vec{r}|\vec{y})$. The sequence of bits $\vec{y}$ can be mapped into a sequence of states $\vec{S} = \{S_1, S_2, \ldots, S_N\}$, which can be visualized as a path in the code's trellis diagram. This means that the solution to the ML decoder is the sequence of bits $\vec{y}$, or the trellis path, that results in the maximum conditional probability for the received bits $\vec{r}$. To simplify the analysis, we assume that the channel is memoryless, which means that the channel effect on one transmitted bit is independent of the effect on any preceding or succeeding bit. In this case, we can write

$$p(\vec{r}|\vec{y}) = \prod_{i=1}^{N}\prod_{j=1}^{K} p(r_i^{(j)}|y_i). \tag{3.13}$$

In implementing this computation, it is more convenient to accumulate partial results rather than to multiply them. The process to achieve this relies on the fact that is equivalent to find the maximum of $\ln(p(\vec{r}|\vec{y}))$ or $p(\vec{r}|\vec{y})$. Therefore, taking natural logarithms of both sides of (3.13) results in

$$\ln(p(\vec{r}|\vec{y})) = \sum_{i=1}^{N}\sum_{j=1}^{K} \ln(p(r_i^{(j)}|y_i)). \tag{3.14}$$

The task of the ML decoder is to explore all possible sequences $\vec{r}$, calculating $\ln(p(\vec{r}|\vec{y}))$ for each of them, and find the sequence that results in the maximum such log-probability. In Figure 3.7, each branch corresponds to the received bits at a given stage, so an equivalent way of looking at this task is to explore all possible paths along the code trellis (each corresponding to a different $\vec{r}$) and choose the one that results in the maximum $\ln(p(\vec{r}|\vec{y}))$. This is a potentially computationally intensive task because of the very large number of possible paths to be evaluated. Fortunately, the Viterbi algorithm provides the means to perform ML decoding efficiently.

3.2.3 The Viterbi Algorithm

Examining (3.14), we see that $\sum_{j=1}^{K} \ln(p(r_i^{(j)}|y_i))$ reflects the probability distribution conditioned on the received encoder output for the ith encoded bit (ith stage or ith branch of the trellis) for a choice of possible received bit $r_i^{(j)}$. This is the basic computation in the Viterbi algorithm and receives the name of "ith branch metric." Denoting the ith branch metric as $M(r_i, y_i) = \sum_{j=1}^{K} \ln(p(r_i^{(j)}|y_i))$, (3.14) can be written as:

$$\ln(p(\vec{r}|\vec{y})) = \sum_{i=1}^{N} M(r_i, y_i). \tag{3.15}$$

In an analogous way, it is possible to define the "path metric" $M(\vec{r}|\vec{y})$ as the computation of $\ln(p(\vec{r}|\vec{y}))$ along a complete trellis path, $M(\vec{r}|\vec{y}) = \ln(p(\vec{r}|\vec{y})) = \sum_{i=1}^{N} M(r_i, y_i)$, and the "$k$th partial path metric" as the computation of a partial result for $\ln(p(\vec{r}|\vec{y}))$ consisting of the accumulation of the ith branch metrics along the first k stages of the trellis path, $M^k(\vec{r}|\vec{y}) = \sum_{i=1}^{k} M(r_i, y_i)$. Figure 3.8 illustrates the calculation of these metrics as well as their relation to the trellis diagram.

Having introduced the various path metrics, we can now present how the Viterbi algorithm works. Assuming that the initial state is {0,0} because the encoder started with its internal memories cleared, a kth partial path metric is calculated for each possible ending state along the kth stage in the trellis. Because the initial state is {0,0}, for the first partial path metric there are only two possible ending states: {0,0} and {1,0}. Therefore, the first partial path metric will be $M^1(\vec{r}|\vec{y}) = M(r_1 = 00|y_1 = 0)$ for the ending state for the first stage {0,0}, and $M^1(\vec{r}|\vec{y}) = M(r_1 = 11|y_1 = 1)$ for the ending state for the first stage {1,0}. Calculating the kth partial path metrics for the first few trellis stages is straightforward because fixing the initial state means that only a subset of all the states can happen. Nevertheless, as Figure 3.8 illustrates, eventually all states become possible ending states for a stage and all the branches are possible transitions for a trellis stage. The measure of the likelihood of a branch is given by the branch metric, with more likely transitions (branches) having a larger metric. As Figure 3.8 illustrates, after the few first stages, there are two paths joining at any given state. The kth partial path metric at these states equals that of the kth partial path with largest partial path metric. Therefore, at each state along the trellis, every time that the corresponding kth partial path metric is computed, one partial path is eliminated from the metric (that with smallest kth partial path metric) and the other, called the *surviving partial path*, is the one that makes the kth partial path metric. In this way, the Viterbi algorithm addresses the issue of large computational complexity by eliminating one partial path at each trellis node (code state). Following this procedure,

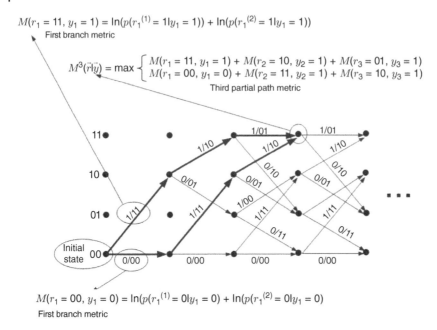

$M(r_1 = 11, y_1 = 1) = \ln(p(r_1^{(1)} = 1|y_1 = 1)) + \ln(p(r_1^{(2)} = 1|y_1 = 1))$
First branch metric

$$M^3(\bar{r}|\bar{y}) = \max \left\{ \begin{array}{l} M(r_1 = 11, y_1 = 1) + M(r_2 = 10, y_2 = 1) + M(r_3 = 01, y_3 = 1) \\ M(r_1 = 00, y_1 = 0) + M(r_2 = 11, y_2 = 1) + M(r_3 = 10, y_3 = 1) \end{array} \right.$$
Third partial path metric

$M(r_1 = 00, y_1 = 0) = \ln(p(r_1^{(1)} = 0|y_1 = 0) + \ln(p(r_1^{(2)} = 0|y_1 = 0)$
First branch metric

Figure 3.8 Example Viterbi decoding algorithm metrics from the trellis in Figure 3.5.

each end state of the last stage will have one associated path metric. This is used to identify the path associated with the most likely sequence of states.

How to use the path metrics depends on whether the convolutional code is terminated or not. A terminated convolutional code is one where, after processing all information bits at the encoder, additional bits equal to zero are input to the encoder to drive the encoder back to the initial state {0,0}. The number of bits equal to zero that are required to flush out the encoder is equal to the number of encoder memory units. In this case, the surviving path along the trellis is the one ending at state {0,0}. A non-terminated convolutional encoder is one where there are no extra bits input at the end, and thus the encoder is equally likely to end in any state. In this case, the surviving path is the one with the largest path metric across all states at the end of the trellis. Just as it is possible for the encoder to end processing in any state, so it is also possible for the encoder to start at a random, unknown state. This occurs when the input to the encoder is an infinite (in practice, a very long) bit stream. In this case, the decoder constrains the decoding complexity by breaking down the complete trellis into sections. In an intermediate section of the trellis, both the initial and final states are random with a typically uniform distribution. In this case, the decoding process progresses as if the section was a window focusing into one part of what would be a very large trellis. The decoder would first output the estimated information bits corresponding to the first stage of the window for the path that yields the largest partial path metric at the last stage of the window, then it would shift the window by one stage, calculate the partial path metrics for the last stage of the window (the only one without results for the partial path metrics), check which path across the window yields the largest partial path metric, and continue repeating the same process.

Lastly, we address how branch metrics are computed. Options vary in complexity and in estimation precision. The branch metric computation also depends on the channel model. We consider the canonical case where the channel is assumed to be AWGN and the metric is what is known as a soft metric (as opposed to a hard metric that is quantized to only two possible values).

We start by considering that $r_i^{(j)}$ is a random variable with a Gaussian probability distribution inherited from the AWGN, with mean $y_i \sqrt{E_b}$ and variance N_0, where y_i is the estimated transmitted bit (± 1), E_b is the received energy per bit, and N_0 is the noise variance. Then, the probability density $p(r_i^{(j)}|y_i)$ is:

$$p(r_i^{(j)}|y_i) = \frac{1}{\sqrt{2\pi N_0}} e^{-(r_i^{(j)} - y_i \sqrt{E_b})^2/(2N_0)},$$

and the ith branch metric is

$$
\begin{aligned}
M(r_i, y_i) &= \sum_{j=1}^{K} \ln(p(r_i^{(j)}|y_i)) \\
&= \sum_{j=1}^{K} \ln\left(\frac{1}{\sqrt{2\pi N_0}} e^{-(r_i^{(j)} - y_i \sqrt{E_b})^2/(2N_0)}\right) \\
&= \sum_{j=1}^{K} \left(-\ln(\sqrt{2\pi N_0}) - \frac{(r_i^{(j)} - y_i \sqrt{E_b})^2}{2N_0}\right) \\
&= \frac{1}{2N_0} \sum_{j=1}^{K} \left(r_i^{(j)} y_i \sqrt{E_b} - \left(r_i^{(j)}\right)^2 - E_b y_i^2 - 2N_0 \ln(\sqrt{2\pi N_0})\right) \\
&= \frac{1}{2N_0} \sum_{j=1}^{K} \left(r_i^{(j)} y_i \sqrt{E_b} - 1 - E_b - 2N_0 \ln(\sqrt{2\pi N_0})\right) \\
&= \frac{\sqrt{E_b}}{2N_0} \sum_{j=1}^{K} r_i^{(j)} y_i - \frac{K}{2N_0}\left(1 + E_b + 2N_0 \ln(\sqrt{2\pi N_0})\right),
\end{aligned}
$$

where we have used the fact that since $r_i^{(j)} = \pm 1$ and $y_i = \pm 1$, it follows that $(r_i^{(j)})^2 = 1$ and $y_i^2 = 1$. Now, note that the metric $M(r_i, y_i)$ is used by accumulating it to other similar metrics and then choosing the maximum of these aggregates, all with the same structure where the term $-\frac{K}{2N_0}\left(1 + E_b + 2N_0 \ln(\sqrt{2\pi N_0})\right)$ and the factor $\sqrt{E_b}/(2N_0)$ act as constants that will equally affect all metric calculations and will have no effect on the final result, other than adding repetitive unnecessary calculations. Therefore, the ith branch metric is calculated as:

$$M(r_i, y_i) = \sum_{j=1}^{K} r_i^{(j)} y_i.$$

3.2.4 Error Correction Performance

The performance of convolutional codes can be measured in terms of the probability of decoding failure (decoding error); this is the event where the decoding result is different from the bits that were input into the encoder. Analyzing the error correction performance of convolutional codes depends on the channel model and the decoding procedure. We will focus on transmission over an AWGN channel and Viterbi decoding with soft metric computation. The technique to analyze convolutional code performance relies on the property of convolutional codes being linear codes. This means that the probability of decoding failure can be calculated by assuming that the input to the encoder is the all-zero sequence. The decoding error probability can then be calculated as the probability the decoder chooses a path through the trellis different from the one for the all-zero input sequence (which, following standard notation, usually corresponds to always staying in state zero). Since the convolutional encoding and decoding process is not constrained to a fixed length, the performance of the convolutional code is measured by calculating the probability that a path that separates from and then rejoins the all-zero path for the first time has a path metric larger than the all-zero path. We specify "for the first time" because if the path separates again from the all-zero path after rejoining it, this is considered a different error case. This calculation can be related to the probability that the two paths differ by d bits. Assuming BPSK modulation and using the trellis metric expressions presented previously, this probability can be shown to be

$$P_e(d|\gamma_b) = \frac{1}{2} \operatorname{erfc}\left(\sqrt{dr\gamma_b}\right), \tag{3.16}$$

where $\gamma_b = E_b/N_0$ is the received signal-to-noise ratio (SNR) per bit, r is the channel code rate, and $\operatorname{erfc}(\gamma)$ is the complementary error function: $\operatorname{erfc}(\gamma) = 2/\pi \int_\gamma^\infty e^{-u^2} du$. This expression for $P_e(d)$ is the probability of an error event resulting in a surviving path along the trellis with d bits different from zero. Inspecting a trellis would show that there exist many such error paths that separate from the all-zero path and rejoin without separating again later. These paths have different numbers d of nonzero bits. The minimum possible number d is the free distance for the code and is denoted d_f. Using the union bound on probability, it is possible to derive an upper bound on the probability of decoding error that takes into account all the error paths:

$$P(\gamma_b) \leq \sum_{d=d_f}^{\infty} a(d) P_e(d|\gamma_b), \tag{3.17}$$

where $a(d)$ is a function that returns the number of paths with d nonzero bits separating from the all-zero path and then rejoining again for the first time. There exist analytical methods to calculate d_f and the values $a(d)$, but it is also possible to know these values from tables published for the best-known convolutional codes. In the aforementioned expression for $P(\gamma_b)$, the sum is from d_f to infinity because it is assumed that the code is not terminated. If the code was terminated, this is, it operates on blocks of input bits, the upper limit of the sum becomes equal to the number of nonzero bits in the longest possible nonzero path in the terminated trellis. Nevertheless, because the probability of nonzero paths usually decreases quite rapidly with growing d, the difference in $P(\gamma_b)$ between the terminated/non-terminated cases is frequently negligible. Furthermore, to simplify the analysis of systems with convolutional codes, it is frequently possible to approximate $P(\gamma_b)$ with only the first term (with $d = d_f$) without a significant sacrifice in accuracy.

Similarly to (3.17), it is possible to derive an upper bound for the bit error probability, [5], which is

$$P(\gamma_b) \leq \sum_{d=d_f}^{\infty} c(d) P_e(d|\gamma_b), \tag{3.18}$$

where $c(d)$ indicates the total number of information errors in all paths with Hamming distance d.

3.3 Modified Linear Codes (Puncturing, Shortening, Expurgating, Extending, Augmenting, and Lengthening)

It is frequently useful in joint source–channel coding (JSCC) applications to be able to modify the characteristics of a linear channel code, often called the *mother code*. These modifications allow one to control, for example, the codeword size or the channel code rate. In the next section, we will focus on puncturing. The different modifications to a linear code are:

Puncturing: This refers to the deletion from the codeword of one or more of its parity bits. In cases where the code is not systematic, this operation refers to the deletion of one or more bits from the codeword.

Extending: This operation is the opposite of puncturing, since it consists of adding one or more parity bits to the codeword.

Lengthening: The codeword is modified by adding one bit to the input message part of the codeword.

Shortening: This is the reverse operation from lengthening, as it consists of deleting one bit from the input message part of the codeword.

Expurgating: The encoding output is modified by deleting one or more codewords in such a way that the result remains a linear code (with a weaker error correction capability).

Augmenting: This is the reverse operation from expurgating. Augmenting consists of adding to the encoder output one or more codewords in such a way that the result remains a linear code.

3.4 Rate-Compatible Channel Codes

Being able to adapt the coding rate, and thereby the error correction strength, of a channel code is useful for JSCC applications. It is even more useful when the channel-encoded bit stream presents an embedded structure where the channel coded bit stream of a higher code rate is a prefix of the bit stream encoded at a lower code rate by the same channel encoder. This means that when comparing the bit streams resulting from encoding at two rates, the complete bit stream resulting from the higher channel coding rate appears at the beginning of the bit stream resulting from the lower channel coding rate. The bit stream resulting from the lower channel coding rate follows the bit stream resulting from the higher channel coding rate with extra redundancy bits. A channel code with this property is called a *rate-compatible channel code*.

Rate-compatible channel codes form a family of codes where all of the member codes are derived from a code with the lowest rate, called the *mother code*. A rate-compatible channel code can be generated by puncturing the encoded bit stream of a code following a strategy where codes of higher rates are obtained by removing from the mother code bit stream every bit already removed from the codes with lower rates and puncturing some other extra bits as needed to obtain the intended higher rate. Rate-compatible channel codes can be generated from any linear code, but due to its straightforward implementation and construction, we will use in the following explanation the example of *rate-compatible punctured convolutional* (RCPC) codes [6, 7].

Figure 3.9 shows an RCPC code example derived from the same convolutional code shown in Figure 3.5. The figure shows three codes with the coding rate increasing from

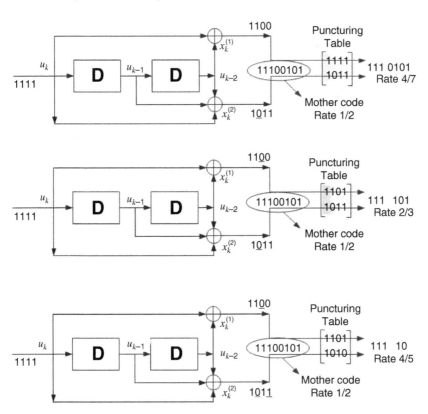

Figure 3.9 Rate-compatible punctured convolutional (RCPC) channel code example.

top to bottom. The mother code in this example is the code from Figure 3.5, which has a coding rate equal to 1/2. All RCPC codes derived from this mother code will have a coding rate larger than 1/2. Of these RCPC codes, the one with smallest coding rate is shown at the top of Figure 3.9. In this case, a code with rate 4/7 is obtained by puncturing one out of every four bits from the bottom output line in the encoder. The deleted bit is shown underlined in the figure. For any RCPC code, the *Puncturing Table* indicates the bits that are to be punctured from the output of the mother code encoder. In the table, each row corresponds to a different encoder output and the columns indicate the order of the bits in the output stream. An entry 0 in the table indicates a bit to be punctured and an entry 1 indicates a bit not to be punctured. For example, in the top RCPC code in Figure 3.9, the puncturing table

$$\begin{bmatrix} 1 & 1 & 1 & 1 \\ 1 & 0 & 1 & 1 \end{bmatrix}, \tag{3.19}$$

indicates that only one bit is to be punctured: the second bit from every group of four bits in the second encoder output. Because the puncturing table has a finite number of columns, the puncturing pattern indicated in each row will repeat periodically. This is, for the puncturing table (3.19), the bits to be punctured are the second, sixth, tenth, etc., from the second output. The length of this pattern (equal to four for Figure 3.9) is called the puncturing period for the RCPC code. Figure 3.9 shows the construction of the rate 2/3 code in the middle encoder. For this case, the puncturing table is

$$\begin{bmatrix} 1 & 1 & 0 & 1 \\ 1 & 0 & 1 & 1 \end{bmatrix}. \tag{3.20}$$

With this puncturing table, the rate 2/3 code is constructed by puncturing exactly the same bits from the mother code bit stream as the rate 4/7 code (the second bit from the second output), and also puncturing a new bit (the third bit in the first output). Similarly, Figure 3.9 shows the puncturing table for the rate 4/5 code:

$$\begin{bmatrix} 1 & 1 & 0 & 1 \\ 1 & 0 & 1 & 0 \end{bmatrix}. \tag{3.21}$$

Here again, the new code is constructed by puncturing the same bits as the code with lower rate (puncturing table (3.20)) and puncturing a new bit (fourth from the second output). This design for the puncturing tables ensures that for two RCPC codes from the same family (i.e. derived from the same mother code), the bit stream at the output of the lower rate code will contain all the bits that are present in the bit stream at the output of the higher rate code. The coded bit stream from the higher rate code will have missing (punctured) some of the bits in the lower rate code. This is the nature of the embedded and rate-compatible property for RCPC codes. Also, note that in all the cases in Figure 3.9, the puncturing period is maintained equal to four. This is not a strict requirement for RCPC code design; however, maintaining the same puncturing period reduces the complexity of encoding and decoding.

Decoding RCPC codes can be done in the same way as regular convolutional codes. For all the RCPC codes in a family, the decoder can be implemented by using the same

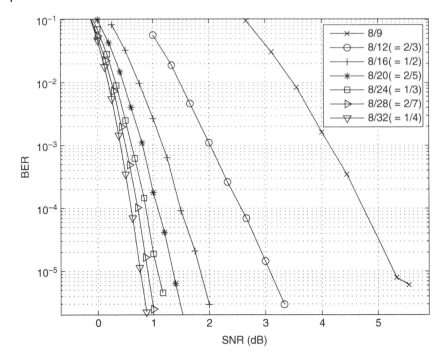

Figure 3.10 Performance of different codes from the same family of RCPC codes with mother code rate 1/4, memory 8 and puncturing period 8. Source: Adapted from [7].

Viterbi decoder based on the same trellis diagram, where the trellis diagram is the one from the mother code. During the decoding of punctured codes, the missing bits are replaced with a value of zero when computing the branch metrics, essentially nulling the contribution of the corresponding branch to the overall path metric. Using the same decoder for all the RCPC codes in the same family has obvious implementation advantages.

Calculating the performance of RCPC codes can also be done in the same way as for regular convolutional codes. The puncturing operation, which results in codes with higher rate, leads to codes with weaker error correcting capability. This means that when evaluating the error performance at a fixed channel SNR (assuming for example an AWGN channel), codes with higher coding rate will exhibit a larger bit error rate (BER). This is illustrated in Figure 3.10, which shows the BER as a function of channel SNR for different RCPC codes from the same family derived from a mother code rate 1/4, memory 8, and puncturing period 8 [7].

Puncturing the mother code affects the parameters that characterize code performance. Specifically, both the free distance, d_f, and the number $a(d_f)$ of trellis paths that separate from the all-zero path and rejoin it after d_f nonzero bits for the first time depend on the code's channel coding rate. Figure 3.11 shows how d_f changes as a function of the channel coding rate r for two representative RCPC code families. The figure shows not only that the free distance decreases with increasing code rate (resulting in weaker error correction

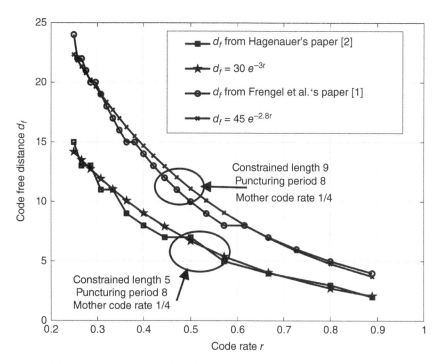

Figure 3.11 Free distance as a function of channel coding rate r for two families of RCPC codes. Source: Adapted from [6] and [7].

capability for codes with higher coding rate), but it also shows that it can be approximated by a function of the form:

$$d_f \approx \kappa e^{-cr}, \tag{3.22}$$

where κ and c are constants. At the same time, Figure 3.12 shows that there is not a clear function that can be used to approximate $a(d_f)$ as a function of channel coding rate. These observations on d_f and $a(d_f)$ will be useful in modeling the performance of RCPC code families.

Channel codes related by a rate-compatibility property are useful in JSCC because they provide a simple method to adapt the channel coding rate, while still using the same decoder in all cases. As will be seen in later chapters, rate-compatible channel codes provide a flexible tool to implement unequal error protection of a source-encoded bit stream. Also, when a feedback channel is available, these codes can be used to provide incremental redundancy. Starting with a weak code (sending as little redundancy as possible), if the transmission is successful, the transmitter can stop, and if it is not, the transmitter sends only the incremental redundancy needed so that, when combined with the weak code redundancy, it will result in a new code of smaller rate, strong enough to achieve a successful transmission. The transmitted incremental redundancy are the bits punctured from the strong code to obtain the weak code used in the first transmission attempt. These hybrid automatic repeat query (ARQ)/Forward Error Correction (FEC) techniques will be discussed in more detail in Chapter 6.

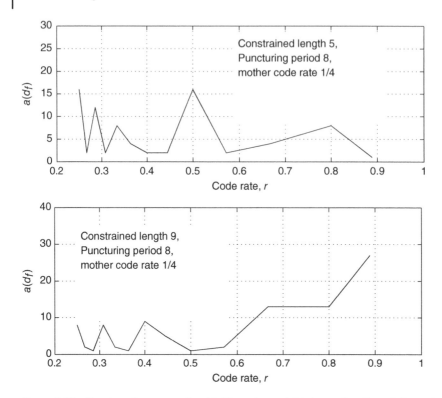

Figure 3.12 Number of error events with Hamming weight d_f as a function of channel coding rate r for two families of RCPC codes.

References

1 Wicker, S. (1995). *Error Control Systems for Digital Communication and Storage*. Prentice Hall.

2 Lin, S. and Costello, D.J. (2004). *Error Control Coding*, 2e. Upper Saddle River, NJ: Prentice-Hall, Inc.

3 Moreira, J.C. and Farrell, P.G. (2006). *Essentials of Error-Control Coding*. Wiley.

4 Ryan, W. and Lin, S. (2009). *Channel Codes: Classical and Modern*. Cambridge University Press.

5 Viterbi, A. (1971). Convolutional codes and their performance in communication systems. *IEEE Transactions on Communication Technology* 19 (5): 751–772.

6 Hagenauer, J. (1988). Rate compatible punctured convolutional (RCPC) codes and their applications. *IEEE Transactions on Communications* 36 (4): 389–399.

7 Frenger, P., Orten, P., Ottosson, T., and Svensson, A.B. (1999). Rate-compatible convolutional codes for multirate DS-CDMA systems. *IEEE Transactions on Communications* 47: 828–836.

4

Concatenated Joint Source–Channel Coding

The most straightforward approach to joint source–channel coding (JSCC) is to keep the source and channel coding blocks as two separate components that are jointly configured. In this approach, the source and channel coding blocks are connected in a serial configuration. The source signal samples are passed through the source encoder, and the output of the source encoder is fed into the channel encoder. While the source and channel codecs are, in principle, designed independently of each other, with concatenated JSCC the joint coding design is implemented by jointly setting their operating parameters and by choosing codec designs that can easily adapt their parameters and interconnect with each other. In this chapter, we will discuss the configuration, design, and performance of concatenated JSCC schemes.

4.1 Concatenated JSCC Bit Rate Allocation

Figure 4.1 shows the typical configuration for concatenated JSCC. On the transmitter side, a block of source samples is fed into the source encoder. The resulting source-encoded data is fed to the channel encoder. The channel-encoded data is then transmitted and, at the receiver, undergoes channel decoding and source decoding to obtain a reconstructed version of the original block of source samples. The source and channel codecs are usually designed independently of each other, yet their main configuration parameter, the source and channel coding rates, can be adapted. The separate source and channel codecs are often designed so that their rates are easily modified. Good examples for this approach are embedded source codecs and rate-compatible punctured channel codecs.

With concatenated JSCC, the main design goal is to jointly allocate the source and channel coding rates. For simplicity, in this chapter, we will consider that the channel is additive white Gaussian noise (AWGN). The design goal is to jointly allocate source and channel coding rates for a given channel signal-to-noise ratio (SNR). We will also consider that the channel transmit bit rate is fixed, so that during one complete encode-transmit-decode cycle, the number of transmitted bits on the channel is fixed to W bits. More sophisticated communication systems will change the transmit bit rate (e.g. by using adaptive modulation), but this is beyond the scope of the concatenated JSCC bit rate allocation problem.

In joint rate allocation, the problem is to balance the *source coding distortion* introduced during quantization and compression at the source encoder, with the *channel-induced*

Joint Source-Channel Coding, First Edition. Andres Kwasinski and Vinay Chande.
© 2023 John Wiley & Sons Ltd. Published 2023 by John Wiley & Sons Ltd.

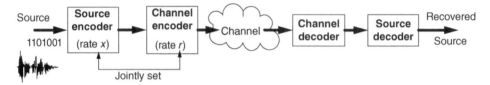

Figure 4.1 Block diagram of a tandem source and channel coding scheme.

distortion associated with channel errors introduced during transmission, so that their combined effect, the *end-to-end distortion*, is minimum. It is useful to have an expression for the end-to-end distortion in terms of source coding and channel-induced distortions. It is often assumed that the end-to-end distortion equals the sum of the distortion due to source coding and the distortion associated with transmission errors. This assumption holds true under some conditions, but is frequently assumed without verifying that the conditions hold. We first summarize the derivation for the end-to-end distortion studied in [1].

Assume that a source signal $s(t)$ is transmitted by first sampling it, followed by quantization and transmission of the samples. At the receiver, the reconstructed signal is denoted $\tilde{s}(t)$. Then, the end-to-end distortion is

$$D = \int_T E\left[(s(t) - \tilde{s}(t))^2\right] dt, \tag{4.1}$$

where T is the length of time over which the source signal is sampled and the distortion is measured. Here the distortion is measured using the mean squared error, as will be the case in this chapter unless otherwise noted. Since $s(t)$ is sampled, we can express it as a reconstruction from the samples s_k, with $k = 1, \ldots, N$. That is, from sampling theory,

$$s(t) = \sum_{k=1}^{N} s_k \phi_k(t),$$

where the functions $\phi_k(t)$ form an orthonormal basis set used to reconstruct the signal $s(t)$ by interpolating the samples s_k. There are multiple possibilities for the choice of basis functions. A common set has the functions

$$\phi_k(t) = \frac{\sin\left(\frac{\pi}{T_s}(t - kT_s)\right)}{\frac{\pi}{T_s}(t - kT_s)},$$

used in the Nyquist–Shannon sampling theorem, where T_s is the sampling period.

For the sampled signal $s(t)$, the samples s_k are quantized and then transmitted through a channel where errors may be introduced. These errors may be corrected with the aid of a forward error correction (FEC) channel code, but the FEC code may be unable to correct some of the error patterns introduced during transmission. The channel code is usually a concatenation of an FEC code and an error detection code that has a very high probability of detecting the errors that could not be corrected by the FEC code. So the source decoder inputs packets of received samples that are marked as being error free or containing an error. As a last line of defense, a well-implemented source decoder has provisions to deal with the unlikely case of a packet containing an error but marked as error free. For packets marked as containing an error, the source decoder will implement a mechanism of *error concealment* to cover up the lost information, for example, by interpolating from neighboring samples that were received, or by using an average value. Error concealment is rarely perfect, so the lost information results in some error when reconstructing the source, which translates into a distortion associated with the transmission errors that could not be corrected through FEC.

The received samples, $\tilde{s}_k$, are used to reconstruct the transmitted signal as:

$$\tilde{s}(t) = \sum_{k=1}^{N} \tilde{s}_k \phi_k(t).$$

Now the end-to-end distortion in (4.1) can be written as

$$D = \int_T E\left[\left(\sum_{j=1}^{N} s_j \phi_j(t) - \sum_{k=1}^{N} \tilde{s}_k \phi_k(t)\right)^2 \right] dt.$$

Because of the orthonormality of the functions $\phi_k(t)$, the end-to-end distortion becomes

$$D = E\left[\sum_{k=1}^{N} (s_k - \tilde{s}_k)^2 \right].$$

Let the quantized sample s_k be denoted as $\hat{s}_k$. Adding and subtracting the quantized samples in the expression for end-to-end distortion yield:

$$D = E\left[\sum_{k=1}^{N} (s_k - \hat{s}_k + \hat{s}_k - \tilde{s}_k)^2 \right]$$

$$= E\left[\sum_{k=1}^{N} (s_k - \hat{s}_k)^2 \right] + E\left[\sum_{k=1}^{N} (\hat{s}_k - \tilde{s}_k)^2 \right] + 2E\left[\sum_{k=1}^{N} (s_k - \hat{s}_k)(\hat{s}_k - \tilde{s}_k) \right]. \tag{4.2}$$

In this result, the first term is the source encoding distortion, D_S, because the orthonormality of the functions $\phi_k(t)$ implies that

$$D_S = \int_T E\left[\left(\sum_{j=1}^{N} s_j \phi_j(t) - \sum_{k=1}^{N} \hat{s}_k \phi_k(t)\right)^2 \right] dt = E\left[\sum_{k=1}^{N} (s_k - \hat{s}_k)^2 \right].$$

The second term is the distortion associated with transmission errors, that is, the mean squared error (MSE) between the transmitted signal (the signal of quantized samples) and the reconstructed signal. Denoting this distortion as D_C, and using the orthonormality of the functions $\phi_k(t)$, we have that

$$D_C = \int_T E\left[\left(\sum_{j=1}^{N} \hat{s}_j \phi_j(t) - \sum_{k=1}^{N} \tilde{s}_k \phi_k(t)\right)^2 \right] dt = E\left[\sum_{k=1}^{N} (\hat{s}_k - \tilde{s}_k)^2 \right].$$

Therefore, (4.2) becomes

$$D = D_S + D_C + 2E\left[\sum_{k=1}^{N} (s_k - \hat{s}_k)(\hat{s}_k - \tilde{s}_k) \right]. \tag{4.3}$$

The value of the last term depends on the interrelation between samples at different locations in the system: s_k, $\tilde{s}_k$, and $\hat{s}_k$. To study this, we first exchange the order of summation and expectation, $E[\sum_{k=1}^{N} (s_k - \hat{s}_k)(\hat{s}_k - \tilde{s}_k)] = \sum_{k=1}^{N} E[(s_k - \hat{s}_k)(\hat{s}_k - \tilde{s}_k)]$, and then consider one term $E[(s - \hat{s})(\hat{s} - \tilde{s})] = E[s\hat{s}] - E[s\tilde{s}] - E[\hat{s}^2] + E[\hat{s}\tilde{s}]$, where the subscripts k are removed for simplicity.

It was shown in [1] that when the quantizer is designed to minimize the average quantization distortion (see (2.3) in Chapter 2), it follows that

$$E[s\hat{s}] = E[\hat{s}^2], \text{ and } E[s\tilde{s}] = E[\hat{s}\tilde{s}], \tag{4.4}$$

which implies that the last term in (4.3) equals zero. Therefore, when using such a quantizer designed to minimize average quantization distortion, the end-to-end distortion equals the sum of the source encoding distortion and the channel-induced distortion:

$$D = D_S + D_C. \tag{4.5}$$

Note that this result makes no assumptions on the independence of the quantization and the channel noises.

It will be useful to apply this result to the frequent case when the error concealment is implemented by replacing the lost samples with their expected value. That is, if we assume that the source has zero mean, the received samples $\tilde{s}_k$ that were lost because of uncorrectable transmission errors are replaced with the value $E[s_k] = 0$. The distortion introduced with this error concealment operation is (for a single sample),

$$D_C = E[(\hat{s} - \tilde{s})^2] = E[\hat{s}^2].$$

Now, by expanding the expression for source encoding distortion we get:

$$D_S = E[(s - \hat{s})^2] = E[s^2] + E[\hat{s}^2] - 2E[s\hat{s}].$$

From this expression, it can be seen that

$$E[\hat{s}^2] = D_S - E[s^2] + 2E[s\hat{s}].$$

At the same time, as it was shown in [1] that $E[s\hat{s}] = E[\hat{s}^2]$, we can write

$$\begin{aligned} E[\hat{s}^2] &= D_S - E[s^2] + 2E[\hat{s}^2] \\ &= E[s^2] - D_S \\ &= \sigma_s^2 - D_S \\ &= D_C. \end{aligned}$$

From this result, replacing the value of D_C into (4.5) yields the following end-to-end distortion when there are uncorrected errors after FEC decoding and the error concealment is implemented by replacing the lost samples with their expected value:

$$D = D_S + \sigma_s^2 - D_S = \sigma_s^2. \tag{4.6}$$

For the rest of this chapter, we will denote as D_F this end-to-end distortion, to differentiate it from the end-to-end distortion resulting from the scenario with no uncorrected transmission errors.

In summary, a transmitted quantized source sample may be received with no errors (in which case the end-to-end distortion is $D = D_S$), or it may be affected by channel errors that cannot be corrected by the channel decoder (in which case the end-to-end distortion is $D_F = D_S + D_C = \sigma_s^2$ for the case of a zero-mean source). Given a probability $P_r(\gamma)$ that the transmission has uncorrectable errors, where γ is the channel SNR, the overall average end-to-end distortion $D(x, r, \gamma)$ is

$$D(x, r, \gamma) = D_F P_r(\gamma) + D_S(x) \left(1 - P_r(\gamma)\right), \tag{4.7}$$

where x is the source encoding rate in bits/sample, and r is the channel coding rate. Equation (4.7) shows that when transmission is accomplished with no errors, an event that has probability $1 - P_r(\gamma)$, the end-to-end distortion equals the source encoding distortion $D_S(x)$. When the channel codec is unable to correct all the channel errors, an event with probability $P_r(\gamma)$, some or all of the source samples need to be discarded and replaced through an estimation or error concealment mechanism. Bear in mind that (4.7) depicts a result that was specialized to the frequent case of a zero-mean source; the general expression is

$$\begin{aligned} D(x, r, \gamma) &= (D_S + D_C)P_r(\gamma) + D_S(x) \left(1 - P_r(\gamma)\right) \\ &= D_C P_r(\gamma) + D_S(x). \end{aligned} \tag{4.8}$$

Note that in all the expressions for overall average end-to-end distortion, the relative weight of the distortion introduced by transmission errors is given by the probability of uncorrectable errors.

The value of D_F may depend on the source statistics, the level of compression, and the algorithm used by the source decoder to conceal errors due to lost data. In most cases, D_F is considered to be much larger than $D_S(x)$. Unless noted otherwise, for the remainder of this chapter we will consider that D_F is a constant. In some cases, it is possible to calculate the value for D_F. One such case is when the source sample statistics follow a normalized Gaussian distribution, and the missing source samples are concealed with the expected value for the source samples, $E[s] = 0$. In this case, D_F becomes $D_F = \sigma_S^2 = 1$.

Since end users perceive the quality of the communication process through the reconstructed source samples, the end-to-end distortion indicates the complete effects affecting the transmission. Therefore, in a concatenated JSCC system, the joint allocation of

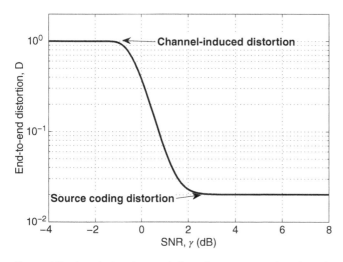

Figure 4.2 A typical end-to-end distortion curve as a function of the channel SNR.

source and channel coding rates will aim to minimize the end-to-end distortion. Figure 4.2 illustrates typical results from plotting the end-to-end distortion as a function of channel SNR when source coding and channel coding rates are fixed. At high SNR, the fixed channel coding rate results in enough protection against channel errors, and the probability of uncorrectable errors $P_r(\gamma)$ is close to zero, so $D(x, r, \text{high } \gamma) \approx D_S(x)$. At low SNR, the opposite happens, as the chosen channel coding rate introduces insufficient redundancy and $P_r(\gamma) \approx 1$. In this case, the end-to-end distortion is approximately equal to the channel-induced distortion: $D(x, r, \text{low } \gamma) \approx D_F$. Between the two regimes of high and low SNR, there is a transition region where for decreasing SNR, the distortion increases from $D_S(x)$ to D_F. Importantly, because source and channel coding rates are fixed, in the transition region the distortion only increases because the rate (or probability) of uncorrected channel errors increases when the channel SNR decreases.

In concatenated JSCC, both the source and the channel coding rates are jointly chosen based on the channel SNR. Typically, and under practical circumstances, both coding rates cannot be chosen from within a continuous range of values. Even when the codecs are designed to adapt the coding rate with very fine granularity, there is a set with a finite number of discrete choices. Let $\mathcal{X} = \{x_1, x_2, \ldots\}$ and $\mathcal{R} = \{r_1, r_2, \ldots\}$ be the sets of possible source encoding rates and channel coding rates, respectively. Each combination of a source and a channel coding rate forms an operating mode Ω_i. We will denote the set of all operating modes by $\Omega = \{\Omega_i\}$. Source and channel coding rates are related to the transmit bit rate, or more precisely the number of transmitted bits per transmission period W, through the relation $W \geq Nx/r$, where N is the number of source samples processed in one transmission period. Although this inequality in principle allows for many possible choices for x given a fixed r, the only reasonable choice is the largest possible one. This is because given a fixed r, W, N, and a channel SNR, the value for $P_r(\gamma)$ gets fixed and, thus, the largest allowed x will result in the smallest possible $D_S(x)$ and end-to-end distortion. As a result, a useful operating mode can be identified through the choice of either a source coding rate from $\mathcal{X}$ or a channel coding rate from $\mathcal{R}$; the other coding rate will be the one that would make $W - Nx/r$ closest to zero but positive. For the rest of the chapter, we choose to define the operating mode based on the channel coding rate: $\Omega = \{\Omega_i\} \equiv \{r_i\}$, and we will adapt the notation accordingly, e.g. $P_{r_i}(\gamma) \equiv P_{\Omega_i}(\gamma)$.

Now, (4.7) can be interpreted as the end-to-end distortion achieved when using one operating mode. We emphasize this by rewriting it as:

$$d_{\Omega_i}(\gamma) = D_F P_{\Omega_i}(\gamma) + D_S(\Omega_i)\left(1 - P_{\Omega_i}(\gamma)\right). \tag{4.9}$$

and we will call (4.9) the *single-mode D-SNR curve*. For each choice of operating mode, there is a different single-mode D-SNR curve. Figure 4.3 shows, on a single plot, all the single-mode D-SNR curves that can be obtained when using a family of rate-compatible punctured convolutional (RCPC) channel codes with mother code rate 1/4, $K = 5$, puncturing period 8 and available channel coding rates {8/9, 4/5, 2/3, 4/7, 1/2, 4/9, 4/10, 4/11, 1/3, 4/13, 2/7, 4/15, 1/4}. In the figure, the source samples are assumed to follow a standard Gaussian distribution and transmission uses binary phase shift keying (BPSK) modulation. The curve in Figure 4.2 is a single-mode D-SNR curve and, in fact, is the sixth single-mode D-SNR curve from the bottom in Figure 4.3 (channel coding rate 4/9).

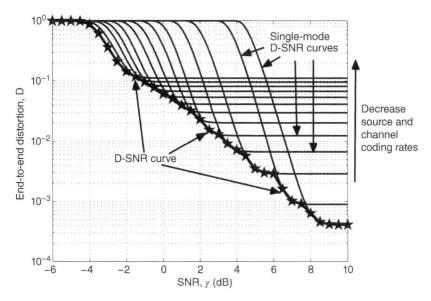

Figure 4.3 The single-mode D-SNR curves and the resulting D-SNR curve.

Figure 4.3 provides insight into the basic trade-offs associated with joint bit rate allocation for concatenated JSCC. The mechanism linking source and channel coding rate implies that choosing a smaller channel coding rate means more redundancy and more channel error protection, but also dictates using a smaller source coding rate. To examine the relation between two single-mode D-SNR curves, we consider that the operating mode indexes are sorted in increasing channel coding rate order: $i > j$ (from Ω_i and Ω_j) if $r_i > r_j$ ($i > j$ also means that $x_i > x_j$). Since, for each single-mode D-SNR curve, the distortion only increases due to more frequent uncorrected channel errors, the single-mode D-SNR curve is practically a constant equal to $D_S(\Omega_i)$ for as long as the SNR is high enough for the channel code to correct most channel errors. Clearly, over the section where two single-mode D-SNR curve are approximately constant, the one with a larger source encoding rate presents lower distortion: $D_S(\Omega_j) > D_S(\Omega_i)$ for $i > j$. Let γ_i^* be the SNR value at which the distortion of the mode-Ω_i single-mode D-SNR curve starts to noticeably increase. For well-behaved channel codes and $i > j$, we have that $\gamma_i^* > \gamma_j^*$ because $r_i > r_j$, i.e. channel errors start to become significant at a channel SNR that is larger for weaker channel codes. The overall effect is that two curves $d_{\Omega_i}(\gamma)$ and $d_{\Omega_{i+1}}(\gamma)$, cross each other at a channel SNR that lies between γ_i^* and γ_{i+1}^*. This can be expressed through the relations $d_{\Omega_{i+1}}(\gamma) < d_{\Omega_i}(\gamma)$ for $\gamma > \gamma_{i+1}^*$ and $d_{\Omega_{i+1}}(\gamma) > d_{\Omega_i}(\gamma)$ for $\gamma < \gamma_i^*$.

The interplay between single-mode D-SNR curves can be observed in practically all systems using concatenated JSCC because it follows simple phenomena that apply to most well-designed source and channel codecs. To illustrate this, Figures 4.4 and 4.5 show the same curves as Figure 4.3 applied to speech and video. Figure 4.4 shows speech source-encoded using the the global system (for mobile) communications (GSM) advance multi-rate (AMR) codec, [2], and then channel-encoded using the RCPC channel codec used for Figure 4.3. For this system, the transmit bit rate was 24.4 Kbps, and there were

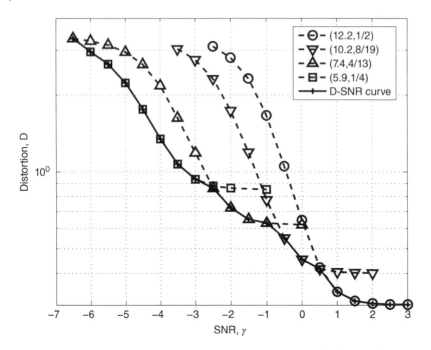

Figure 4.4 The single-mode D-SNR curves and the resulting D-SNR curve for a speech source.

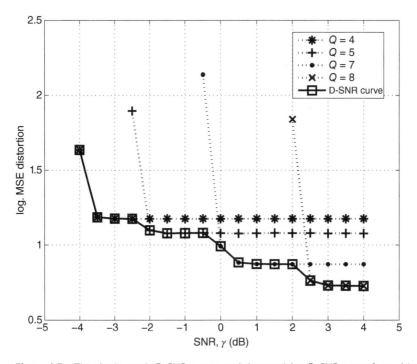

Figure 4.5 The single-mode D-SNR curves and the resulting D-SNR curve for a video source.

four possible (source coding rate, channel coding rate) operating modes: (12.2Kbps, 1/2), (10.2Kbps, 8/19), (7.4Kbps, 4/13), and (5.9Kbps, 1/4). In this figure, the distortion is measured as $4.5 - Q$, where Q is the result of the International Telecommunications Union (ITU)-T perceptual evaluation of speech quality (PESQ) measure standard P.862 [3]. As described in Chapter 1, PESQ is a perceptually based quality measure with output values ranging between 4.5 and 1 [4], where higher values mean better quality. PESQ scores accurately predict those obtained in a typical subjective test involving polling a panel of human listeners (the mean opinion score [MOS] test [5]). Figure 4.5 shows the results for a single frame (frame number 190, the tenth P frame after an I frame) of the 30 frames per second quarter common intermediate format (QCIF) video sequence Foreman encoded with the Moving Picture Experts Group (MPEG)-4 codec, with source coding rate controlled through the choice of quantization factor and protected with an RCPC family of codes with the same characteristics as for Figure 4.3, but with a constraint length $K = 9$. Both the speech and video source codecs implemented error concealment mechanisms. Overall, these figures illustrate that, despite the very different settings and measurement details, the mechanisms dictating the interplay between single-mode D-SNR curves still apply.

Figures 4.3, 4.4, and 4.5 show the nature of the problem of joint bit rate allocation for concatenated JSCC. For each channel SNR, it is necessary to identify the operating mode that results in the minimum end-to-end distortion. The design problem can be written as

$$\min_{\Omega} \left\{ D_F P_{\Omega_i}(\gamma) + D_S(\Omega_i) \left(1 - P_{\Omega_i}(\gamma)\right) \right\}, \tag{4.10}$$

such that $W \geq Nx/r$

The solution to this problem is the operating mode that results in the minimum end-to-end distortion for a given channel SNR. When considering a range of channel SNRs, the solutions of (4.10) form a function relating the minimum end-to-end distortion with the channel SNR. We will call this function the *D-SNR curve*. Figures 4.3, 4.4, and 4.5 illustrate this curve. The D-SNR curve is composed of sections of different single-mode D-SNR curves. These sections correspond to the ranges of channel SNR for which a single-mode D-SNR curve shows less distortion than the rest. The figures also illustrate how concatenated JSCC is implemented in practice: a table at the transmitter stores the operating mode associated with each minimum end-to-end distortion segment of the D-SNR curve, and the channel SNR values delimiting that segment. The operating mode is chosen based on the information provided by a channel estimation algorithm, similar to the implementation of adaptive modulation schemes in modern communication systems. This underscores one of the advantages of concatenated JSCC: its implementation simplicity.

4.2 Performance Characterization

Characterizing the performance of concatenated JSCC means finding expressions for the D-SNR curve, as this represents the solution to the design problem (4.10).

4.2.1 Practical Source and Channel Codecs

As can be inferred from the way concatenated JSCC is implemented, when considering the use of practical source and channel codecs, the result of solving (4.10) does not lend itself to closed form expressions for the D-SNR curve. Nevertheless, the D-SNR curve can be studied by approximating it through a curve determined by a subset of its points. Each of these points, belonging to a section that is part of the solution to (4.10), is selected from a different single-mode D-SNR curve. As illustrated in Figure 4.6, for a point $(\gamma, d_{\Omega_{i+1}}(\gamma))$ from the $i + 1$st single-mode D-SNR curve to belong to the overall D-SNR curve it must follow $D_S(\Omega_{i+1}) \leq d_{\Omega_{i+1}}(\gamma) \leq D_S(\Omega_i)$, approximately. Figure 4.6 shows on top an enlarged section of Figure 4.3, concentrating on showing the overall D-SNR curve and the contribution of two contiguous single-mode D-SNR curves. On the bottom of Figure 4.6, only the bottom single-mode D-SNR curve is shown, divided into three sections: one where the distortion is practically equal to the source encoding distortion as channel errors introduce negligible distortion, a second section that is part of the overall D-SNR curve and a third section,

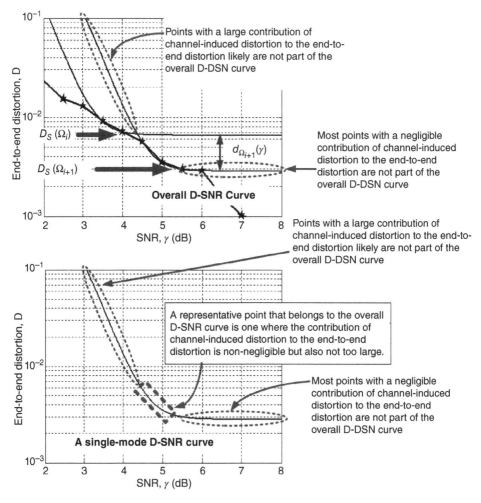

Figure 4.6 Illustration of points from a single-mode D-SNR curve most likely to belong to the overall D-SNR curve.

characterized by a large contribution of channel-induced distortion to the overall distortion. The figure illustrates that the points from a single-mode D-SNR curve that are most likely to be part of the overall D-SNR curve are those where the contribution of channel errors to the end-to-end distortion is small but not so small that the overall distortion nearly equals the source encoding distortion. Points that do not fall in this category, that is, points with a relative contribution of channel errors that is large or negligible, would usually not be part of the overall D-SNR curve. Then, for each single-mode D-SNR curve, the point chosen to represent the D-SNR curve is such that $d_{\Omega_i}(\gamma) = (1 + \Delta)D_S(\Omega_i)$, $\forall \Omega_i$, where Δ is a small number that accounts for the relative contribution of channel errors. Formally, all points such that

$$D(\gamma) = (1 + \Delta)D_S(\Omega_i) = D_F P_{\Omega_i}(\gamma) + D_S(\Omega_i)\left(1 - P_{\Omega_i}(\gamma)\right), \tag{4.11}$$

form the overall D-SNR curve. Equivalently, from (4.11), the D-SNR curve is formed by those points where the probability of post-channel decoding errors is

$$P_{\Omega_i}(\gamma) = \frac{\Delta}{\dfrac{D_F}{D_S(\Omega_i)} - 1}. \tag{4.12}$$

While JSCC schemes often involve a complex interplay of effects, (4.12) simply translates the D-SNR curve characterization problem into a problem in FEC coding. In solving this problem, we could also use the fact that channel-induced errors are relatively few, and thus apply approximations that are accurate in the low post-decoding error rate regime. The expression (4.12) also presents a road map for deriving an approximate expression for the D-SNR curve: For most channel code families, an expression that relates the frame error probability with channel SNR is well known. This relation is normally expressed using some parameters that specify the error correcting capability of the channel code (for example, the minimum distance). This relation can be matched to (4.12), linking end-to-end distortion with channel SNR.

Consequently, we need to first work with the probability of a source block with post-channel decoding errors, which we denote next using the simplified notation $P(\gamma)$. If the source block contains b bits, $P(\gamma)$ is related with the probability of a bit error $P_b(\gamma)$ as $P(\gamma) = 1 - (1 - P_b(\gamma))^b$. Next, when the channel codes are RCPC codes [6, 7], using this relation and following the review in Chapter 3, we can approximate the probability of a source block with post-channel decoding errors as [8]:

$$P(\gamma) \approx 1 - \left(1 - \sum_{d=d_f}^{\infty} a(d)P_e(d|\gamma)\right)^{Nx}, \tag{4.13}$$

where d_f is the free distance of the code and $a(d)$ is the number of error events with Hamming weight d and probability of occurrence $P_e(d|\gamma)$. For BPSK modulation, $P_e(d|\gamma) = .5\,\mathrm{erfc}\sqrt{d\gamma}$ [6], where $\mathrm{erfc}(\gamma)$ is the complementary error function: $\mathrm{erfc}(\gamma) = 2/\pi \int_\gamma^\infty e^{-u^2}\,du$. Although the presentation next assumes the use of BPSK modulation, other popular modulation schemes can be represented by a similar form using a weighted complementary error function with the square root of the SNR as a factor in the argument. Equation (4.13) follows from a bound that is tight at low bit error rate (BER) operation, which is a condition equivalent to our case of Δ being a small number. In this regime, it can be assumed that most error events are those with $d = d_f$, in which case (4.13) can be written as

$$P(\gamma) \approx 1 - \left(1 - \frac{a(d_f)}{2} \, \mathrm{erfc} \sqrt{d_f \gamma}\right)^{Nx}. \tag{4.14}$$

As discussed earlier, the probability of uncorrected channel errors is very dependent on the channel coding rate, or equivalently, on the operating mode (e.g. see (4.10)). This may seem to have been lost in (4.14), but this is not the case because both d_f and $a(d_f)$ depend on the channel coding rate. As shown in Figure 3.11, the free distance d_f can be approximated by (3.22), which for the application discussed here can be further written as

$$d_f \approx \kappa e^{-cr} \approx \kappa e^{-cNx/W},$$

where recall that κ and c are constants. Note that in this equation, we have approximated $r \approx Nx/W$. This follows the earlier discussion; to minimize end-to-end distortion, the source coding rate and channel coding rate need to be chosen so that the inequality $W \geq Nx/r$ is as close as possible to an equality. Also, Figure 3.12 in Chapter 3 shows that there is not a clear function that can be used to approximate $a(d_f)$ as a function of channel coding rate. In what follows, $a(d_f)$ will be approximated through its average value,

$$\bar{a} = \sum_\pi \frac{a_\pi(d_f)}{|\pi|},$$

where π is an index into each of the members of the family of RCPC codes used for error protection and $|\pi|$ is the number of codes in the family. It will be seen later that this approximation is sufficient to yield good results. Using the approximations for d_f and $a(d_f)$, (4.14) becomes

$$P(\gamma) \approx 1 - \left(1 - \frac{\bar{a}}{2} \, \mathrm{erfc} \sqrt{\kappa e^{-cNx/W} \gamma}\right)^{Nx}. \tag{4.15}$$

Equation (4.12) can now be equated with (4.15),

$$\frac{\Delta}{\frac{D_F}{D_S(\Omega_i)} - 1} \approx 1 - \left(1 - \frac{\bar{a}}{2} \, \mathrm{erfc} \sqrt{\kappa e^{-cNx/W} \gamma}\right)^{Nx}, \tag{4.16}$$

and then solved for γ to show the explicit relation between γ and source coding rate x.

$$\gamma(x) \approx \frac{1}{\kappa} e^{cNx/W} \left(\mathrm{erfc}^{-1} \left\{ \frac{2}{\bar{a}} \left[1 - \left(1 - \frac{\Delta}{\frac{D_F}{D_S(x)} - 1}\right)^{1/(Nx)} \right] \right\} \right)^2. \tag{4.17}$$

Equation (4.17) shows the channel SNR as a function of the source coding rate. The purpose of deriving this equation is to later relate the source coding rate with the end-to-end distortion, which will result in an approximation for the D-SNR curve. Before doing this, some more work is needed on (4.17). We first rewrite (4.17) as

$$\gamma(x) \approx \frac{1}{d_f(x)} g(x), \tag{4.18}$$

where

$$
g(x) = \left(\text{erfc}^{-1} \left[\frac{2}{\bar{a}} \left[1 - \left(1 - \frac{\Delta}{\frac{D_F}{D_S(x)} - 1} \right)^{1/(Nx)} \right] \right] \right)^2 .
\tag{4.19}
$$

Using the approximation $\text{erfc}^{-1}(y) \approx \sqrt{-\ln(y)}$, [9], $g(x)$ can be approximated as

$$
g(x) \approx \ln\left(\frac{\bar{a}}{2} \right) - \ln\left[1 - \left(1 - \frac{\Delta}{\frac{D_F}{D_S(x)} - 1} \right)^{1/(Nx)} \right]
\tag{4.20}
$$

$$
\approx \ln\left(\frac{\bar{a}}{2} \right) + \theta ,
$$

where we have defined

$$
\theta = -\ln\left[1 - \left(1 - \frac{\Delta}{\frac{D_F}{D_S(x)} - 1} \right)^{1/(Nx)} \right] .
\tag{4.21}
$$

The approximation $\text{erfc}^{-1}(y) \approx \sqrt{-\ln(y)}$ is used here because it is very good in representing the functional behavior of the inverse error function. The modulation scheme has a probability of error that behaves approximately as e^{-x^2} with respect to SNR. This same behavior is seen with many other modulation schemes. Continuing working with (4.21),

$$
\theta = -\ln\left[1 - \left(1 - \frac{\Delta}{\frac{D_F}{D_S(x)} - 1} \right)^{1/(Nx)} \right]
\tag{4.22}
$$

$$
\Rightarrow \ln\left(1 - e^{-\theta} \right) = \frac{1}{Nx} \ln\left(1 - \frac{\Delta}{\frac{D_F}{D_S(x)} - 1} \right)
\tag{4.23}
$$

$$
\Rightarrow -e^{-\theta} \approx \frac{1}{Nx} \left(-\frac{\Delta}{\frac{D_F}{D_S(x)} - 1} \right)
\tag{4.24}
$$

$$
\Rightarrow \theta \approx \ln\left(\frac{N}{\Delta} x \left(\frac{D_F}{D_S(x)} - 1 \right) \right)
$$

$$
\approx \ln\left(\frac{N}{\Delta} x \right) + \ln\left(\frac{D_F}{D_S(x)} \right) ,
\tag{4.25}
$$

where (4.24) follows from the fact that both logarithms in (4.23) are of the form $\ln(1 - y)$ with $y \ll 1$, which can be seen from (4.22) and (4.12) considering that $P_{\Omega_i}(\gamma)$ is typically small. The approximation in (4.25) follows from recognizing that in general $D_F/D_S(x) - 1 \approx D_F/D_S(x)$, i.e. the reconstructed source error after transmission errors is typically much larger than the quantization distortion.

Recall that for well-designed source codecs, the D–R function is convex and decreasing and is frequently considered to be of the form

$$
D_S(x) = \mu \, 2^{-2vx} ,
\tag{4.26}
$$

since by choosing μ and v it can represent, approximate or bound a wide range of practical and theoretical systems. Following (4.26), it can be observed in (4.25) that θ is the sum of $\ln(Nx/\Delta)$ and a term that is approximately linear in x. Since generally $N/\Delta \gg x$, we can approximate $\ln(Nx/\Delta) \approx \ln(N\bar{x}/\Delta)$, with $\bar{x}$ being the average value of x, and consider that the approximately linear term would determine the overall behavior of $\theta(x)$. Using (4.26) and (4.20), it follows that

$$g(x) \approx \ln\left(\frac{\bar{a}}{2}\right) + \theta,$$

$$\approx \ln\left(\frac{\bar{a}}{2}\right) + \ln\left(\frac{N}{\Delta}x\right) + \ln\left(\frac{D_F}{D_S(x)}\right)$$

$$\approx \ln\left(\frac{\bar{a}}{2}\right) + \ln\left(\frac{N}{\Delta}\bar{x}\right) + \ln\left(\frac{D_F}{\mu}2^{2vx}\right)$$

$$\approx \ln\left(\frac{\bar{a}ND_F\bar{x}}{2\Delta\mu}\right) + vx\ln(4). \tag{4.27}$$

Going back to (4.18), and rewriting the channel SNR in decibels we get

$$\gamma_{dB}(x) \approx 10\log\left(\frac{1}{d_f(x)}g(x)\right)$$

$$\approx -10\log(d_f) + 10\log(g(x))$$

$$\approx -10\log(\kappa e^{-cNx/W}) + 10\log(g(x))$$

$$\approx \frac{10cNx}{W}\log(e) - 10\log(\kappa) + 10\log(g(x)), \tag{4.28}$$

Through algebraic operations, it can be shown that the coefficients of Taylor's expansion of $\log(g(x))$ around $\bar{x}$ are of the form $t^i/i!$, with

$$t = \frac{v\ln(4)}{\ln\left(\frac{\bar{a}ND_F\bar{x}}{2\Delta D_S(\bar{x})}\right)}$$

being much smaller than 1 for typical system parameters. This means that the weight of coefficients in Taylor's expansion falls quite rapidly and $\log(g(x))$ can be accurately approximated through a first-order expansion. Combining these facts with (4.17) and (4.20) and denoting

$$\Psi = \ln\left(\frac{\bar{a}ND_F\bar{x}}{2\Delta D_S(\bar{x})}\right),$$

the channel SNR in decibels can be approximated from (4.28) as

$$\gamma_{dB}(x) \approx \frac{10cNx}{W}\log(e) - 10\log(\kappa) + 10\left(\log(\Psi) + \frac{v\log(4)}{\Psi}(x - \bar{x})\right). \tag{4.29}$$

The important outcome of this result is that the channel SNR in decibels can be approximated as a linear function of the source coding rate x, i.e. $\gamma_{dB}(x) \approx A_1 x + B_1$, where

$$A_1 = 10\left(\frac{cN}{W}\log(e) + \frac{v}{\Psi}\log(4)\right),$$

$$B_1 = 10\log\left(\frac{\Psi D_F^{(1/\Psi)}}{\kappa D_S(\bar{x})^{(1/\Psi)}}\right).$$

Equivalently, we can write $x \approx A_2 \gamma_{dB} + B_2$, with $A_2 = 1/A_1$ and $B_2 = -B_1/A_1$, and combining (4.11) with (4.26) get

$$D(\gamma) = (1+\Delta)D_S = \mu(1+\Delta)2^{-2vx},$$

which, in turn, leads to

$$D(\gamma) = \mu(1+\Delta)10^{-v(\log 4)x}$$
$$\approx \mu(1+\Delta)10^{-v(\log 4)(A_2 \gamma_{dB}+B_2)}$$
$$= \mu(1+\Delta)10^{-vB_2(\log 4)}\gamma^{-10vA_2 \log 4}.$$

In this way, we arrive at the final expression approximating the D-SNR curve:

$$D(\gamma) \approx (G_c\gamma)^{-10m}, \tag{4.30}$$

where, from (4.29),

$$m = \frac{v \log(4)}{A_1}$$
$$= \frac{v}{10}\left(\frac{cN}{W \ln(4)} + \frac{v}{\Psi}\right)^{-1}, \tag{4.31}$$

$$G_c = [v(1+\Delta)]^{-1/(10m)}10^{-B_1/10}$$
$$= \frac{\kappa}{\Psi[v(1+\Delta)]^{1/(10m)}}\left(\frac{D_S(\bar{x})}{D_F}\right)^{1/\Psi}, \tag{4.32}$$

The result in (4.30) is the approximation of the D-SNR curve assuming the use of RCPC codes. The result was obtained by first choosing a subset of points to represent the D-SNR curve. These points, one from each single-mode D-SNR curve, were chosen as those where the contribution of channel-induced distortion to the end-to-end distortion is a small fraction Δ of the source encoding distortion, (4.11). Typical of performance analysis for channel codes, this choice translated the problem into one relating channel SNR with the probability of uncorrected errors, (4.12). From this point, it is possible to use channel coding performance analysis techniques to arrive at an expression relating the channel SNR to the channel coding rate, or equivalently, the source encoding rate. Applying this to the original expression (4.11) results in the approximate expression for the D-SNR curve.

This approach can be applied to other channel code families, such as Reed-Solomon (RS) codes. We now consider a family of RS codes operating on b-bit symbols and with parameters (n, k), i.e. the encoder operates at a rate $r = k/n$, encoding k symbols into an n-symbol codeword. The channel code rate is controlled through the choice of k. Let $L = \frac{W}{nb}$ denote the number of codewords in the frame at the output of the channel encoder. For block codes such as RS codes, the probability of having a source block with post-channel decoding errors is $P(\gamma) = 1 - (1 - q(\gamma))^L$, where $q(\gamma)$ is the probability of channel decoder failure. Assuming that the use of a bounded distance decoder and denoting $l = \lfloor \frac{n-k}{2} \rfloor$, we have $q(\gamma)$ approximated as [10],

$$q(\gamma) = P\text{ (codeword symbols in error)} > l] = \sum_{j=l+1}^{n}\binom{n}{j}P_s^j(1 - P_s)^{n-j}, \tag{4.33}$$

where P_s is the probability of a symbol error. For b-bit symbols, $P_s(\gamma) = 1 - \left(1 - P_b(\gamma)\right)^b$, where P_b is the bit error probability, which depends on the modulation and the channel conditions. Considering that P_s and $q(\gamma)$ are small numbers because we assume operation in the low BER regime (Δ is a small number), we can use a first order Taylor's series to approximate $P_s(\gamma) \approx bP_b(\gamma)$, $1 - (1 - q(\gamma))^L \approx Lq(\gamma)$ and assume that the first term in (4.33) accounts for most of q's magnitude, which means that

$$P(\gamma) \approx L \binom{n}{l+1} P_s^{l+1}(1 - P_s)^{n-l-1}. \tag{4.34}$$

Following procedures similar to those illustrated above for RCPC codes [11], it can be shown that in the case of RS codes, the channel SNR when measured in decibels also changes with the source coding rate following mainly a linear relation, i.e. we can again write a relation of the form $\gamma_{dB} \approx A_1 x + B_1$. Following the same steps as for RCPC codes, we reach the conclusion that

$$D(\gamma) \approx (G_c\gamma)^{-10m},$$

where

$$m = \frac{W \log(4)}{t_1(0.5)N}, \tag{4.35}$$

$$G_c = (1 + \Delta)^{-1/(10m)} \frac{10^{t_1(0.5)/20}}{\ln(f(0.5))}, \tag{4.36}$$

and

$$t_1(r) = \frac{10f(r)'}{f(r)\ln(f(r))\ln(10)}$$

$$= \frac{10}{\log(f(r))} \left[\frac{2}{1 - r^2} + \frac{1 - \frac{1}{u(r)}}{\sqrt{u(r)}[n(1 - r) + 2]} \left(\frac{n\ln(h(r))}{n(1 - r) + 2} + \frac{n}{r + 1} - \frac{W\ln(4)}{N} \right) \right].$$

$$f(r) = \frac{b^{\frac{n-l-1}{l+1}}}{1 - \sqrt{u(r)}} \approx \frac{b^{\frac{n-l}{l}}}{1 - \sqrt{u(r)}} \approx \frac{b^{\frac{1+r}{1-r}}}{1 - \sqrt{u(r)}}.$$

$$u(r) = 1 - 4h(r)^{\frac{2}{n(1-r)+2}},$$

$$h(r) = \left(\frac{1+r}{2}\right)^n \sqrt{\frac{\pi}{2}n\frac{\Delta}{L}} 2^{-2vWr/N}.$$

Up to here, the important conclusion is that the D-SNR curve for concatenated JSCC schemes with practical source and channel codecs can be approximated with the function $D(\gamma) \approx (G_c\gamma)^{-10m}$. This is linear on a log–log scale (i.e. when measuring distortion on a logarithmic scale and the channel SNR in decibels). In this case, m is the slope, or the rate of decrease of distortion as the channel SNR increases. Also, G_c is the SNR value for which the distortion reaches a reference value of 1. Figure 4.7 illustrates these observations by showing the D-SNR curve and its characterization for systems with different RS and RCPC codes. The setting $\mu = v = 1$ was used for the figure, which corresponds to assuming that the input signal samples follow a standard Gaussian distribution and that long block source codes are used. For this source model, and assuming that the error concealment replaces each lost sample by its expected value, we explained earlier that $D_F = 1$ because the mean squared

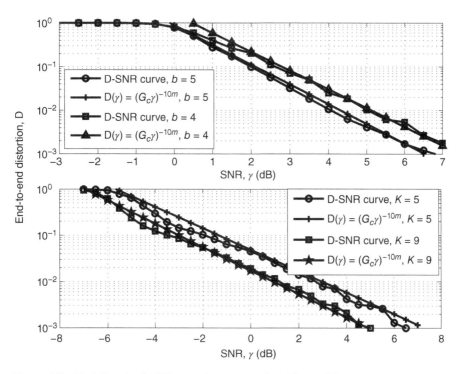

Figure 4.7 Modeling the D-SNR curve in systems with different RS (top) and RCPC (bottom) codes.

error equals the sample variance. In the figure, the linear relation between distortion and channel SNR on the log–log scale shows clearly, but the figure also shows a region for the D-SNR curve where the linear behavior does not hold. This region corresponds to the case where there is no further adaptive choice of operating mode (because there is no other channel code with smaller rate to choose from), and so the distortion follows the increase in channel errors with a fixed channel code.

The procedure used to approximate the D-SNR curve is based on the mechanics of the concatenated JSCC bit rate allocation procedure and principles that are quite general, and the approximation is applicable in various concatenated JSCC schemes. Figures 4.8 and 4.9 show the curves from the earlier Figures 4.4 and 4.5, but now including the characterization of the D-SNR curve as the linear curve on the log–log scale $D(\gamma) = (G_c\gamma)^{-10m}$. This approximation is still pertinent despite the different settings, distortion measure, and reduced number of operating mode choices.

A powerful consequence of the linear behavior is that the performance of a concatenated JSCC scheme can be fully described with only two parameters, m and G_c. These depend on variables linked to the source codec, the channel codec, and the system setup. Figure 4.10 uses the D-SNR curve approximation to compare the performance of concatenated JSCC schemes using different channel codes. This is useful for design, when choosing source and channel codes. The comparison among different D-SNR curves also allows one to visualize the differences between asymptotic and idealized results, obtained from information theory methods. This can be seen in Figure 4.10, which also shows the information theory

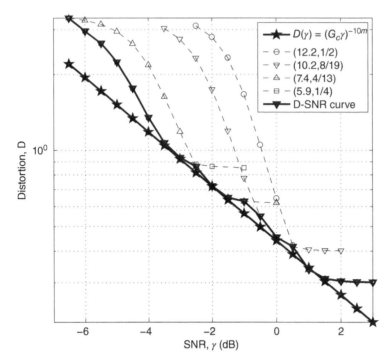

Figure 4.8 The single-mode D-SNR curves, the D-SNR curve, and the characterization of the D-SNR curve as a linear function on a log–log scale for a speech source.

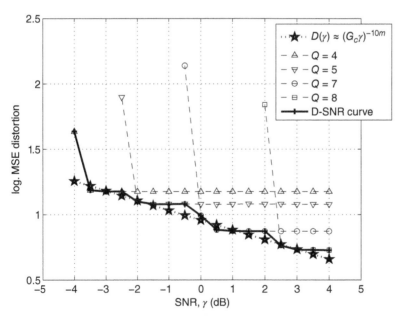

Figure 4.9 The single-mode D-SNR curves, the D-SNR curve, and the characterization of the D-SNR curve as a linear function on a log–log scale for a video source.

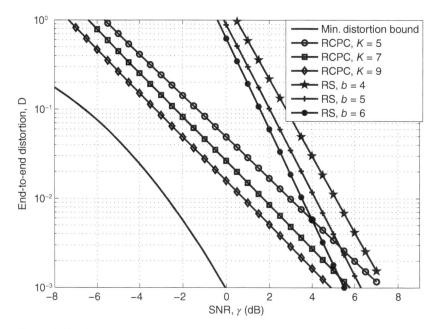

Figure 4.10 Comparison among different concatenated JSCC schemes and the minimum distortion bound.

function known as the *minimum distortion bound*. The minimum distortion bound is the D-SNR curve for a system where the conditions for Shannon's separation principle hold and where the channel is used at its capacity. Following Goblick's approach [12], the minimum distortion bound is taken as the optimal performance theoretically attainable (OPTA) curve by simply equating channel capacity to coding rate and applying this value in (4.26). Figure 4.10 illustrates how the use of stronger error correcting codes, with larger complexity, achieves a performance closer to the minimum distortion bound. The figure also shows the performance difference between the bound and systems with practical constraints such as delay and system complexity. Also, the figure allows one to compare performance between different channel code families. For example, the use of RS codes results in a larger m (rate of distortion reduction) but RCPC codes show lower distortion at lower SNRs. The performance for the code families becomes similar at high SNR.

The designer of a concatenated JSCC scheme needs to make a few key decisions, including picking the source and channel codecs. As Figure 4.10 shows for channel codes, the designer needs to balance implementation complexity with performance. A similar decision arises with the source codec. For practical source codecs, implementation complexity relates to coding efficiency. The modeling of the D-SNR curve allows one to also study the effects of the source codec efficiency on the performance of a concatenated JSCC scheme. The results presented so far have assumed for simplicity that $\mu = \nu = 1$ in (4.26). This implicitly assumes that the source encoder is able to compress the input samples perfectly. In reality, different source codecs exhibit different compression efficiencies, so the same distortion values are achieved at different encoding rates. The efficiency of the

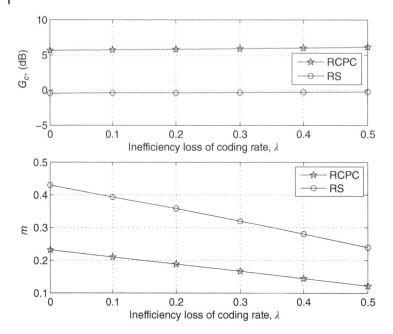

Figure 4.11 Comparing m and G_c for systems using source codecs with different compression efficiency.

source codec can be explicitly incorporated into the formulation by modifying the D-R function (4.26) and writing

$$D_S(x) = 2^{-2(1-\lambda)x} = 2^{-\hat{\lambda}x}, \tag{4.37}$$

where $\hat{\lambda} \triangleq 2(1 - \lambda)$ and λ introduces the inefficiency as a loss in coding rate. That is, with the choice of λ, one controls how much larger the source coding rate needs to be in order to achieve the same distortion as the ideal system $D_S(x) = 2^{-2x}$. By modifying (4.26) according to (4.37), it is possible to see the effects of source codec efficiency on the performance, as shown in Figure 4.11. As can be seen, the source coding efficiency has little effect on G_c. This is due to λ having little effect on γ_{dB} (this is more clear for RCPC codes). In contrast, larger source encoder efficiencies translate into greater values for m. In fact, changes in m are approximately proportional to $\hat{\lambda}$, as seen in Figure 4.11. This implies that when using source codecs with different efficiencies in the concatenated JSCC scheme, the D-SNR curve will change m but not G_c. Graphically, this will translate to the D-SNR line "pivoting" around the approximately fixed point where distortion equals 1 and changing slope (the rate of distortion decrease). The slope will become more steep with more efficient source codecs. From the observations in Figure 4.10, it follows that when a designer wishes to translate the D-SNR curve along the channel SNR axis, one needs to change the complexity or the family of channel codes used.

The choice of Δ in (4.11) determines the subset of points used to approximate the D-SNR curve. By examining Figure 4.3, it is clear that Δ cannot be chosen to be large or very small, since the corresponding distortion points may not belong to the D-SNR curve. The previous study used Δ equal to 0.1. As long as Δ is picked reasonably small, the results in terms of

D-SNR curves, and values for G_c and m, show little sensitivity to the value of Δ [11]. In terms of quantifiable values, a choice for Δ in the range between 0.05 and 0.25 has been seen to work well; the aim is to pick points in the "elbow" region of the single-mode D-SNR curves where distortion starts to rapidly increase due to channel-induced errors. In this section of the single-mode D-SNR curve, relatively large changes in distortion translate to small changes in γ_{dB}. This corresponds to small changes in the points used to represent the D-SNR curve and is one reason why the subset of points chosen to approximate the D-SNR curve is effective.

4.3 Application Cases

Due to its simplicity, concatenated JSCC can be used in many applications. This section will focus on their use in a direct sequence code-division multiple access (DS-CDMA) system. With DS-CDMA, adding each new network user increases the interference to the rest of the users, so DS-CDMA is said to have a "soft" capacity limit based on the limit of interference that is acceptable. As the limit on interference is related to variables such as acceptable error rates and transmit powers, the use of concatenated JSCC allows for linking interference control with the parameter settings affecting end-to-end distortion.

Consider the uplink of a DS-CDMA system, with ideal power control, BPSK modulation, and an AWGN channel. A matched filter is used at the receiver. Then power assignment to transmit a frame and interference from other users are related to the target signal-to-interference plus noise ratio (SINR) required by each of the U users by [13],

$$\gamma_i \geq \frac{(B/R_i)\, P_i}{\sigma^2 + \sum_{j \neq i} P_j}, \qquad i = 1, 2, \cdots, U, \tag{4.38}$$

where P_i is the power assigned to user i, as measured at the receiver, necessary to obtain the target SINR γ_i, B is the system bandwidth, R_i is the transmit bit rate, and σ^2 is the channel's noise variance. Given a channel coding rate, the target SINR is set to ensure that the error probability remains below a limit determined by subjective and/or objective measures of quality. As a consequence, the value for γ_i is directly related to a channel coding rate setting.

In (4.38), the ratio between system bandwidth and transmit bit rate is called the *Processing Gain* and will be denoted $G_{Pi} = B/R_i$. It can be shown that, at the receiver, P_i is related to γ_i by [13, 14],

$$P_i = \frac{\Upsilon_i \sigma^2}{1 - \sum_{j=1}^{U} \Upsilon_j}, \qquad i = 1, 2, \cdots, U. \tag{4.39}$$

where

$$\Upsilon_i = \left(1 + \frac{G_{Pi}}{\gamma_i}\right)^{-1}. \tag{4.40}$$

In order to achieve better end-to-end transmission, Υ_i should be as large as possible. Indeed, transmission using higher bit rates, thus higher Υ_i, naturally allows for lower source compression ratios and use of stronger error protection. Also, it is desirable to use a channel coding rate that is as large as possible, so as to compress the source as little as possible. This translates into seeking a setting for γ_i as large as possible and, thus, larger Υ_i. From (4.39), a first constraint on channel coding rate and transmit bit rate assignments should result in positive power assignments by meeting the condition $\sum_{j=1}^{U} \Upsilon_j < 1$. Yet, as the value for $\sum_{j=1}^{U} \Upsilon_j$ grows closer to 1, power assignments grow rapidly and the system approaches a region of instability for power assignments. Therefore, the constraint that is used for channel coding rate and transmit bit rate assignments is

$$\sum_{i=1}^{U} \left(1 + \frac{G_{Pi}}{\gamma_i}\right)^{-1} = 1 - \epsilon, \tag{4.41}$$

where ϵ is a small positive number set during design. The choice of ϵ implicitly allows for controlling maximum power settings. When all active users share the same source encoder and the same channel encoder, it is easy (and intuitive) to see that it is optimal to assign the same target SINRs $\gamma_i = \gamma$ and transmit bit rates $R_i = R$ to all calls (this is, assign the same Υ_i to all calls, thus everybody gets an equal share of resources). Therefore, denoting $G_P = B/R$, (4.41) can now be written as

$$U \left(1 + \frac{G_P}{\gamma}\right)^{-1} = 1 - \epsilon,$$

and the target SINR and the number of voice calls are related by

$$\gamma = \frac{G_P}{\frac{U}{1-\epsilon} - 1}. \tag{4.42}$$

Now, from (4.30), the joint assignment of source coding rate and channel coding rate should be such that

$$\gamma \approx \frac{D^{-1/10m}}{G_c}.$$

Combining this result with (4.42) and performing some algebraic operations yields

$$U = (1 - \epsilon) \left[1 + \left(D^{1/m} 10^{(G_{C_{dB}} + G_{P_{dB}})}\right)^{1/10}\right], \tag{4.43}$$

where, for convenience, $G_{C_{dB}}$ and $G_{P_{dB}}$ denote, respectively, the magnitudes G_C and G_P measured in decibels. This expression relates the number of calls to the end-to-end distortion in a simple way that uses four system parameters: ϵ and $G_{P_{dB}}$ from the CDMA network design, and $G_{C_{dB}}$ and m from the JSCC bit rate allocation design. This can be used to design call admission control protocols. Also, the approximate D-SNR expression (4.30) allows straightforward analysis of how the CDMA network behavior would change with different source or channel coding. For example, (4.43) reveals that $G_{C_{dB}}$ and $G_{P_{dB}}$ both have the same effect on the CDMA system performance, so if the parameters that affect $G_{P_{dB}}$ (system bandwidth and transmit bit rate) cannot be changed, a system designer could equally control performance by changing $G_{C_{dB}}$. From the discussion in previous sections, and especially from Figure 4.10, it can be seen that controlling $G_{C_{dB}}$ can be achieved best by using channel codes with different strengths without needing to redesign the source codec.

References

1 Totty, R. and Clark, G. Jr. (1967). Reconstruction error in waveform transmission (corresp.). *IEEE Transactions on Information Theory* 13 (2): 336–338.

2 ETSI/GSM (1998). Digital cellular telecommunications system (Phase 2+); Adaptive Multi-Rate (AMR) speech transcoding (GSM 06.90 version 7.2.1). *Document ETSI EN 301 704 V7.2.1 (2000-04)*.

3 ITU-T (2001). Recommendation P.862: Perceptual evaluation of speech quality (PESQ): an objective method for end-to-end speech quality assessment of narrow-band telephone networks and speech codecs.

4 ITU-T (2003). Recommendation P.862.1: mapping function for transforming P.862 raw result scores to MOS-LQO.

5 ITU-T (1996). Recommendation P.800: methods for subjective determination of transmission quality.

6 Hagenauer, J. (1988). Rate compatible punctured convolutional (RCPC) codes and their applications. *IEEE Transactions on Communications* 36 (4): 389–399.

7 Frenger, P., Orten, P., Ottosson, T., and Svensson, A.B. (1999). Rate-compatible convolutional codes for multirate DS-CDMA systems. *IEEE Transactions on Communications* 47: 828–836.

8 Malkamaki, E. and Leib, H. (1999). Evaluating the performance of convolutional codes over block fading channels. *IEEE Transactions on Information Theory* 45 (5): 1643–1646.

9 Chiani, M., Dardari, D., and Simon, M.K. (2003). New exponential bounds and approximations for the computation of error probability in fading channels. *IEEE Transactions on Wireless Communications* 2 (4): 840–845.

10 Wicker, S. (1995). *Error Control Systems for Digital Communication and Storage*. Prentice Hall.

11 Kwasinski, A. and Liu, K.J.R. (2008). Toward a unified framework for modeling and analysis of diversity in joint source-channel coding. *IEEE Transactions on Communications* 56 90–101.

12 Goblick, T. Jr. (1965). Theoretical limitations on the transmission of data from analog sources. *IEEE Transactions on Information Theory* 11 (4): 558–567.

13 Sampath, A., Mandayam, N.B., and Holtzman, J.M. (1995). Power control and resource management for a multimedia CDMA wireless system. *PIMRC'95*, Toronto, Canada.

14 Kwasinski, A. and Farvardin, N. (2003). Resource allocation for CDMA networks based on real-time source rate adaptation. *IEEE International Conference on Communications, 2003. ICC '03*, Volume 5 (11–15 2003), pp. 3307–3311.

5

Unequal Error Protection Source–Channel Coding

Practical source encoders frequently need to combine multiple techniques to achieve good compression performance. So, the bit stream at the output of a source encoder may be formed by different components generated by different compression techniques. For example, in a video encoded bit stream, there is data related to the motion vectors obtained from using motion compensation, data related to image texture that has been differentially encoded and entropy coded, and headers carrying important metadata about the bit stream. These have different importance for the source reconstruction, so channel errors will have different impact on the source reconstruction quality for different parts of the bit stream. While an error in one part of the bit stream may add little distortion to the source reconstruction, an error of the same magnitude may have a devastating effect on a different part. This is a case where following Shannon's source–channel coding theorem would result in a performance penalty. A sensible approach is to apply error protection of different strength to the different parts of the bitstream, based on each part's impact on distortion. This approach exemplifies a class of joint source–channel coding (JSCC) techniques known as unequal error protection (UEP) source–channel coding, which is the subject of this chapter.

5.1 Effect of Channel Errors on Source Encoded Data

Channel error effects on source reconstruction quality may differ depending on the importance or sensitivity of the parts affected. Headers within the source encoded bitstream are highly important; they usually contain information that is indispensable to configure the decoder. If this information is corrupted, then it may not be possible to decode large portions of transmitted data. Certain parts of the bit stream may contain information that has been encoded with techniques that are more sensitive to channel errors. For example, if using variable-length entropy coding, channel errors will have a more significant effect on earlier bits because all succeeding bits will need to be discarded and cannot be used to improve reconstruction quality. As another example, in differential encoding, one source encoded sample may depend on several previous samples, so an error in a sample's reconstruction would affect several successive samples. We illustrate this point with the following example.

Joint Source-Channel Coding, First Edition. Andres Kwasinski and Vinay Chande.
© 2023 John Wiley & Sons Ltd. Published 2023 by John Wiley & Sons Ltd.

Assume that a portion of a bit stream encodes samples from a first-order autoregressive Gaussian source x_n (n being the discrete-time index) that behaves according to

$$x_n = \rho x_{n-1} + w_n, \qquad n = 1, 2, \ldots,$$

where w_n is the noise sequence, modeled as a discrete-time, zero-mean Gaussian random process with unit variance, and ρ controls the correlation between two successive samples. Since any given sample is correlated with the previous sample, it is natural to use differential encoding to compactly represent the samples x_n. In this case, using a linear predictor (Section 2.2.2) only requires a one-tap prediction filter, with its only coefficient being $a_0 = \rho$, [1]. Recall that with differential encoding, the transmitted data consist of a sequence of words, each equal to the quantized error between the input sample and its predicted value. Figure 5.1 illustrates the channel effects on the differentially encoded source when the error between the input sample and its predicted value is quantized as eight-bits words and only the tenth word in the transmitted data sequence is in error. Before this tenth word, the source reconstruction faithfully follows the source. When error is introduced in the tenth transmitted word, the difference between the source and its reconstruction becomes large as expected. The sensitivity of differential encoding becomes evident as the error in the tenth transmitted word affects the other ones that follow through the prediction and differential methods used in the kernel of the decoder. The difference between the input source and its reconstruction decays slowly following the damped time constant given by the coefficient of the first-order linear prediction filter. If there were many errors in the transmitted data sequence, not just the tenth word, the difference between the input source and its reconstruction might be significant for all samples (see the example in Figure 5.2 with all transmissions going through a binary symmetric channel (BSC) with a bit error rate (BER) of 0.001).

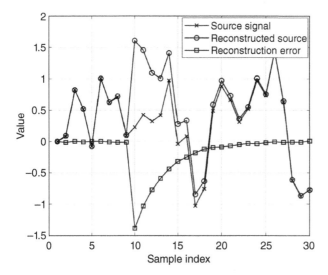

Figure 5.1 The effect of channel errors on a differentially encoded first-order autoregressive Gaussian source. Channel errors are introduced only in the tenth transmitted eight-bit word.

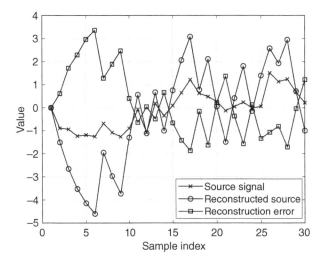

Figure 5.2 The effect of channel errors on a differentially encoded first-order autoregressive Gaussian source when all the transmitted data is subject to a BSC with bit error rate 0.001.

A good example of the effect of errors when using differential encoding is block-based motion-compensated video coding. Figure 5.3 shows a block diagram, modified from Figure 2.26 in Section 2.5.3, that shows the motion compensation loop where temporal differential encoding is applied during encoding. The effects of channel error propagation on the recovered video sequence have been studied by many research groups. The work in [2] modeled the distortion due to error propagation and arrived at a simple modeling expression that reinforces the observations made above. The study considered the sequence of frames between two Intra-coded frames. The channel-introduced error was modeled as a process with zero mean and variance σ_e^2 that is introduced in the decoder after the inverse discrete cosine transform (DCT) block. It was assumed that in terms of average distortion, all frames are affected by the same error variance σ_e^2. For a given video sequence and decoder implementation (e.g. given error concealment scheme), it was seen in [2] that the error variance is directly proportional to the transmission packet loss rate P_L: $\sigma_e^2 = \sigma_{e0}^2 P_L$. To model the average channel-induced distortion, it is necessary to see how the error process with variance σ_e^2 propagates over the sequence of P-frames and results in a difference signal (or video frame), v, between the originally encoded P-frame x and the corresponding decoded frame y (the analysis is only on Inter- predicted frames because there is no temporal error propagation in Intra-predicted frames).[1] Assuming that the error signal is introduced at frame index 0, the variance of the error frame $v(n)$ at time index n is, [2],

$$\sigma_v^2(n) = \sigma_e^2 \frac{1 - \beta n}{1 + \gamma n},$$

where β is the Intra refresh rate (equal to the inverse of the number of frames between two I-frames) and the *leakage* γ describes the effect of the filter implied in the motion compensation processing on damping the error propagation. Typical values for γ are between 0 and 1.

---

1 Inter- and Intra-predicted frames are described in Section 2.5.3.

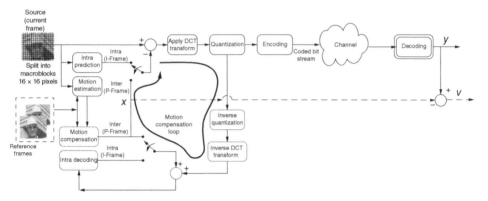

Figure 5.3 Simplified block diagram for the H.264 video encoder, showing the motion compensation loop.

Figure 5.4a illustrates how the variance of the error frame σ_v^2 decreases from frame to frame after an error frame with variance $\sigma_e^2 = 1$ is introduced in frame 0. In the figure, $\beta = 1/15$.

While for simplicity, the Intra refresh rate is frequently thought of as relating to frames that are completely Intra coded, with modern video codecs, Intra coding is not necessarily applied over a complete single frame. Instead, it is applied over a portion of a frame. The following frame will apply Intra coding over a different part of the frame and so on until Intra coding is applied over all regions of a frame over a time period of several frames.[2] After a number of frames, Intra coding would be applied to the same portion of the frame that started the sequence of Intra coding. The number of frames that elapse between the Intra coding of the same portion of the frame is the Intra refresh period, and its inverse is β, the Intra refresh rate. As frames are usually divided into equal size parts, the Intra refresh rate is also the fraction of any given frame that is Intra coded. This implies that any given error will propagate over at most $1/\beta$ frames. Then, since the filtering operation affecting error propagation is linear, it is possible to use the superposition principle, and assuming that the superimposed error signals are uncorrelated from frame to frame (an assumption that is valid as long as the channel error rate is small), the mean channel-induced distortion can be calculated as [2],

$$D_c = \sigma_{e0}^2 \sum_{n=0}^{1/\beta-1} \sigma_e^2 \frac{1 - \beta n}{1 + \gamma n}.$$

Figure 5.4b illustrates this result by showing D_c as a function of the leakage γ. The value of γ depends on the strength of the filtering operation in the loop and on the spectral shape of the introduced error.

Sometimes channel errors of the same magnitude can have different effects on the source reconstruction distortion because a source is encoded using samples that are not equally important. This unequal importance may be due to human subjective perception of distortion. We illustrate this with an example of image coding using a simplified JPEG scheme.

---

2 Note that periodic Intra refresh is not the only approach for Intra refresh. Video codecs may implement random Intra refresh, or Intra blocks may be chosen based on frame content, motion, or residual error from Inter coding. However, for simplicity, we assume a periodic Intra refresh here.

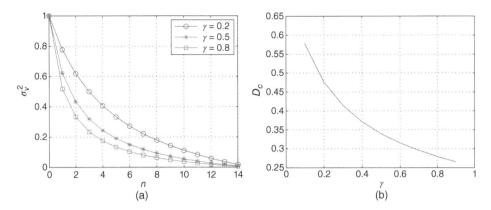

Figure 5.4 **(a)** Variance of difference frame $v(n)$ as a single error event is introduced in frame 0. **(b)** Average channel-induced distortion as a function of the leakage.

Figure 5.5 Original image to be encoded and decoded using a simplified JPEG technique. Source: fibPhoto/Shutterstock.

We consider the image in Figure 5.5 encoded using an eight-by-eight block DCT followed by quantization using the quantization matrix (2.16), as explained in Section 2.5.1.

The result of encoding and decoding without channel errors is shown in Figure 5.6, with the reconstructed image on the left and the corresponding pixel-by-pixel squared error on the right. The measured peak signal-to-noise ratio (PSNR) is 35.26 dB.

In block-based DCT techniques (e.g.in JPEG), the DC coefficient is more important both in terms of distortion associated with errors in this coefficient, and in terms of human perception of image quality. After subjecting the discrete cosine (DC) coefficients to a BSC with a huge BER of 50%, Figure 5.7 shows the reconstructed image and squared error. The measured PSNR is 25.53 dB, and the image is badly distorted.

The other coefficients are less important than the DC coefficient. After affecting the DCT coefficients with coordinates (1,1) with a BER of 50%, Figure 5.8 shows the reconstructed image and corresponding pixel-by-pixel squared error. The measured PSNR is 34.71 dB, only

Figure 5.6 (Left) Recovered image from Figure 5.5, after encoding and decoding, (Right) corresponding squared error. Source: fibPhoto/Shutterstock.

Figure 5.7 The recovered image from Figure 5.5, after encoding and decoding when the DC coefficients of the block DCT transforms are transmitted through a BSC with bit error rate 50%. On the left is the reconstructed image and on the right is the corresponding squared error. Source: fibPhoto/Shutterstock.

Figure 5.8 The recovered image from Figure 5.5, after encoding and decoding when the coefficients in position (1,1) of the block DCT transforms are transmitted through a BSC with BER 50%. On the left is the reconstructed image and on the right is the corresponding squared error. Source: fibPhoto/Shutterstock.

about 0.5 dB less than the case with no channel errors. The errors are difficult to perceive in the reconstructed image, but can be seen in the squared error image; the biggest distortions are in edge areas, for example, in the opera house roof.

These examples illustrate that different parts of the source encoded bit stream result in different source reconstruction distortion when affected by channel errors. It is often necessary to add forward error correction (FEC) to combat these errors before transmission. Adding redundant bits will help identify and correct some errors, but since these added bits increase the transmitted bit rate, one must address how much error protection redundancy to add. Adding redundancy matched to the average error-related distortion performance across all the parts is possibly both excessive for the unimportant parts and insufficient for the most important parts. The best trade-off between distortion performance and added error control redundancy comes from assigning different levels of error protection to different parts of the source encoded bit stream, according to the impact that channel errors on that part would have on the end-to-end distortion. This assignment of different error protection to different parts of the source encoded bit stream is called *UEP*.

To implement UEP, it is necessary for the channel coding component to know the relative importance of different parts of the source encoder output. Figure 5.9 shows the standard block diagram for a digital communication system with source and channel codecs. The figure shows with big arrows the flow of source data from source encoding, through channel encoding, the communication channel, channel decoding, and ending at the source decoder. It also shows, with thinner arrows, possible exchange of information between different blocks. These exchanges allow for implementation of different JSCC techniques. Most of the information that is exchanged feeds into one of the decoder blocks (shown with dashed lines), allowing for *joint source–channel decoding* techniques, which will be discussed in Chapter 10. Figure 5.9 shows where UEP fits into this larger picture. Going from the source to the channel encoders, *Source Significance Information* indicates the relative importance of different parts of the source encoder output, allowing for the design of UEP schemes. UEP has received significant attention from the digital communications research

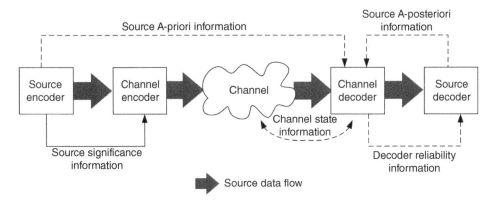

Figure 5.9 Main processing blocks in a typical digital communication system, showing the flow of source data (indicated with the bold arrow) and possible exchange of information between source and channel codecs that allow for JSCC schemes.

and development community, and there exist many techniques. In the rest of this chapter, we will describe some of the salient ones.

5.2 Priority Encoding Transmission Schemes for Unequal Loss Protection

The design of the error correction subsystem depends on the nature of the errors and how the errors are dealt with once they cannot be completely removed. Once a packet, frame, or block is identified as having errors, one possibility is to eliminate it from the receiver processing stream. There exist very efficient codes, such as the family of cyclic redundancy check (CRC) codes, capable of accurately detecting the presence of error over many bytes at the cost of transmitting just a few extra bits. While these codes are powerful in detecting the presence of errors, they cannot correct them. Thus, the usual option is to discard the complete packet where errors have been detected. From the viewpoint of later decoding stages where packets need to be processed, packet deletion appears as if the communication channel loses packets. The Internet also behaves as a packet loss channel. Packets traveling through a router (or any store-and-forward intermediate node) that is congested may be dropped from the processing queue, that is, the packets are deleted inside the communication channel. As a consequence, the Internet behaves as a packet loss channel, where the statistics of the loss process depend on a myriad of factors and variables. We discuss next techniques designed to provide UEP when the effect of the channel is modeled as a packet loss process.

A number of techniques that apply UEP can be seen as derived from the concept of *priority encoding transmission* (PET), [3]. In PET, a message is divided into parts of different priority or importance. Each part is encapsulated into one or more packets and receives FEC of different strength based on an assigned priority. The application of FEC results in extra redundancy packets that are combined with the data packets to make a *segment*. As a result, the original message is transmitted as a sequence of segments composed of data packets and redundancy packets. In PET, the priority for a segment corresponds to the fraction of packets from the segment that needs to be successfully received so that the part of the message transmitted within the segment can be recovered. A high priority segment is encoded in such a way that it can be successfully decoded even if a relatively small fraction of packets are received.

A main task in designing a PET scheme is defining the priorities for the message parts. The channel coding strength will be chosen to correspond to the importance of each source packet. A simple application of PET was described in [4], where PET was applied to Moving Picture Experts Group (MPEG)-coded video. Similar in its basic elements to H.264 coding, MPEG follows the DCT-transform, motion compensation and entropy coding techniques summarized in Chapter 2. The encoder output presents I-frames, P-frames, and B-frames. These have different importance, with I-frames needing more error protection, followed by P-frames, and B-frames needing the least protection. Because of this, the scheme in [4] assigns priorities based on the frame type. The Group of Pictures (GoP) is the sequence of frames starting with an I-frame and ending with the frame preceding the next I-frame. The GoP is split into three types of segments: segment type 1 is used to send I-frames and the FEC will be chosen such that if 60% of its packets are received, the segment will be decodable. Segment type 2 is used to send P-frames and is set to require 80% of packets to be received for decoding, while segment type 3 has B-frames and needs 95% of packets.

How should the FEC scheme be configured to ensure that the transmitted data in a segment can be recovered from any set fraction of the transmitted packets? When reviewing the erasure correction capabilities of block codes in Chapter 3, we saw that for an (n, k) block code exhibiting the maximum-distance separable (MDS) property (e.g. a Reed–Solomon code), it is possible to recover the k original encoded packets from any k received packets out of the n transmitted ones. In other words, an entire transmitted encoded message of length n packets can sustain the loss of any $n - k$ of the transmitted packets. This property can be directly applied in PET by using an MDS block code encoding k packets (a value that is known) into $n = k/p$ packets, where p is the fraction of packets that is required to be correctly received.

When using block codes, such as Reed–Solomon codes, in a configuration to correct packet erasures, the assignment of priorities in PET is essentially the problem of designing the amount of redundancy to be added to each message part or, equivalently, the channel coding rate assigned to each message part. The approach taken for MPEG video coding in [4] was based on the heuristics that I-frames need to be more protected than P-frames and that B-frames require the least protection. Ideally, the priority scheme (i.e. channel coding rate assignment) should be based on a metric conveying end-to-end distortion, which depends on all elements determining the system.

In [5] the idea of PET is applied to protect images encoded using an embedded source encoder, such as the Set Partitioning in Hierarchical Trees (SPIHT) encoder discussed in Section 2.5.2. The bit stream generated by embedded encoders can also be described as *progressive*, because the reconstructed source quality progressively improves as more bits are received and used in the source decoder. This property directly provides the heuristics to define the priority for segment parts. The earliest bytes are the most important, not only because they provide the baseline description of the source that reduces distortion the most (as compared with not receiving any source information at all) but also because the embedded bit stream structure makes unusable any data received after the first uncorrectable error has occurred in the received bit stream. In [5], PET was applied to an image encoded with an embedded coder, where the source encoder output is a message with M bytes. As the encoder produces a progressive representation of the source, the first byte within the message is the most important and the last byte is the least important. The transmission used N packets. The transmitted message is divided into L segments, each with N bytes, for a total of $L \times N = M$ transmitted bytes. When FEC is added to the transmitted stream, in a segment i, there are m_i source coded bytes and $N - m_i$ added redundancy bytes. Figure 5.10 illustrates this assignment with $M = 30$, $N = 5$, and $L = 6$. The most important byte in the message is B_1 and the least important is B_{30}. In this example, the message is truncated at byte B_{20}, making B_1 through B_{20} the 20 bytes from the message at the output of the source encoder that are transmitted, and B_{21} through B_{30} the bytes that are not transmitted. In the figure, B_F are the redundancy bytes added for FEC.

Figure 5.10 illustrates how this PET-derived scheme provides different levels of packet loss protection to the segments. The FEC subsystem is a Reed–Solomon (n, k_i) coder, where n is always equal to N, the number of packets, and k_i equals m_i and needs to be determined

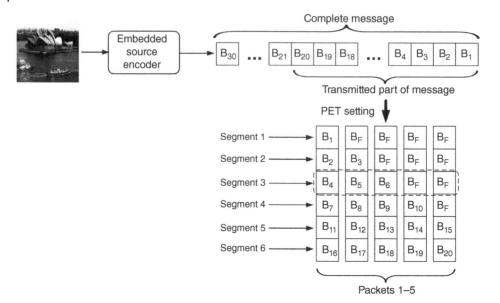

Figure 5.10 Priority Encoding Transmission configured for transmission of an embedded-coded image using six segments and five packets (each of size 6 bytes).

at design time using the criteria explained below. Consider the case shown in Figure 5.11, where the third packet is lost during transmission (marked with all bytes crossed out). The data byte B_1 in segment 1 will be successfully recovered because for this segment $k_1 = 1$, which means that the message data in the segment would be recovered as long as at least one byte of the segment is received. Similarly, bytes B_2 and B_3 will be successfully recovered from segment 2 because the setting $k_2 = 2$ implies that the source encoded bytes can be recovered if any two packets are received. For segment 3, the setting is $k_3 = 3$, which means that the source encoded bytes can be recovered as long as any three packets are received. Consequently, even though byte B_6 was lost as part of the third packet, its value would be recovered from the FEC bytes received in other packets. Bytes B_7 through B_{10} in the fourth segment would still be recovered because for this segment $k_4 = 4$, which is the number of useful packets that were received. The setting for segment 5 is $k_5 = 5$, which means that all

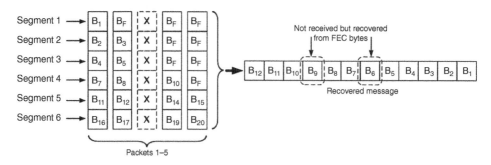

Figure 5.11 Effect of losing the third packet during transmission of the PET example in Figure 5.10.

the five transmitted packets need to be received in order to recover all the source encoded bytes in the segment. Consequently, it would not be possible to recover the lost byte B_{13}. Due to the progressive nature of the source encoded stream, all bytes after B_{13} would have to be discarded and could not be used in reconstructing the image, even if received correctly. These correctly received but unusable bytes are B_{14}, B_{15}, B_{16}, B_{17}, B_{18}, B_{19}, and B_{20}. Nevertheless, bytes B_{11} and B_{12} from segment 5, which were received as part of packets 1 and 2, could be used in the reconstruction of the image because they precede byte B_{13}.

The examples in Figures 5.10 and 5.11 illustrate how PET applied to an embedded-coded image provides different levels of protection against packet losses. The more important bytes can be recovered from the added FEC after packet losses have occurred. How does one choose the values $k_i = m_i$ of source encoded bytes allocated to each segment? In [5], the total number M of transmitted bytes and the number N of transmitted packets are predetermined. The source encoder produces a number of bytes equal to M, so if no FEC were used, all source encoded bytes could be transmitted. The choice of adding FEC bytes both determines priorities for the source encoded bytes and their related levels of FEC and determines the number of transmitted source encoded bytes, $M - \sum_{i=1}^{L}(N - m_i)$. Choosing values m_i means finding the trade-off between source encoding rate and overall channel coding rate. Since N and M are fixed at the start, the number of segments $L = M/N$ is also determined (ignoring possible zero padding in the packets).

The broad idea in [5] to determine the values m_i is to consider the improvement in quality resulting from receiving an incremental part of the progressively encoded source data. This improvement in quality is balanced against the FEC that must be added so as to reliably receive the source encoded data. Let $f_i = N - m_i$ and let $p_d(f_i)$ be the probability that the receiver can decode segment i, which is equivalent to the probability that f_i or fewer packets are lost, and is calculated based on the transmitter-channel-receiver characteristics. Then, the goal of the design is to find a vector $\mathbf{f} = (f_1, f_2, \dots, f_L)$ with the FEC redundancy allocation in each segment. We define the incremental quality (measured as PSNR for the current example of image transmission) resulting from receiving segment i when all the segments 1 through $i - 1$ have been successfully received:

$$g_i(\mathbf{f}) = Q[F(i, \mathbf{f})] - Q[F(i - 1, \mathbf{f})],$$

where $F(i, \mathbf{f})$ is the sequence of bytes from the source encoded byte stream starting with the first byte and ending with all the source encoded bytes contained in segment i, and $Q[F(i, \mathbf{f})]$ is the source reconstructed quality when using the fragment $F(i, \mathbf{f})$ as the only input to the source decoder. To complete the definition of the incremental quality, $g_i(\mathbf{f})$, the case $g_1(\mathbf{f})$ is the increase in quality when the first segment is successfully received when compared to the source representation when no useful data is received (in the case of image encoding, this would be an image with a uniform mid-gray color, for example). It is now possible to state the design problem as

$$\max_{\mathbf{f}} G(\mathbf{f}) = \sum_{i=1}^{L} g_i(\mathbf{f}) p_d(f_i), \tag{5.1}$$

$$\text{subject to} \quad m_i + f_i = N, \qquad i = 1, 2, \dots L$$

$$f_i \geq f_{i+1}, \qquad i = 1, 2, \dots L - 1.$$

The expression $G(\mathbf{f})$, (5.1), is the expected quality (PSNR in the case of an image) for the reconstructed source. The problem statement includes the condition $f_i \geq f_{i+1}$, $i = 1,2,\dots L-1$ that formalizes the segment priority order inherent to PET, but it is also a condition that recognizes the progressive property in the source encoded byte stream since it implies that when segment i can be decoded, all previous segments can also be decoded.

The design problem (5.1) was solved in [5] through a suboptimal heuristic algorithm of low complexity that evaluated the expected quality $G(\mathbf{f})$ obtained after each iteration in which 1 to Q bytes of FEC were added or subtracted to each segment while also satisfying the condition $f_i \geq f_{i+1}$. The algorithm was applied to the 512×512 grayscale Lenna image compressed with SPIHT operating at an encoding rate of 0.2 bits per pixel. Aiming at asynchronous transfer mode (ATM) packet transmission, the packet size was chosen as 47 bytes plus an extra byte to indicate sequence number. Therefore, the number of segments was 47. The total number of transmitted packets was 137. The channel was modeled as having a packet loss process following an exponential distribution with a mean loss rate of 20%.

Table 5.1 shows, from [5], the fraction of packets needed for a segment to be correctly received. The most important segments (those identified with a smaller index) are allocated a larger number of FEC bytes, and so are decodable with a smaller fraction of packets being received. As shown in [5], the operation of this unequal loss protection scheme results in a quality of the reconstructed image that gracefully degrades as the fraction of packets lost increases. This is not the case for an equivalent scheme with FEC allocated based

Table 5.1 Fraction of packets needed for segment reception as reported in [5] for a 512×512 grayscale Lenna image.

	Segment number							
	1	2	3	4	5	6	7	8
k_i/N	0.25	0.29	0.31	0.36	0.40	0.43	0.45	0.47
	Segment number							
	9	10	11	12	13	14	15	16
k_i/N	0.48	0.51	0.52	0.53	0.53	0.54	0.54	0.54
	Segment number							
	17	18	19	20	21	22	23	24
k_i/N	0.54	0.56	0.58	0.59	0.59	0.60	0.60	0.60
	Segment number							
	25	26	27	28	29	30	31	32
k_i/N	0.60	0.60	0.60	0.62	0.62	0.62	0.64	0.64
	Segment number							
	33	34	35	36	37	38	39	40
k_i/N	0.64	0.64	0.64	0.67	0.67	0.67	0.67	0.67
	Segment number							
	41	42	43	44	45	46	47	
k_i/N	0.68	0.68	0.70	0.70	0.70	0.70	0.70	

on equal loss protection. In this case, as the fraction of lost packets increases, the quality first remains unchanged but once the fraction of lost packets reaches a certain value, the quality decreases sharply to unacceptable levels. The unequal loss protection scheme shows a smooth decrease and outperforms the equivalent equal loss protection scheme over most of the range of packet loss rates.

5.3 Dynamic Programming Algorithm for Optimal UEP

We continue with the discussion of UEP for embedded sources, describing the work in [6] that studied a very general setup and an optimal solution for UEP allocation. A challenge in transmitting embedded source encoded data is to minimize the damage caused by error propagation. Consider the transmission over a memoryless channel of N_s source samples encoded together with an embedded source coder. The bits at the output of the source encoder are organized into fixed-length source packets of k_s bits each. As each additional packet arrives, the source rate is some increasing multiple of k_s/N_s. The progressive nature of the encoded bit stream implies that representations of the source at rates that are multiples of k_s/N_s can be obtained from various prefixes of the overall stream of source packets. While the source packets are of fixed length, the scheme studied next allows for source packets to have a variable-length error correction. When using families of rate compatible error correction codes, the use of variable-length error correction has the advantages that the family of codes can be implemented with a single channel coder–decoder pair and that the variable codeword lengths of the rate compatible families are usually multiples of a smaller fixed-length channel block, which can be used for synchronization.

The scheme to be studied next proceeds with transmission as follows. Each source packet output by the source coder is encoded with a potentially different channel code from a single family of channel codes, and the resulting channel coded bits are transmitted over a noisy memoryless channel. The receiver attempts to recover the source packets from the received channel codewords that may contain errors. The channel decoder either correctly decodes a source packet or detects an error and declares a *source-packet decoding failure*. The probability of undetected errors is usually very small and will be assumed to be zero. As progressive source packets that follow a packet decoding failure cannot improve the reconstructed quality and need to be discarded, at any stage in transmission, the source is reconstructed only from the decoded bit stream up to the first source packet that contains a detectable error or irrecoverable erasure.

We denote the family of error correction-detection channel codes by $C = \{c_1, c_2, \ldots c_J\}$, and the code rates by $r_c(c_i), i = 1, \ldots, J$. A codeword for a source packet of length k_s bits, protected by code c_i, has length $k_s/r_c(c_i)$ bits. For $c_i \in C$, let the probability of packet decoding failure for the channel be $P_e(c_i)$. If the first i source packets are received and available to the decoder, the source can be reconstructed with a source coding rate ik_s/N_s bits per source sample, where N_s is the number of source samples. Let $r_s \overset{\Delta}{=} k_s/N_s$ be the rate in bits per sample that can be obtained from a source packet for source reconstruction.

The design of UEP for the source packets aims at assigning a channel code to each source packet before transmission. This assignment is specified through a *code-allocation policy*, π, which allocates channel code $c_\pi^i \in C$ to the ith source packet out of the source coder. A policy

π is described by the number of source packets to be transmitted ($N(\pi)$) and by a sequence of channel codes $\{c_\pi^1, c_\pi^2, \ldots, c_\pi^{N(\pi)}\}$ to be used with the sequence of source packets. Note that $N(\pi)$ can also be thought of as the index of the terminating source packet for the policy.

The normalized transmission rate (in channel bits per source sample) for a policy π is

$$R_{T_\pi} \triangleq \sum_{i=1}^{N(\pi)} \frac{r_s}{r_c(c_\pi^i)}. \tag{5.2}$$

For transmitting an image, there are several single-parameter criteria that can be used to measure the performance of a code-allocation policy, such as the *expected distortion* at the receiver using a policy π (at transmission rate $R_T(\pi)$). As the source is reconstructed only from the source packets received prior to a source packet decoding failure, the expected distortion can be expressed as

$$\overline{D}_\pi \triangleq \sum_{i=0}^{N(\pi)} D(ir_s) P_{i|0}(\pi), \tag{5.3}$$

where $D(r)$ is the operational distortion-rate performance of the source coder with rate r measured in bits per sample. Equation (5.3) depends on the probability $P_{i|0}(\pi)$, which can be introduced in more general terms as $P_{i|k-1}(\pi)$: the conditional probability that exactly the first i source packets are decoded correctly given that the first $k-1$ packets are decoded correctly, while using the policy π. For integers $k = 1, 2, \ldots, N(\pi)$ and $i = k-1, k, k+1, \ldots, N(\pi)$, $P_{i|k-1}(\pi)$ can be computed as

$$P_{i|k-1}(\pi) \triangleq \begin{cases} P_e(c_\pi^k) & i = k-1, \\ \prod_{j=k}^{i} (1 - P_e(c_\pi^j)) P_e(c_\pi^{i+1}) & i = k, k+1, \ldots, N(\pi) - 1, \\ \prod_{j=k}^{N(\pi)} (1 - P_e(c_\pi^j)) & i = N(\pi). \end{cases} \tag{5.4}$$

Note that $\sum_{i=k-1}^{N(\pi)} P_{i|k-1}(\pi) = 1$ for $k = 1, 2 \ldots, N(\pi)$.

A second criterion that can be used to measure the performance of a code-allocation policy is the quality-rate performance. For image encoding, the quality can be measured through the PSNR and, thus, the criterion becomes the PSNR-rate performance of the source-coder $PSNR(r)$, where

$$PSNR(r) \triangleq 10\log_{10} \frac{255^2}{D(r)} \text{ (dB)}. \tag{5.5}$$

Then the *expected PSNR* for the policy π is given by

$$\overline{PSNR}_\pi \triangleq \sum_{i=0}^{N(\pi)} PSNR(ir_s) P_{i|0}(\pi). \tag{5.6}$$

For transmitting bit streams with the progressive decoding property, a third performance criterion is the average number of source encoder bits per sample received before a source packet decoding failure (which is the beginning of a possible error propagation). We call this criterion the *average useful source coding rate*. This criterion is motivated by the fact that a longer error-free prefix means a better source reconstruction. For a policy π, the average number of source packets received before a source packet decoding failure can be written analogously to (5.6) as

$$V_\pi \overset{\Delta}{=} \sum_{i=0}^{N(\pi)} i P_{i|0}(\pi). \tag{5.7}$$

Note that the average useful source coding rate is given by $r_s V_\pi$.

Using the three performance criteria just introduced, the channel code-allocation problems for the UEP scheme under the constraint of total transmission rate R bits per source sample can be expressed in terms of the following optimization problems.

1. **Problem A:** For minimization of the average distortion the problem is

$$\min_\pi \overline{D}_\pi \text{ subject to } R_{T\pi} \leq R. \tag{5.8}$$

2. **Problem B:** For maximization of the average PSNR, the problem is

$$\max_\pi \overline{PSNR}_\pi \text{ subject to } R_{T\pi} \leq R. \tag{5.9}$$

3. **Problem C:** Finally, as r_s is a constant, to maximize the average useful source coding rate, the problem is

$$\max_\pi V_\pi \text{ subject to } R_{T\pi} \leq R. \tag{5.10}$$

The design criterion (5.7) does not involve the source statistics or the source-coder performance. As long as transmitter and receiver share the same estimate for the probabilities $P_{i|0}(\pi)$, the receiver can also carry out this optimization and hence the UEP policy can be available at the receiver without needing to transmit it as side information. The assumption that the probabilities $P_{i|0}(\pi)$ are known by the transmitter and the receiver is not far-fetched, as they only require knowledge of the performance models for the different system components (channel code families, modulation scheme, etc.) and of the channel characteristics, which are frequently estimated at the receiver and sent to the transmitter.

While any of the above optimization criteria can be chosen to suit the application, choosing to maximize the average PSNR or minimize the average distortion should be done when it is possible to robustly transmit the UEP policy that is being used, as there is no inherent way for the receiver to obtain this information. The design criterion (5.7) (and hence Problem C) has the advantages that it does not need the UEP policy to be transmitted, it is simpler than the other two criteria, and it is also useful in situations where the source coder is not embedded but error propagation is still an issue. For example, in variable-length coded macroblocks with synchronization symbols, error propagation within a macro-block can be prevented or maximally delayed by UEP design based on maximizing (5.7). In addition, the optimal policies for criterion (5.7) allow provably optimal progressive transmission at intermediate rates. This point will be addressed later in this chapter.

We are now set to study the solution to optimization problems A, B, and C. These are not conventional rate allocation problems because the associated cost functions (5.3), (5.6), and (5.7) are not additive. Nevertheless, it can be shown that the three problems can be solved exactly by a framework based on dynamic programming, a mathematical optimization approach that transforms a complex problem into a sequence of simpler problems. The simpler problems are all essentially of the same nature, leading to a multistage optimization procedure. Dynamic programming is a tool with a rich history of applications for communications and networking such as the Viterbi algorithm (to decode convolutional codes and many other applications) and Dijkstra's algorithm to solve routing problems.

The principal idea of the solution to the optimization problems A, B, and C is to write the objective function in the absence of noise (distortion, PSNR, or number of source packets) as a sum of incremental rewards, which are accumulated as each source packet is successfully decoded by the receiver. It is the use of incremental rewards that leads to solving simpler problems, which can be used to obtain the overall solution in a multistage procedure. Let δ_i denote the incremental reward when the ith source packet is successfully received. Hence, if the task is to minimize the average distortion, δ_i is defined as

$$\delta_i \stackrel{\Delta}{=} D((i-1)r_s) - D(ir_s), \quad i = 1,2,\dots. \tag{5.11}$$

Similarly for average PSNR maximization, δ_i is defined as

$$\delta_i \stackrel{\Delta}{=} PSNR(ir_s) - PSNR((i-1)r_s), \quad i = 1,2,\dots. \tag{5.12}$$

And, for maximization of the average useful source coding rate, δ_i is defined as

$$\delta_i \stackrel{\Delta}{=} 1, \quad i = 1,2,\dots. \tag{5.13}$$

The objective functions in (5.3), (5.6), and (5.7) are related to these incremental rewards as follows. For a code-allocation policy $\pi = \{c_\pi^1, c_\pi^2, \dots c_\pi^{N(\pi)}\}$ and for integers $k, 1 \le k \le N(\pi)$, define,

$$\Delta(k,\pi) \stackrel{\Delta}{=} \sum_{i=k}^{N(\pi)} \left(\sum_{j=k}^{i} \delta_j \right) P_{i|k-1}(\pi). \tag{5.14}$$

Note that for $k = 1$ we can write

$$\Delta(1,\pi) = \sum_{i=1}^{N(\pi)} \left(\sum_{j=1}^{i} \delta_j \right) P_{i|0}(\pi).$$

For Problem A, replacing the value of δ_j defined in (5.11) and operating we get,

$$\Delta(1,\pi) = \sum_{i=1}^{N(\pi)} \left(\sum_{j=1}^{i} D((j-1)r_s) - D(jr_s) \right) P_{i|0}(\pi)$$

$$= \sum_{i=1}^{N(\pi)} \left(\sum_{k=0}^{i-1} D(kr_s) - \sum_{j=1}^{i} D(jr_s) \right) P_{i|0}(\pi) \tag{5.15}$$

$$= \sum_{i=1}^{N(\pi)} \left(D(0) - D(ir_s) \right) P_{i|0}(\pi)$$

$$= D(0) \sum_{i=1}^{N(\pi)} P_{i|0}(\pi) - \sum_{i=1}^{N(\pi)} D(ir_s) P_{i|0}(\pi)$$

$$= D(0) - \overline{D}_\pi,$$

where step (5.15) follows from the change of variables $k = j - 1$ and the last step from $\sum_{i=1}^{N(\pi)} P_{i|0}(\pi) = 1$ and from using (5.3). Similarly, from the values of δ_i defined in (5.12) and

(5.13), and using (5.6) and (5.7) it can be verified that

$$
\Delta(1, \pi) = \begin{cases} \overline{PSNR}_{\pi} - PSNR(0), & \text{for Problem B,} \\ \overline{V}_{\pi}, & \text{for Problem C.} \end{cases} \tag{5.16}
$$

Hence, Problems A, B, and C defined in (5.8), (5.9) and (5.10) reduce to the following problem:

$$
\max_{\pi} \Delta(k, \pi) \text{ subject to } R_T(k, \pi) \stackrel{\Delta}{=} \sum_{i=k}^{N(\pi)} \frac{r_s}{r_c(c_{\pi}^i)} \le R, \quad \text{for } k = 1. \tag{5.17}
$$

Now, from (5.4) for $k = 1, 2, \dots, N(\pi)$ and $i = k, k+1, \dots, N(\pi)$, it can be seen that

$$
P_{i|k}(\pi) \stackrel{\Delta}{=} \begin{cases} \prod_{j=k+1}^{i} \left(1 - P_e(c_{\pi}^j)\right) P_e(c_{\pi}^{i+1}) & i = k, k+1, \dots, N(\pi) - 1, \\ \prod_{j=k+1}^{N(\pi)} \left(1 - P_e(c_{\pi}^j)\right) & i = N(\pi). \end{cases}
$$

Consequently, the following holds:

$$
P_{i|k-1}(\pi) = \left(1 - P_e(c_{\pi}^k)\right) P_{i|k}(\pi). \tag{5.18}
$$

From (5.14) and (5.18), and considering that $P_{k|k}(\pi) = 1$ by definition, notice that $\Delta(k, \pi)$ satisfies the following recursion.

$$
\Delta(k, \pi) = \begin{cases} \left(1 - P_e(c_{\pi}^{N(\pi)})\right) \delta_{N(\pi)}, & \text{for } k = N(\pi), \\ \left(1 - P_e(c_{\pi}^k)\right) (\delta_k + \Delta(k+1, \pi)), & \text{for } k = 1, 2, \dots, N(\pi) - 1. \end{cases} \tag{5.19}
$$

Also, the normalized transmission rate (in channel bits per source sample) defined in (5.2) can be equivalently expressed as

$$
R_{T_{\pi}} \stackrel{\Delta}{=} R_T(1, \pi),
$$

where

$$
R_T(k, \pi) \stackrel{\Delta}{=} \sum_{i=k}^{N(\pi)} \frac{r_s}{r_c(c_{\pi}^i)}
$$
$$
= \frac{r_s}{r_c(c_{\pi}^k)} + R_T(k+1, \pi).
$$

Notice that, for a policy π, $\Delta(k, \pi)$ and $R_T(k, \pi)$ do not depend on $c_{\pi}^1, c_{\pi}^2, \dots, c_{\pi}^{k-1}$. Hence, the solution to the maximization problem in (5.17) needs to be specified only over a subsequence of channel codes, namely, $c_{\pi}^k, c_{\pi}^{k+1}, \dots$. Moreover, for any policy π, $\Delta(k+1, \pi)$ and $R_T(k+1, \pi)$ do not depend on the kth channel code c_{π}^k. Consequently, (5.19) leads to the following dynamic programming result for solving (5.17).

Let $\{c_*^k, c_*^{k+1}, c_*^{k+2}, \dots, c_*^{N^*(k,R)}\}$ be the solution for the maximization problem in (5.17). That is, it is the subsequence of channel codes achieving the maximum in (5.17) for starting source packet index k and rate constraint R. Let

$$
r_{min} \stackrel{\Delta}{=} \min_{c \in C} \frac{r_s}{r_c(c)}.
$$

Then the following results determine the dynamic programming solution.

1. For notational convenience, let $\Delta^*(k, R)$ denote the optimal value of the objective function in (5.17) (the total reward). Then, $\Delta^*(k, R)$ satisfies the dynamic programming equation,

$$\Delta^*(k, R) = \begin{cases} 0, & \text{if } R < r_{min}, \\ \max_{c \in C} (1 - P_e(c))(\delta_k + \Delta^*(k + 1, R - \frac{r_s}{r_c(c)})), & \text{otherwise} \end{cases} \quad (5.20)$$

2. The channel code c_*^k is the channel code achieving the maximum in (5.20).
3. The subsequence $\{c_*^{k+1}, c_*^{k+2}, \ldots, c_*^{N^*(k,R)}\}$ solves (5.17) for starting source packet index $k + 1$ and rate constraint $R - \frac{r_s}{r_c(c_*^k)}$.
4. Finally, the terminating source packet index is found by

$$N^*(k, R) = N^*(k + 1, R - \frac{r_s}{r_c(c_*^k)}) \text{ if } \Delta^*\left(k + 1, R - \frac{r_s}{r_c(c_*^k)}\right) > 0$$

$$= k \text{ otherwise.}$$

For arbitrary incremental rewards δ_i, these results can be translated into the following algorithm.

Algorithm 5.1 (optimal UEP) $\Delta^*(k, r)$ *is computed as a recursive function call.*

$$\Delta^*(k, r) := 0 \text{ if } r < r_{min}$$

$$:= \max_{c \in C} (1 - P_e(c)) \left(\delta_k + \Delta^* \left(k + 1, r - \frac{r_s}{r_c(c)} \right) \right). \quad (5.21)$$

The channel code achieving the maximum in Eq. (5.21) is used for encoding the kth source packet.

Notice that the channel code obtained by the algorithm for the kth source packet depends on all the δ_i as well as the target transmission rate.

The computation of $\Delta^*(1, R)$ depends on the computation of $\Delta^*(2, r)$ for a finite number of values of r, all of which are strictly smaller than R. The computation of $\Delta^*(k, r)$ in turn depends on the computation of $\Delta^*(k + 1, r')$ for even smaller values of r'. The recursion terminates by returning a value of 0 when k is sufficiently large so that the target transmission rate falls below r_{min}. It may appear that the number of calls to the recursion grows exponentially, but the computation load can be reduced by storing the computed values $\Delta^*(k, r)$ in memory. Figure 5.12 illustrates how the values of $\Delta^*(k, r)$ can be computed using a time-varying trellis.[3]

The above algorithm has a complexity proportional to R^2, where complexity is considered to be the number of calls to the recursive function in which maximization in (5.21) needs to be performed. This can be seen as follows. Let ρ be the smallest grain of rate-increment per sample in the code family, i.e.

$$\rho = \frac{1}{N_s} \gcd \left(\left\{ \frac{k_s}{r_c(c)}, c \in C \right\} \right).$$

---

3 The use of a trellis is common when solving problems using dynamic programming.

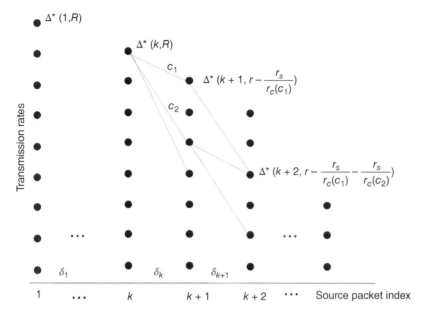

Figure 5.12 Trellis for maximizing the performance for arbitrary δ_i.

Then all achievable transmission rates, i.e. those in the collection $G(R) \stackrel{\Delta}{=} \{R_T(\pi): R_T(\pi) \le R\}$, must be multiples of ρ. $|G(R)|$ grows linearly with R. Now from Algorithm 1, the computation of $\Delta^*(1, r)$ for all values of $r \in G(R)$, requires computation of $\Delta^*(2, r)$ for all values of $r \in G(R - r_{min})$, which, in turn, requires the computation of $\Delta^*(3, r)$ for all values of $r \in G(R - 2r_{min})$ and so on. Hence, the total number of function calls needed to compute $\Delta^*(1, r)$ for all $r \in G(R)$ is upper bounded by

$$|G(R)| + |G(R - r_{min})| + |G(R - 2r_{min})| + \dots + |G(R - (\lfloor R/r_{min} \rfloor - 1)r_{min})|.$$

This is an arithmetic progression, upper bounded by αR^2, for some α.

We discussed earlier the advantages of Problem C that maximizes the average useful source coding rate. It is easy to see that, if $\delta_i = $ constant $\forall i$, then the optimization of (5.17) does not depend on the starting source packet index k. Hence, for such a case, we have

$$\Delta^*(k, R) = \Delta^*(1, R) \quad \forall R \text{ for } k = 1, 2, \dots . \tag{5.22}$$

Further, in such a case, the channel code obtained by the algorithm for the kth source packet depends only on the target transmission rate. Hence, to solve (5.10), Algorithm 1 can be rewritten as follows.

Algorithm 5.2 (maximization of average useful source coding rate) *If the incremental rewards are constant, i.e. $\delta_i = 1 \;\; \forall i$, then $\Delta^*(1, r)$ is computed as a recursive function call.*

$$\Delta^*(1, r) := 0 \text{ if } r < r_{min}$$

$$:= \max_{c \in C}(1 - P_e(c)) \left(1 + \Delta^* \left(1, r - \frac{r_s}{r_c(c)} \right) \right). \tag{5.23}$$

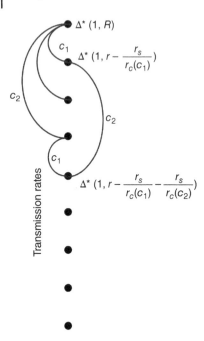

Figure 5.13 Trellis for maximizing the average useful source coding rate.

Figure 5.13 illustrates the trellis used for the computation of $\Delta^*(1, r)$ for the maximization of average useful source coding rate.

Following a complexity analysis similar to the one for Algorithm 1, it can be seen that for Algorithm 2, the number of function calls needed to compute $\pi^*(1, R)$ varies linearly with R. Hence, $|G(R)|$ computations of $\Delta^*(1, r)$ are sufficient to compute $\Delta^*(1, R)$.

Theoretically, the constraints added in the design of embedded source coders result in a coding performance that is not as good as that of the best nonembedded source coders. Nevertheless, good embedded source coders, such as SPIHT, by design have very good performance at all rates, a performance almost as good as that of comparable nonembedded source coders. For embedded source coders, it is desirable to perform JSCC such that, in addition to having the best end-to-end performance for a given target transmission rate, the coder also exploits the progressive characteristic of the encoded bit stream by achieving good performance at intermediate rates. In terms of notation, we say that two policies π_1 and π_2 with transmission rates R_1 and R_2, $R_2 > R_1$, allow progressive transmission, if the output at rate R_2 can be obtained by appending $(R_2 - R_1)$ bits per source sample to the bit stream at rate R_1. In other words, the bit stream for target rate R_1 can be obtained as a prefix of the bit stream for target rate R_2.

A powerful and flexible channel coding technique that is especially apt for progressive transmission is rate compatible channel coding. Recall from Section 3.4 in Chapter 3 that rate compatible codes are a family of channel codes in which the codewords of a low rate code can be obtained by adding some extra parity bits to the codeword of a high rate code. Popular examples of such codes are rate compatible punctured convolutional (RCPC) codes [7, 8]. These codes, when combined with an outer error detection code, such as a CRC encoding fixed-length source packets, provide good error correction and detection capabilities. Similarly, Reed–Solomon codes and their punctured versions are used for erasure channels. We assume next that the channel code family C is rate compatible.

Consider two policies $\pi_1 = \{c_{\pi_1}^1, c_{\pi_1}^2, \ldots, c_{\pi_1}^{N(\pi_1)}\}$ and $\pi_2 = \{c_{\pi_2}^1, c_{\pi_2}^2, \ldots, c_{\pi_2}^{N(\pi_2)}\}$ designed by some scheme for target rates R_1 and $R_2, R_2 > R_1$. Then, if the channel code family is rate compatible, progressive transmission at rates R_2 and R_1, for $R_2 > R_1$, is possible using the two policies π_2 and π_1, if and only if, $N(\pi_2) \geq N(\pi_1)$ and

$$\frac{r_s}{r_c(c_{\pi_1}^i)} \leq \frac{r_s}{r_c(c_{\pi_2}^i)} \quad \text{for } 1 \leq i \leq N(\pi_1). \tag{5.24}$$

To prove this statement, consider that if condition (5.24) is satisfied, progressive transmission is accomplished by first transmitting the bitstream corresponding to policy π_1 followed by the extra redundancy bits needed to obtain the lower rate codes for policy π_2, i.e.

$$\frac{r_s}{r_c(c_{\pi_2}^i)} - \frac{r_s}{r_c(c_{\pi_1}^i)}$$

bits per source sample for packet i, $i = 1,2, \ldots$. Clearly, this cannot be done if (5.24) is not satisfied. Figure 5.14 illustrates the sequence of transmission for two policies. Consider a source bitstream divided into packets containing equal number of source bits. The picture represents a depiction of channel code-word length (bit allocation of source and parity bits) to source packets, such that more bits are allocated to source packets of higher importance and hence lower source packet index. This unequal allocation holds true for both policies. The height of a column for a given source-packet index is proportional to the inverse of channel code rate (i.e. if the channel code rate is k/n, the height of the column represents n/k). To achieve progressive transmission of the two policies, first, all source and parity bits for policy 1 (bits represented by the area below the curve labeled Policy 1) are transmitted. Then the remaining bits, the area corresponding to the shaded region between Policy 2 and Policy 1, are transmitted. They may correspond to new source bits for source packets with higher indices, as well as additional parity bits for source packets with lower indices.

If we require the transmission to be progressive with optimal UEP, then the policies $\pi^*(1, R)$ obtained by solving (5.8), (5.9), or (5.10) for different values of R must allow progressive transmission. As such, it is necessary to verify that those policies satisfy the conditions in (5.24). To do so, let us consider first the optimization criterion (5.7), where $\delta_i \stackrel{\Delta}{=} 1$,

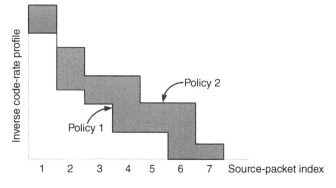

Figure 5.14 Progressive transmission with two policies. The shaded area represents source and parity bits transmitted second.

$i = 1, 2, \ldots$. Let $\pi^*(1, R) = \{c^1, c^2, \ldots, c^{N(\pi^*(1,R))}\}$ be the optimal policy solving (5.10) for rate R. Then by the results that determine the dynamic programming solution, the subsequence $\{c^2, \ldots, c^{N(\pi^*(1,R))}\}$ solves the corresponding version of (5.17) for the starting index 2 and rate constraint

$$R - \frac{r_s}{r_c(c^1)}.$$

Now, if the δ_i's are constant, then as discussed earlier, the optimization (5.17) does not depend on the starting index k. Hence, a policy that assigns c^2 to the *first* source packet, c^3 to the second source packet and similarly assigns $c^{N(\pi^*(1,R))}$ to the $N(\pi^*(1, R)) - 1^{st}$ source packet is the optimal policy

$$\pi^*\left(1, R - \frac{r_s}{r_c(c^1)}\right)$$

for starting index 1 and rate constraint

$$R - \frac{r_s}{r_c(c^1)}.$$

Consider now arbitrary reward sequences δ_i. For all transmission rates R, if π is an optimal policy then it can be shown through a simple interchange argument that $P_e(c^i_\pi) \le P_e(c^j_\pi)$ for $1 \le i \le j \le N(\pi)$. Consequently, we must have the property that $r_s/r_c(c^i_\pi) \ge r_s/r_c(c^j_\pi)$ for $1 \le i \le j \le N(\pi)$ for an optimal policy π, i.e. the optimal policies are *code rate increasing*. Therefore, we have,

$$\frac{r_s}{r_c\left(c^i_{\pi^*(1,R)}\right)} \ge \frac{r_s}{r_c\left(c^{i+1}_{\pi^*(1,R)}\right)} = \frac{r_s}{r_c\left(c^i_{\pi^*(1,R-\frac{r_s}{r_c(c^1)})}\right)} \quad \text{for } i = 1, 2, \ldots, N(\pi^*(1, R)) - 1.$$

(5.25)

This implies that the conditions for progressive transmission in (5.24) are satisfied. These thoughts lead to the property that when implementing UEP with a channel code family consisting of rate compatible codes, if $\pi^*(1, R)$ is the optimal policy solving (5.10) for target rate R, then $\pi^*(1, R)$ and

$$\pi^*\left(1, R - \frac{r_s}{r_c\left(c^1_{\pi^*(1,R)}\right)}\right)$$

allow optimal *progressive* transmission at rates R and

$$R - \frac{r_s}{r_c\left(c^1_{\pi^*(1,R)}\right)}.$$

(5.26)

Continuing in the same way, this property can now be applied to the optimal policy at the rate given by (5.26), to obtain another lower intermediate transmission rate where the optimal policy can be executed. In the same manner, a sequence of intermediate transmission rates can be obtained, at which progressive transmission is possible with provably optimal UEP. The sequence of bits transmitted follows the scheme discussed in Figure 5.14. First, bits corresponding to the policy for a low target bitrate are transmitted. Then the extra parity

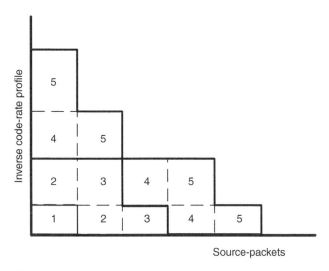

Figure 5.15 Optimal progressive transmission of five source packets; the numbers indicate the sequence in which bits are transmitted.

check bits and new source packets needed to achieve a higher target bit rate are transmitted. Figure 5.15 sketches the inverse code-rate profile of an optimal policy consisting of five source packets. Similarly to Figure 5.14, each column in the figure represents the codeword for one source packet, i.e. its source bits and parity bits. Therefore, the height of the column represents an inverse code-rate n/k for a codeword with channel coderate k/n. In this figure, the labels $1, 2, 3, 4, 5$ indicate the order in which bits corresponding to the source packets are transmitted, to achieve a progressive transmission of five policies at five target bitrates. For example, the bits labeled 1 are followed by bits labeled 2. Bits labeled 2 have two components in this example. Some of the bits labeled 2 are coming from the codeword for source packet 1, and some bits coming from the codeword for source packet 2. Similarly, bits labeled 3 have bits from two source packets, while bits labeled 5 have bits coming from four source packets.

The resulting bit stream has some interesting properties. The bits for a single channel codeword are not necessarily contiguous. This deferred transmission of redundancy creates the possibility that a source packet decoding failure at one target rate can be overcome and the source packet recovered if the target rate is increased and more bits for that packet are received. In that sense, the bit stream is always progressive. We also note that it can be shown by counterexamples that, even for the criterion of average useful source coding rate, progressive transmission using optimal policies may not be possible at *arbitrary* rate pairs R_1 and R_2.

Having described dynamic programming for UEP allocation, we now examine its performance. The work in [6] presented simulation results for the 512×512 grayscale Lenna image compressed with the SPIHT algorithm with arithmetic coding. The source bit stream was divided into source packets of length 32 bytes, UEP was added using the above dynamic programming algorithm, and the resulting bit stream was transmitted over a BSC. The FEC used for UEP consisted of three different families of RCPC codes concatenated with a 2-byte CRC outer error detection code. Channel code family A is a collection of RCPC codes derived from a 64-state, rate 1/3 convolutional mother code taken from [7]. The codes were decoded with a list-Viterbi decoding algorithm with a search depth of 100 paths. Code family B was a relatively weaker RCPC code family derived from a 16-state, rate 1/4 mother code also taken from [7]. These codes were decoded with a list-Viterbi decoding algorithm with a search depth of only 10 paths. Code family C is the same as code family B but used without list decoding (search depth = 1). To run the dynamic programming algorithm, it is necessary to know the parameters P_e of the code families, which were obtained from simulations.

The results of using the dynamic programming algorithm for a BSC with BER of 0.01 are shown in Figures 5.16–5.21. Figures 5.16, 5.18 and 5.20 show the average PSNR performance of the scheme optimizing the PSNR for channel BER 0.01 for code families A, B, and C, respectively. The figures also show the performance of equal error protection (EEP) schemes using the channel codes from the same family. The code rates in the legends do not include the (fixed) code rate of the outer CRC code.

For clarity, Figures 5.17, 5.19, and 5.21 depict the difference in the average PSNR of the optimized scheme and that of different EEP schemes, vs. the total transmission rate. Figures 5.17, 5.19, and 5.21 also include the difference in PSNR of the scheme maximizing the expected PSNR and the scheme maximizing the average useful source coding rate. The figures show that the optimized UEP schemes always perform as well as or better than any EEP scheme from the same family, for all transmission rates.

A second observation is that the improvement of the optimal UEP scheme over any fixed EEP scheme depends on the transmission rate. For example, in Figure 5.19, the loss of the EEP scheme with code rate 4/7 varies from about 0.15 to 0.5 dB depending on the transmission rate. This variability highlights the nature of how UEP schemes address

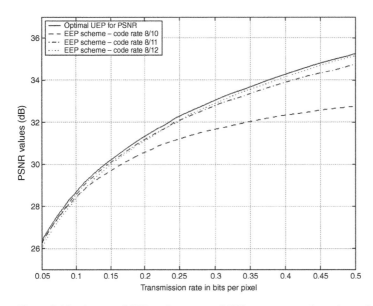

Figure 5.16 Average PSNR performance of UEP over memoryless channels for the image Lenna. Code family A, BER = 0.01.

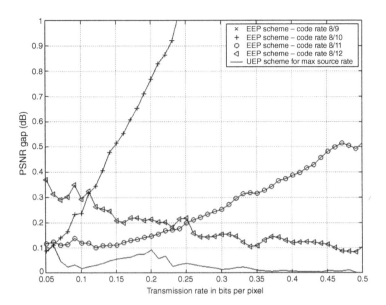

Figure 5.17 The loss of PSNR in EEP schemes and optimal UEP scheme maximizing average useful source coding rate compared to the optimal UEP scheme maximizing PSNR for the image Lenna. Code family A. BER = 0.01.

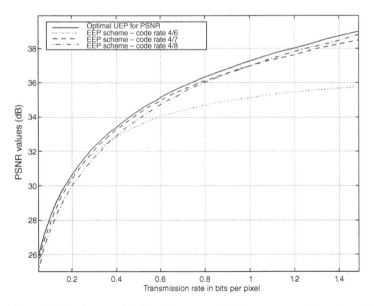

Figure 5.18 Average PSNR performance of UEP over memoryless channels for the image Lenna. Code family B, BER = 0.01.

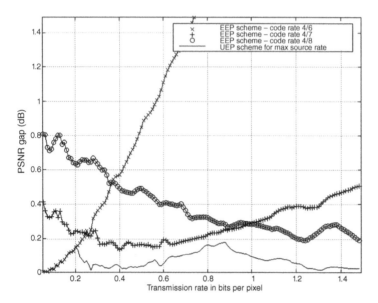

Figure 5.19 The loss of PSNR in EEP schemes and optimal UEP scheme maximizing average useful source coding rate compared to the optimal UEP scheme maximizing PSNR for the image Lenna. Code family B, BER = 0.01.

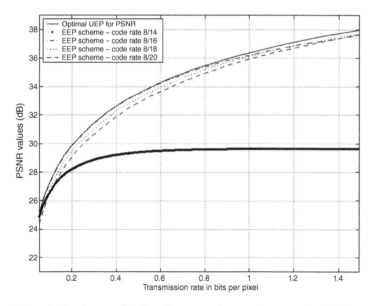

Figure 5.20 Average PSNR performance of unequal error protection for memoryless channels for the image Lenna. Code family C, BER = 0.01.

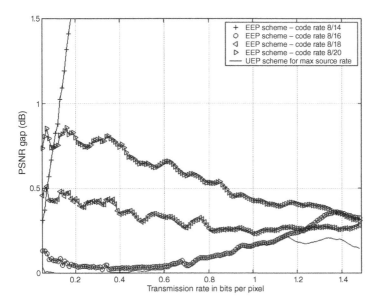

Figure 5.21 The loss of PSNR in EEP schemes and optimal UEP scheme maximizing average useful source coding rate compared to optimal UEP scheme maximizing PSNR, for the image Lenna. Code family C, BER = 0.01.

EEP weaknesses. A low code rate EEP scheme that performs well (close to optimal) at high transmission rates is overprotective at low transmission rates. A higher code rate EEP scheme may be efficient at low transmission rates but, as the transmission rate increases, the average PSNR may saturate as the probability of source packet decoding failure somewhere in the image increases with the target transmission rate. It is not possible to "switch" between two EEP schemes at the crossover points during a progressive transmission. The performance loss of the UEP scheme maximizing the average useful source coding rate also depends on the transmission rate. But the loss is smaller than that for any EEP scheme and hence the UEP scheme maximizing the average source coding rate will also perform better than any EEP scheme at all transmission rates.

The UEP scheme is more effective when the available channel code family is weak. With the strong code family A, the performance of a single channel code is fairly close to the optimal for a significant portion of the range of transmission rates. In such cases, the benefit of UEP is more limited. When the channel code family is weak, any EEP scheme performs close to the optimal only for a small range of transmission rates. At other rates, its performance may be substantially suboptimal compared to the UEP scheme. This behavior can also be observed for channels with different BERs.

Finally, we note that the UEP approach with the optimal dynamic programming algorithm is general enough that it can be applied to memoryless packet erasure channels. In this case, the code family is replaced by blocks codes such as Reed–Solomon codes and their punctured versions. The performance for this application obtained from simulations can be seen in Figures 5.22 and 5.23. The figures show results for transmission over a memoryless packet erasure channel of packet size 8 bytes. The family of channel codes was derived from a (255, 32) Reed–Solomon code with single-byte size symbols. Eight consecutive 1-byte

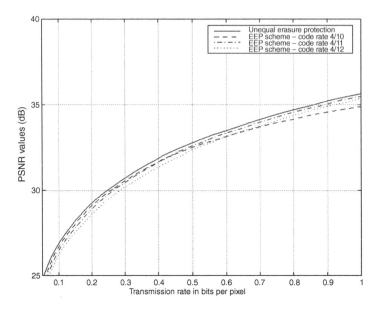

Figure 5.22 Average PSNR performance of EEP and the optimal UEP scheme for the Lenna image for memoryless packet erasure channels: packet size 8 bytes, erasure rate 20%.

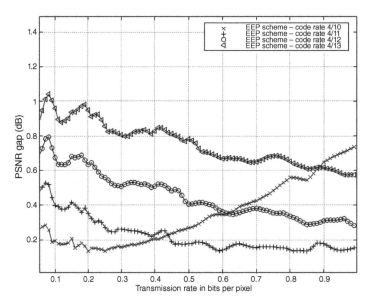

Figure 5.23 Average PSNR gain of the optimal UEP scheme over equal erasure protection schemes: memoryless erasure channels: packet size 8 bytes, erasure rate 20%.

symbols were arranged in one packet to yield a mother packet erasure correcting (PEC) code of parameters (31,4). The family consisted of $(n, 4)$ PEC codes, for $4 \leq n \leq 31$, obtained as punctured versions of the mother code. Recall that a $(n, 4)$ PEC code is capable of correcting up to $n - 4$ packet erasures. The results in Figures 5.22 and 5.23, for a packet erasure channel with a packet loss rate of 20%, show again that no single EEP policy performs closer to the optimal at all transmission rates. Depending on the target rate, gains up to 0.5 dB can be obtained over any EEP scheme chosen from the family.

5.4 Unequal Error Protection Using Digital Fountain Codes

Digital fountain (DF) codes, also known as *rateless codes*, allow for low complexity encoding and decoding and require almost no feedback. First implemented as the Luby transform code, [9], fountain codes work similarly to other block codes (such as Reed–Solomon and /Bose–Chaudhuri–Hocquenghem [BCH] codes) in that the source data can be recovered from any sufficiently large subset of coded symbols. At the same time, fountain codes exhibit a high erasure correction performance for large block lengths. The most salient property of a fountain code is that the encoder can generate a potentially unlimited number of coded symbols. The transmitter could keep sending coded symbols until enough are received to recover the source data, at which time the transmission can be stopped by sending a feedback command.

A fountain code is defined by specifying a *degree distribution* (DD) and a *selection distribution* (SD). The degree value specifies the number of information symbols that determine the corresponding coded symbol. A DD $\rho(d)$ indicates the probability that the degree value would be d. For a given degree d, the coded symbol is computed from a subset of d information symbols chosen according to the SD function.

A simple example of fountain code encoding is given by the original Luby transform (LT) code implementation. The encoder inputs a block of k binary information symbols $\mathbf{x}^T = \{x_1, x_2, \ldots, x_k\}$ and y_m denotes the mth coded symbol. To generate y_m the encoder first chooses the corresponding degree $d_m \in \{1, 2, \ldots, k\}$ by sampling a random variable with distribution equal to the DD $\rho(d)$. Using the generated d_m, the coded symbol y_m is calculated by single-bit modulo 2 addition (bitwise exclusive-OR) of a subset of d_m information symbols from $\mathbf{x}$ chosen according to the SD. In the original LT code implementation, the selection distribution is uniform, so the coded symbol y_m is calculated by single-bit modulo 2 addition of d_m information bits picked at random from the k information symbols in $\mathbf{x}$. This selection has the advantage of simplicity, but does not take into account the different importance of the information symbols. Note also that in this encoding procedure, each coded symbol is generated independently of the others.

During transmission over an erasure channel, some coded symbols will be lost, and only n of them will be received and used during decoding. Fountain codes can be decoded with low complexity using a belief propagation algorithm. The algorithm progresses from degree-one coded symbols to increasing degrees. Figure 5.24 illustrates the progress of the decoding algorithm to recover $k = 4$ information symbols from $n = 5$ coded symbols. Using a graph that connects information and coded symbols, the algorithm starts decoding from the degree-one coded symbols. In Figure 5.24a, the only degree-one coded symbol is the

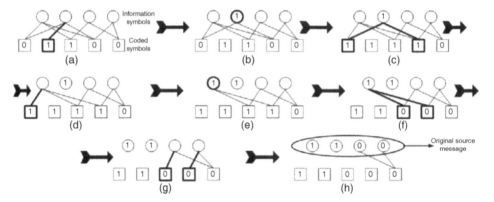

Figure 5.24 The decoding steps of a fountain code with $n = 5, k = 4$ using a belief propagation algorithm.

second one. Because the degree is one, the coded symbol can directly be used to estimate the corresponding information symbol. This is shown in Figure 5.24b, where the value 1 has propagated from the second code symbol to the connected information symbol, and the edge connecting them is removed from the graph. Next, the decoded information symbol is added modulo 2 to the coded symbols to which it has connecting edges in the graph (Figure 5.24c). Because of the modulo 2 addition properties, this operation removes the contribution of the information symbol just decoded from the coded symbols that remain to be used in the decoding process, so the corresponding connecting edges are deleted after this operation (Figure 5.24d). The algorithm continues, as shown in Figure 5.24d, by identifying for each cycle a new degree-one coded symbol (after edge deletion) and repeating the same steps. Figures 5.24e–h illustrate the rest of the decoding operation.

The original LT code implementation of fountain codes uses a uniform SD, implicitly assigning the same importance to all information symbols, so it does not provide UEP. Also, note that the decoding algorithm can progress only if there are degree-one coded symbols to work with at the start of each iteration. In the graph, the set of coded symbols that have one connected edge at a given iteration is called the *ripple*. The decoder stops when the ripple is an empty set. Moreover, because lower degree coded symbols are processed earlier by the decoding algorithm, a fountain code designed for UEP should aim at having coded symbols with lower degree connected in the code graph with more important information symbols. In this way, the chance of decoding the more important information bits is higher if the decoder stops the decoding process at some early iteration.

Fountain code designs for UEP were classified in [10] into two categories: the *weighted approach* [11, 12] and the *expanding window fountain codes* [13]. In the weighted approach, the uniform SD is transformed through the use of a set of weights. This is implemented by partitioning the k-bit information block into r disjoint subsets $s_1, s_2, \ldots, s_r$, where the bits for each subset are a sequence within the information block. The sizes of the r subsets may be different and are the design parameters for the partition. For encoding, the degree d_m for an encoded symbol is chosen as usual. Then, a subset s_i is chosen according to a predetermined probability w_i ($\sum_{i=1}^{r} w_i = 1$), and the coded symbol y_m is calculated by modulo 2 addition of d_m information symbols selected following a uniform distribution only from s_i. When designing a UEP code, the probabilities w_i for selecting each subset are chosen so that the probability of recovering more important bits is increased with respect to the other bits.

This weighted concept was expanded in [10] as a *generalized weighted approach* where the probability distributions for choosing the disjoint subsets $s_1, s_2, \ldots, s_r$ depend on the degree of the coded symbol. Then, the code is defined through the $r \times k$ matrix

$$\mathbf{P} = \begin{bmatrix} p_{1,1} & p_{1,2} & \cdots & p_{1,k} \\ p_{2,1} & p_{2,2} & \cdots & p_{2,k} \\ \vdots & \vdots & \cdots & \vdots \\ p_{r-1,1} & p_{r-1,2} & \cdots & p_{r-1,k} \\ p_{r,1} & p_{r,2} & \cdots & p_{r,k} \end{bmatrix},$$

where $p_{i,j}$ is the probability of choosing subset s_i given that the degree of the coded symbol is j, and the DD is expressed as a k-element vector $\vec{\lambda}$, where the ith vector entry is the probability that a coded symbol has degree i.

Note that because the subsets s_i form a partition, each column of $\mathbf{P}$ sums to 1. Then, the design of the code involves finding k parameters from $\vec{\lambda}$ and $(r-1) \times k$ parameters from $\mathbf{P}$.

In contrast to the weighted approach, the expanding window fountain codes divide the block of k information symbols into r successively larger windows. The windows are nested, with larger windows including the content of all smaller windows. This can be realized by first partitioning the k-bit information block into r variable-size disjoint subsets $s_1, s_2, \ldots, s_r$, followed by defining each of the r embedded windows $\{W_j\}_{j=1}^r$ as $W_j = \bigcup_{l=1}^j s_l$. To generate a coded symbol from the expanding window fountain code, a window is selected according to a discrete probability distribution, then the degree for the coded symbol is chosen using a DD that depends on the selected window. Finally, the coded symbol is calculated by picking from the selected window a number of information symbols equal to the degree and adding those symbols using modulo 2 arithmetic.

In [10], a *generalized expanding window fountain* (GEWF) code was introduced. First, the degree of the coded symbol is chosen, and this degree is used to condition the probability distribution for selecting one of the windows. Consequently, the GEWF code is specified by a k-element vector $\vec{\lambda}$, where the ith vector element is the probability that a coded symbol has degree i, and an $r \times k$ matrix

$$\Gamma = \begin{bmatrix} \gamma_{1,1} & \gamma_{1,2} & \cdots & \gamma_{1,k} \\ \gamma_{2,1} & \gamma_{2,2} & \cdots & \gamma_{2,k} \\ \vdots & \vdots & \cdots & \vdots \\ \gamma_{r-1,1} & \gamma_{r-1,2} & \cdots & \gamma_{r-1,k} \\ \gamma_{r,1} & \gamma_{r,2} & \cdots & \gamma_{r,k} \end{bmatrix},$$

where $\gamma_{i,j}$ is the probability of choosing window W_i given that the degree of the coded symbol is j. The columns of the matrix Γ sum to 1; thus, the matrix has $(r-1) \times k$ independent parameters.

The choice for the design parameters $\vec{\lambda}$ and $\mathbf{P}$, or $\vec{\lambda}$ and Γ, depends on the particular application that requires UEP. The characteristic that UEP fountain codes can be designed to maximize the number of information symbols that can be recovered before the belief propagation decoding algorithm needs to stop (because there are no more degree-one coded symbols) is useful for protecting a progressive multimedia source. Because of this, we will explain next the application of UEP fountain codes to progressive transmission.

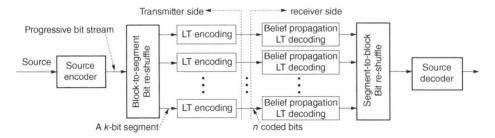

Figure 5.25 A system using UEP fountain codes for transmitting a multimedia source using a progressive source encoder.

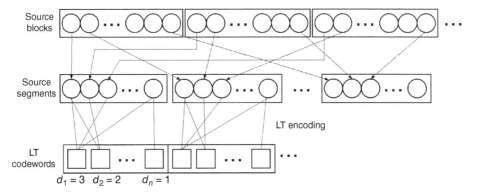

Figure 5.26 Reorganization of source bits from blocks into equal size segments, followed by encoding of each segment using the same UEP fountain code applied to each.

The UEP fountain code approaches presented need to be extended to operate on blocks of symbols, rather than on individual information symbols, because this is how multimedia sources are typically transmitted. Figure 5.25 shows the block diagram for a system that uses UEP fountain codes for the transmission of a progressive multimedia source. For a progressive stream partitioned into blocks, the importance of bits within a block tends to be about the same as they are from nearby positions in the source encoded bit stream, or are encoding the same type of source information (e.g. motion vectors for video). Then, the first step in the UEP scheme is to reorganize the source bits into equal size blocks, which will be called *segments*, such that the bits in each segment are of different importance. This operation, illustrated in Figure 5.26, will increase the effectiveness of the UEP scheme. In the figure, a progressive source encoded bit stream with B bits exists as source blocks. After re-shuffling, the bits are organized into $\lfloor B/k \rfloor$ segments with k bits that ideally belong to a diverse set of locations within the progressive source encoded bit stream.

After the bit reorganization, the same realization of a UEP fountain code is applied to each segment. The coded bits are then transmitted over an erasure channel. Assuming that all streams of coded symbols go through the same erasure channel, the receiver will collect for decoding the same number of bits, denoted n, for each transmitted k-bit segment. If, following decoding, m useful information bits can be recovered from one k-bit segment, the same number of useful information bits will be recovered from each of the other segments because the same UEP fountain code realization was applied to all segments. As a consequence, the UEP fountain code decoding procedure will yield $\lfloor B/k \rfloor m$ information bits from the progressive stream that will be used for source decoding.

Designing the UEP fountain code involves finding the parameters $\vec{\lambda}$ and $\mathbf{P}$, or $\vec{\lambda}$ and $\boldsymbol{\Gamma}$, in a smart way that considers the transmission of a progressive stream. The belief propagation decoding algorithm has a nonzero probability of stopping at any iteration (depending on the realization of channel erasures and the encoding structure). For this type of algorithm, a common practice is to have a negligible probability that the decoding iterations would exceed a limit. This threshold is chosen based on the number of information symbols being encoded. After limiting the number of iterations for the decoding algorithm, the design goal becomes choosing $\vec{\lambda}$ and $\mathbf{P}$ (or $\boldsymbol{\Gamma}$) so that the average distortion is minimized at each number of decoding iterations given that n coded symbols are received from the erasure channel for each segment. Formally, this design goal can be written for the generalized weighted approach as,

$$\min_{\vec{\lambda},\mathbf{P}} \overline{D}_M \text{ such that } n \text{ coded symbols are received unerased,} \tag{5.27}$$

and for the GEWF as,

$$\min_{\vec{\lambda},\boldsymbol{\Gamma}} \overline{D}_M \text{ such that } n \text{ coded symbols are received unerased,} \tag{5.28}$$

where $\overline{D}_M$ is the average source reconstructed distortion after completing M fountain code decoding iterations. It is challenging to solve this design optimization problem as is. Consequently, the approach followed in [10] was to first reduce the number of parameters to be found. The first step is to have the SD be a function that can be characterized by a few parameters. One choice is to have the SD follow an exponential function. This choice is natural because low-degree coded symbols will make, on average, more connections with the more important information sets. For the weighted fountain code approach, the exponential SD is introduced by defining the parameters $p_{i,j}$, the probability of choosing subset s_i given that the degree of the coded symbol is j, as an exponential function of the degree number i,

$$p_{i,j} = A_j + B_j e^{-(i-1)/C_j}, \tag{5.29}$$

for $i = 1, 2, \ldots, k$, and where the design parameters are now $A_j \geq 0$, $B_j \geq 0$ and $C_j \geq 0$ for $j = 1, 2, \ldots, r - 1$. Since the $p_{i,j}$ are probabilities, the parameters have the extra constraint that $\sum_{j=1}^{r} p_{i,j} = 1$ for all i. Introducing the exponential SD reduces the number of design parameters from $rk - 1$ to $3(r - 1) + k - 1$. If, for example, the information bits for each segment are divided into $r = 2$ sets and segments have $k = 100$ bits, the number of design parameters is reduced from 199 to 102.

The number of design parameters can be further reduced by using a standard DD function and a predetermined partition set $s_1, s_2, \ldots, s_r$. A possible standard DD choice is the robust soliton distribution, derived from the ideal soliton distribution, which has ideal behavior in terms of the expected number of encoding symbols needed to recover the block of information symbols. Recalling that the DDs, $\rho(d)$, characterize the probabilities for the degrees $d \in \{1, 2, \dots, k\}$, the ideal soliton distribution is given by, [9],

$$\rho(d) = \begin{cases} 1/k & \text{for } d = 1, \\ \frac{1}{d(d-1)} & \text{for } d = 2, 3, \dots, k. \end{cases}$$

The ideal soliton distribution has the drawback that it generates codes with an expected ripple size that is too small. If used in a practical implementation, the decoder would likely stop decoding early in the process. The robust soliton distribution addresses this weakness by resulting in codes that have high probability of keeping a large enough ripple size at each decoding step. The robust soliton probability distribution for the degrees $d \in \{1, 2, \dots, k\}$ is given by, [9],

$$\mu(d) = \frac{\rho(d) + \tau(d)}{\beta}, \text{ for } d = 1, 2, \dots k,$$

where $\beta = \sum_{d=1}^{k} (\rho(d) + \tau(d))$ is a normalization constant and $\tau(d)$ is defined as

$$\tau(d) = \begin{cases} R/dk, & \text{for } d = 1, \dots, k/R - 1, \\ R \ln(R/\delta)/k, & \text{for } d = k/R, \\ 0 & \text{for } d = k/R + 1, \dots, k. \end{cases}$$

with $R = c\sqrt{k}\ln(k/\delta)$ for a suitable constant $c > 0$ and δ is the acceptable failure decoding probability for the given number of coded symbols.

When choosing the exponential SD, a standard DD function and a predetermined partition set $s_1, s_2, \dots, s_r$, the number of design parameters that needs to be optimized reduces to only $3(r - 1)$. If an additional optimization is run to determine the partition set $s_1, s_2, \dots, s_r$, the number of design parameters increases to $4(r - 1)$.

For GEWF codes, the selection of the degree for a coded symbol is conditioned on the chosen window. This is done through the definition of a compound DD function given by

$$\Lambda^c(x) = \sum_{i=1}^{k} \Lambda_i^c x^i,$$

$$\Lambda_i^c(x) \overset{\Delta}{=} \sum_{j=1}^{k} \rho_j \Phi_i^{(j)},$$

where Λ_i^c is the probability of choosing degree i, $0 \leq \{\rho_j\}_{j=1}^r \leq 1$ such that $\sum_{j=1}^r \rho_j = 1$, and $\Phi_i^{(j)}$ is the conditional probability of choosing a degree i given that the embedded window W_j was selected for the coded symbol (and, as such, $\Phi_i^{(j)} \overset{\Delta}{=} 0$ if $i > |W_j|$).

Similar to the weighted approach, the number of design parameters that needs to be optimized can be reduced by choosing $\gamma_{j,i}$ to follow an exponential distribution tailored to the size of each embedded window. The definition of $\gamma_{j,i}$ is in this case, for $i = 1, 2 \dots, k$,

$$\gamma_{j,i} = \begin{cases} \bar{A}_j + \bar{B}_j e^{-(i-1)/\bar{C}_j}, & \text{if } i \leq |W_j|, \\ 0 & \text{if } i > |W_j|, \end{cases}$$

where the choice of the design parameters $\{\bar{A}_j, \bar{B}_j, \bar{C}_j\}_{j=1}^{r-1}$ needs to satisfy $\sum_{j=1}^r \gamma_{j,i} = 1, \forall i$.

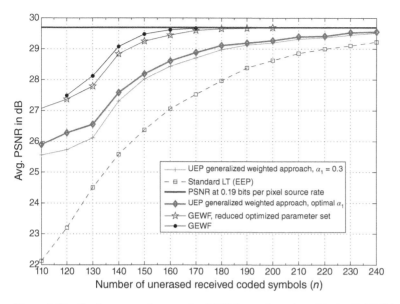

Figure 5.27 Performance of generalized UEP fountain codes with the *Goldhill* image and $k = 100$.

Using these approaches to reduce the number of optimization parameters, the work in [10] found the optimum solution to the fountain code designs by numerical experimentation and exhaustive search. Figure 5.27 illustrates the performance obtained with the generalized UEP fountain codes when protecting the progressive transmission of the 512×512 grayscale *Goldhill* image (seen in Figure 5.28). The progressive stream to be protected contains 50,000 bits, resulting in a source rate of approximately 0.19 bits per pixel. In the results shown in Figure 5.27, information blocks had $k = 100$ bits and each was divided into $r = 2$ blocks. The optimization of the UEP fountain codes followed (5.27) or (5.28), where the distortion is measured by the mean squared error when the belief propagation decoding algorithm runs for a maximum of 70 iterations. Figure 5.27 shows the reconstructed image quality measured by average peak signal-to-noise ratio (PSNR) as a function of the number of unerased received coded symbols (n). The average of the measured quality was calculated based on 10,000 realizations of the complete system. The curves shown in Figure 5.27 are

- PSNR at 0.19 bits per pixel source rate: This curve shows the average PSNR obtained from the complete progressive source coded bit stream.
- Standard LT (EEP): This curve shows the performance achieved with an LT fountain code with EEP.
- UEP generalized weighted approach, $\alpha_1 = 0.3$: This curve shows the result when using the weighted approach. Because $r = 2$, the information block is divided into two disjoint sets. The first set has size $|s_1| = \alpha_1 k = 30$, and the second set has size $|s_2| = (1 - \alpha_1)k = 70$. This code uses the exponential SD but, to further reduce the number of design parameters, $\bar{A}_1 + \bar{B}_1 = 1$, so the design parameters are $\bar{A}_1$ and $\bar{C}_1$. For DD, this code uses the robust soliton distribution with $\delta = c = 0.01$.

Figure 5.28 The Goldhill test image. Source: Robin Weaver/Alamy Images.

- UEP generalized weighted approach, optimal α_1: This curve shows the result when using the weighted approach as above, but where α_1 is one additional design parameter subject to optimization.
- GEWF, reduced optimized parameter set: This curve shows the result when using the generalized expanding window fountain UEP code. The settings for this code share similarities with the "UEP Generalized Weighted Approach, optimal α_1" curve in that the sizes of the two expanding windows are optimized and the exponential SD is used with $\bar{A}_1 + \bar{B}_1 = 1$ so the design parameters are also $\bar{A}_1$ and $\bar{C}_1$. This code uses the compound DD previously explained. Since $r = 2$, the probability of choosing degree i for any coded symbol is given by $\Lambda_i^c = \rho_1 \Phi_i^{(1)} + (1 - \rho_1)\Phi_i^{(2)}$, where $\Phi_i^{(1)} = 0$ for $i > |W_1|$ and ρ_1 is another design parameter subject to optimization.
- GEWF: This curve shows the result when using the GEWF UEP code but, unlike the "GEWF, reduced optimized parameter set," now the three parameters $\bar{A}_1, \bar{B}_1$, and $\bar{C}_1$ from the exponential SD are optimized.

The results show the improvement in reconstructed image quality when using a UEP scheme. The EEP scheme shows a degradation in average quality of at least 3.5 dB when the number of unerased received coded symbols is 110. Figure 5.27 also shows that the

GEWF UEP code outperforms the generalized weighted UEP approach by approximately 1dB.

Finally, it is worth noticing that the performance of the UEP fountain codes changes with the number of iterations performed by the belief propagation decoding algorithm. As expected, the quality of the reconstructed source is noticeably reduced if the number of iterations is much less than the number used during the optimization of the design parameters. This establishes a trade-off between the performance of the UEP scheme and the complexity of the decoding procedure. Yet, the performance loss observed when reducing the number of decoding iterations can be mitigated to a great degree by choosing code design parameters that are optimized for the number of decoding iterations. In this way, and depending on the many variables involved in a particular system, it may be a more sensible approach to design the code for a small number of decoding iterations, avoiding excessive performance loss and operating with lower decoding complexity, and as a trade-off accept a relatively small loss relative to the performance that could be achieved with the maximum number of decoding iterations.

References

1 Pearlman, W.A. and Said, A. (2011). *Digital Signal Compression, Principles and Practice.* Cambridge University Press.

2 Stuhlmuller, K., Farber, N., Link, M., and Girod, B. (2000). Analysis of video transmission over lossy channels. *IEEE Journal on Selected Areas in Communications* 18 (6): 1012–1032.

3 Albanese, A., Blomer, J., Edmonds, J. et al. (1996). Priority encoding transmission. *IEEE Transactions on Information Theory* 42 (6): 1737–1744.

4 Leicher, C. and Technische Universitt Munchen (1994). Hierarchical encoding of MPEG sequences using priority encoding transmission (PET). *Technical report.* Technische Universitat Munchen.

5 Mohr, A.E., Riskin, E.A., and Ladner, R.E. (1999). Unequal loss protection: graceful degradation of image quality over packet erasure channels through forward error correction. *IEEE Journal on Selected Areas in Communications* 18: 819–828.

6 Chande, V. and Farvardin, N. (2000). Progressive transmission of images over memoryless noisy channels. *IEEE Journal on Selected Areas in Communications* 18 (6): 850–860.

7 Hagenauer, J. (1988). Rate compatible punctured convolutional (RCPC) codes and their applications. *IEEE Transactions on Communications* 36 (4): 389–399.

8 Frenger, P., Orten, P., Ottosson, T., and Svensson, A.B. (1999). Rate-compatible convolutional codes for multirate DS-CDMA systems. *IEEE Transactions on Communications* 47: 828–836.

9 Luby, M. (2002). LT codes. *The 43rd Annual IEEE Symposium on Foundations of Computer Science, 2002. Proceedings*, pp. 271–280.

10 Arslan, S.S., Cosman, P.C., and Milstein, L.B. (2012). Generalized unequal error protection LT codes for progressive data transmission. *IEEE Transactions on Image Processing* 21 (8): 3586–3597.

11 Rahnavard, N. and Fekri, F. (2005). Finite-length unequal error protection rateless codes: design and analysis. *Global Telecommunications Conference, 2005. GLOBECOM '05. IEEE*, Volume 3, p. 5.

12 Chang, S.-K., Yang, K.-C., and Wang, J.-S. (2008). Unequal-protected LT code for layered video streaming. *IEEE International Conference on Communications, 2008. ICC '08*, pp. 500–504, May 2008.

13 Sejdinovic, D., Vukobratovic, D., Doufexi, A. et al. (2009). Expanding window fountain codes for unequal error protection. *IEEE Transactions on Communications* 57 (9): 2510–2516.

6

Source–Channel Coding with Feedback

Decision feedback in the form of an acknowledgment (ACK) signaling of a successful reception and/or a negative acknowledgment (NACK) signal to indicate failed transmission has been used extensively in communication situations where there is a feedback channel available from the receiver to the transmitter. Link layer protocols based on *automatic repeat query* (ARQ) and combined ARQ and forward error correction (FEC), also called *Hybrid ARQ*, are used for data communication in a wireless environment. Feedback and retransmission are also used at the transport layer for end-to-end error recovery, e.g. in the TCP/IP protocol. Conventionally, these protocols are designed for reliable transmission of data. The ACK/NACK generation is accomplished by an error detection mechanism such as a cyclic redundancy check (CRC) or bounded distance decoding. The purpose of the error detection mechanism is to evaluate whether decoding has been successful or not in correcting errors introduced during transmission. Protocols designed for data transmission, attempt to trade off the probability of undetected bit errors with the average code rate or throughput. In this chapter, we investigate how a *distortion metric can be incorporated into the design* of a communication system for a loss tolerant source that incorporates an ACK/NACK feedback channel, thus allowing for the use of an ARQ or Hybrid ARQ protocol.

In this chapter we revert to first principles in the formulation and design methodology. This approach involves viewing source encoding as *quantization* followed by *index assignment* and decoding as reproduction of the source from the received, possibly corrupted, information. This chapter is a partly theoretical and partly experimental investigation of the effective use of ACK/NACK feedback, primarily on the receiver side, when the objective is to obtain the best trade-off between the transmission rate and the distortion at the receiver.

6.1 Joint Source–Channel Coding Formulation for a System with ACK/NACK Feedback

A general point-to-point discrete time communication system for transmission of a loss tolerant source using ACK/NACK feedback can be described by the block diagram in Figure 6.1. The source produces a random vector $\mathcal{X}$ of fixed dimension and known statistics taking values in a finite-dimensional real-valued space R^k. The encoding

Joint Source-Channel Coding, First Edition. Andres Kwasinski and Vinay Chande.
© 2023 John Wiley & Sons Ltd. Published 2023 by John Wiley & Sons Ltd.

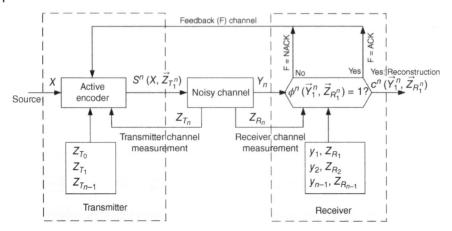

Figure 6.1 General JSCC system with ACK/NACK feedback at the nth step in transmission.

and transmission of this vector from the transmitter to the receiver takes place in several *steps*. In each step, some channel symbols are transmitted from the transmitter to the receiver over the noisy channel. A general discrete time noisy channel takes channel symbols from input alphabet $\mathcal{I}$ and generates received symbols from output alphabet $\mathcal{Y}$.

At the end of every step, a feedback $F \in \{ACK, NACK\}$ is transmitted from the receiver to the transmitter over the feedback channel. The source encoder's transmission for step $n + 1$ may take place only if the feedback after step n was a NACK. If it was an ACK, the packet in question is already received successfully, and hence no further transmission is necessary. Therefore, for any packet undergoing transmission, the sequence of feedbacks is a series of NACKs with an ACK at the end.

The random variables Z_{Tn} and Z_{Rn}, taking values in alphabet $\mathcal{Z}$, represent the *transmitter channel measurement* and the *receiver channel measurement*, respectively, which may be available at the two ends as additional information about the channel. This information can be assumed to be uncorrelated with the source X. Note that, for analysis, the receiver channel measurement Z_{Rn} *can be omitted without loss of generality* as it can be included with the received noisy symbols Y_n as a combined received information.

The transmitter can be described mathematically by an *encoding rule* **S** which is a sequence of integers $l_n \geq 0$, $n = 1, 2, \ldots$, and a sequence of encoder maps $S^n : \mathcal{X} \times \mathcal{Z}^n \rightarrow \mathcal{I}^{l_n}, n = 1, 2, \ldots$. The first transmission of length l_1 is always carried out. On receiving NACK at the $n - 1$th step, the transmitter sends l_n symbols given by computing $S^n(X, Z_{T1}, Z_{T2}, \ldots, Z_{Tn})$ over the noisy channel at the nth step. This vector of channel symbols (also called a *channel codeword, transmit codeword, or nth step codeword*) is corrupted by the channel and is received as random vector Y_n taking values in $\mathcal{Y}^{l_n}$. We will say that Y_n is the *received codeword* at the nth step. For simplicity, with no loss of generality, we shall assume that $l_1 = l_2 = \ldots = l_n = \ldots = L$, i.e. exactly L symbols, are transmitted over the noisy channel between two feedbacks. L is the *packet length* or the *codeword length*.

The receiver is described by the *feedback generation rule* and the *reproduction rule*. The feedback generation rule, ϕ, is a sequence of *feedback generation maps* $\phi^n : \mathcal{Y}^{nL} \rightarrow \{0, 1\}, n = 1, 2, \ldots$. At the nth step, let the realizations of the received codewords be $y_1, y_2, \ldots, y_n$ for $y_i \in \mathcal{Y}^L$. Then an ACK is transmitted over the feedback channel if $\phi^n(y_1, y_2, \ldots, y_n) = 1$, which indicates that at this step decoding was successful in correcting errors introduced during transmission over the noisy channel, while decoding was unsuccessful in all the preceding steps. A NACK is transmitted if $\phi^n(y_1, y_2, \ldots, y_n) = 0$, which indicates that decoding was unsuccessful in this and all preceding steps. For mathematical convenience, we also define the constant "function" ϕ^0 which is either 0 or 1. We assume here that the ACK/NACK feedback is instantaneous and error-free.[1]

The *reproduction rule* **c** is a sequence of *reproduction maps* $c^n : \mathcal{Y}^{nL} \to C \subset \mathcal{X}$ for $n = 1, 2, \ldots$, C is the *reproduction codebook*. For mathematical convenience, constant function $c^0 \in C$ should be defined. If an ACK is generated at the nth step, i.e. if $\phi^n(y_1, y_2, \ldots, y_n) = 1$, then the source is reconstructed as $c^n(y_1, y_2, \ldots, y_n)$. It is not necessary for C to be discrete. We will use $\vec{y}_1^n$ and $\vec{Y}_1^n$ to denote the sequences $y_1, y_2, \ldots, y_n$ and random vectors $Y_1, Y_2, \ldots, Y_n$, respectively. Similarly, $\vec{Z}_{T_1^n}$ denotes the sequence $Z_{T1}, Z_{T2}, \ldots, Z_{Tn}$.

The noisy channel is assumed to be independent of the source vector. The channel can be described by: (i) the joint distributions of transmitter channel measurement $F_{\vec{Z}_{T_1^n}}(\vec{Z}_{T_1^n})$, $n = 1, 2, \ldots$, and (ii) transition probabilities, which are conditional probability density functions of vectors $f_{\vec{Y}_1^n | \vec{Z}_{T_1^n}, \vec{I}_1^n}(\vec{y}_1^n | \vec{z}_1^n, \vec{i}_1^n)$, $n = 1, 2, \ldots$ for $y_n \in \mathcal{Y}^L$ and $i_n \in \mathcal{I}^L$, satisfying appropriate consistency conditions on marginal distributions.

6.1.1 Performance Measurement

The simplest performance measures for loss tolerant systems are the distortion and the transmission rate. The transmission rate is the average channel uses (i.e. channel symbols) per source sample. Unless noted otherwise, we shall assume that distortion between the original source vector X and its reproduction at the receiver $\hat{X}$ is quantified using the squared error distortion measure, i.e. $d(X, \hat{X}) = \| X - \hat{X} \|^2$.

As discussed earlier, the receiver channel measurement need not be explicitly mentioned and will be omitted in the rest of the discussion. Note that the source is reproduced at the nth step only if the current step resulted in an ACK feedback and the previous $n - 1$ steps resulted in a NACK, i.e. $\phi^i(\vec{y}_1^i) = 0$ for $i = 1, 2, \ldots n - 1$ and $\phi^n(\vec{y}_1^n) = 1$. The function can be defined as:

$$\psi^n(\vec{y}_1^n) \overset{\Delta}{=} \prod_{i=0}^{n-1} (1 - \phi^i(\vec{y}_1^i)) \phi^n(\vec{y}_1^n). \tag{6.1}$$

It is straightforward to show that the average distortion for a given transmitter, receiver, and channel can be computed as:

$$D(\phi, \mathbf{c}) = E \left[\sum_{n=0}^{\infty} d(X, c^n(\vec{Y}_1^n)) \psi^n(\vec{Y}_1^n) \right]. \tag{6.2}$$

---

[1] Though this assumption is limiting, it simplifies the analysis of the design and evaluation of feedback-based joint source–channel coding (JSCC) systems. Some effect of delay can be mitigated by buffers at the transmitter and the receiver along with a "selective-repeat" strategy. As ACK/NACK feedback requires very low data rate on the feedback channel, it can be protected by strong error correction and can be reasonably assumed to be error-free.

For clarity, let us write down the expectation calculations explicitly:

$$D(\mathbf{S}, \phi, \mathbf{c}) = \int_{\mathcal{X}} f_X(x) \left(\sum_{n=0}^{\infty} \left(\int_{\mathcal{Y}^{nL}} d(x, c^n(\vec{\mathcal{Y}}_1^n)) \psi^n(\vec{\mathcal{Y}}_1^n) f_{\vec{Y}_1^n|X,S}(\vec{\mathcal{Y}}_1^n|x) d\vec{\mathcal{Y}}_1^n \right) \right) dx, \qquad (6.3)$$

where

$$f_{\vec{Y}_1^n|X,S}(\vec{\mathcal{Y}}_1^n|x) \stackrel{\Delta}{=}$$

$$\int_{\mathcal{Z}^n} f_{Z_T^n}(\vec{z}_{T_1^n}) f_{\vec{Y}_1^n|\vec{Z}_{T_1^n},\vec{I}_1^n}(\vec{\mathcal{Y}}_1^n|\vec{z}_{T_1^n}, S_1(\vec{z}_{T_1^1}, x), S_2(\vec{z}_{T_1^2}, x), \ldots, S_n(\vec{z}_{T_1^n}, x)) d\vec{z}_{T_1^n}. \qquad (6.4)$$

define the "effective" transition probabilities as seen by the receiver. Similarly, the expected transmission rate, which is proportional to the expected value of the stopping time, is given by:

$$R(\mathbf{S}, \phi) = E \left[\sum_{n=0}^{\infty} nL \psi^n(\vec{Y}_1^n) \right]. \qquad (6.5)$$

6.1.2 Classification of the Transmitters

The transmitter, or more specifically, each encoder map S^n can be conceived as a composition of two maps, namely a *quantizer* $Q^n : \mathcal{X} \to \mathcal{N}$ and an *index assignment* $b^n : \mathcal{N} \to \mathcal{I}^L$. The quantizer divides the source space $\mathcal{X}$ into a finite number of partitions, and the index assignment map assigns a unique vector of channel symbols to each partition. The index assignment may include explicit or implicit redundancy for the purpose of error control coding. In a system with ACK/NACK feedback with transmission over multiple steps, the transmitter can be classified into three categories based on how the quantizer and the index assignment map change at each step.

1. *Active encoder (embedded source coding/multiple description-based source coding + Hybrid ARQ)*: We say that the encoder at the transmitter is an *"active encoder"* (Figure 6.2), if both the quantizer and the index assignment are time-varying, i.e. they are allowed to vary at each step in transmission. A quantizer changing with n can be thought of as an embedded source coding, because the partition of $\mathcal{X}$ after the nth step is a refinement of the partition obtained up to step $n - 1$. It can also be conceived as multiple descriptions as the individual quantizers $Q^n, n = 1, 2, 3\ldots$ are different descriptions of the source transmitted at different times. Clearly, this kind of encoding allows the *source distortion to diminish to an arbitrarily small value*.
2. *Incremental redundancy transmission or general Hybrid ARQ*: When the quantizer map is fixed (i.e. time-invariant) and only the index assignment map varies with

Figure 6.2 Active encoder at nth step.

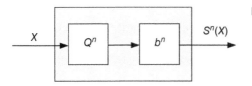

Figure 6.3 System with incremental redundancy transmission, e.g. using RCPC codes.

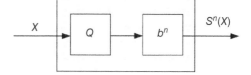

Figure 6.4 Passive Encoder or Type I hybrid ARQ transmitter for any step.

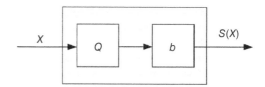

each step, the encoder implements incremental redundancy transmission or Type III Hybrid ARQ (Figure 6.3). The advantage of this configuration is that the source coding can be separated from the transmission protocol. On the other hand, the drawback over the more general encoder is that the *distortion at the receiver is limited by the quantizer-induced distortion*, and it cannot be driven to zero no matter how well the channel behaves or how efficient the error control scheme is.

3. *Passive encoder/pure retransmission encoder/Type I hybrid ARQ Transmitter*: The simplest system using ACK/NACK feedback is one in which the source coding and the index assignment are time-invariant. On receiving a NACK, the transmitter retransmits a copy of the same codeword. In such a case, we say that the encoder (or the transmitter) is *"passive"*. This is attractive because it is simple. But it does not make use of the feedback channel in the best possible way at the transmitter side. The passive encoder with pure retransmissions also represents the transmitter of a Type I Hybrid ARQ scheme.

6.1.3 Decoder Structure and Design

The receiver or the decoder can be classified analogously based on degrees of freedom, complexity, and memory usage. Note that the decoder consists of the feedback generation rule and the reproduction rule. The simplest form of decoder, the *Type I hybrid ARQ decoder with discard*, uses only the current observation for generating a feedback i.e. the feedback generation map ϕ^n does not depend on $\vec{y}_1^{n-1}$. A Type I hybrid ARQ decoder has low computational and memory requirements, but it does not make use of the full potential of ACK/NACK feedback.

The low complexity of a Type I hybrid ARQ decoder with discard follows from its operation based on retransmitting a codeword until it is successfully received and an ACK is sent over the feedback channel. The reception process ignores and discards the retransmitted copies of the codeword that were unsuccessfully received because it was not possible to correct all channel-induced errors during decoding. However, the unsuccessfully received codewords contain valuable information because usually only a very small part of the codeword would be affected by errors after decoding. This leads to the idea of a more general decoder where the unsuccessfully received codewords are not discarded but, instead, are combined together gradually building what amounts to an error-correcting code of increasing strength. With a general structure as shown in Figure 6.5, at the nth step, this more

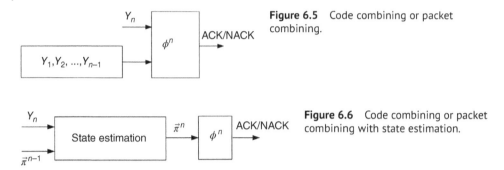

Figure 6.5 Code combining or packet combining.

Figure 6.6 Code combining or packet combining with state estimation.

general decoder can use all the received codewords up to step n in generating ACK/NACK feedback. If the encoder is active, such a decoder is said to be doing *code combining*. If the encoder is passive, the decoder is said to be doing *packet combining*. We will use the term code combining for both decoders.

Clearly, a code combining decoder is more complex and has larger and variable memory requirements. The memory requirements can be reduced if the decoder, instead of storing all the received codewords, stores only a "state" estimated from the past received code-words. The decoder structure with this decomposition is depicted in Figure 6.6. Note that the "state" need not be a sufficient statistic. It may be used only to impose additional struc-ture on the feedback generation maps.

We now consider some key points related to the decoder design. The design of the decoder amounts to the definition of the feedback generation rules and the reproduction rules. It can be argued that the design of the decoder must precede the design of the encoder. We shall see that systems which use ACK/NACK feedback are primarily receiver-driven. Even in scenarios involving a passive transmitter, by letting the feedback generation maps change, the receiver can exercise considerable control over the end-to-end performance of the sys-tem. Nevertheless, design of the transmitter side to use feedback well is an interesting and harder problem that we shall visit briefly later in this chapter.

Notice that the decoder performance (Eqs. (6.2) and (6.5) as functions of $(\phi, \mathbf{c})$) depends on the encoding rule only through the effective transition probabilities (6.4). We shall assume in the rest of this chapter that the effective transition probabilities given by (6.4) are known at the receiver.

In Section 6.2 we restrict our attention to the design of an optimal decoder for *a pas-sive encoder* and *a memoryless noisy channel*, when the number of steps is allowed to be unbounded. The optimality criterion is the trade-off between end-to-end distortion and transmission rate. We obtain the optimal design and also describe some suboptimal but competitive, computationally simpler decoder designs.

Later in this chapter, we will focus on decoder design when the encoder is active but predesigned, under the constraint that the maximum number of steps is bounded. We draw parallels between JSCC with ACK/NACK feedback and pruned tree-structured vector quan-tization. We also analyze the decoder structure and show that the optimal feedback genera-tion rules are embedded in a special sense. This property of embeddedness has applications in progressive transmission.

6.2 Packet Combining for Joint Source–Channel ARQ over Memoryless Channels

In this section we restrict our attention to a passive transmitter scenario, where, on receiving a NACK, *the transmitter can only do a retransmission of the codeword* transmitted earlier (i.e. the scenario of Figure 6.4). On the other hand, we look at an active receiver system in which the receiver retains all the (noisy) copies of the received codeword and can use them for generation of the next feedback or reproduction of the source. This is analogous to *packet combining* or *diversity combining* in the context of data transmission [1]. Clearly, as the encoder is passive, any transmitter channel measurement is not used, and we shall assume in these sections that the *transmitter channel measurement is absent*.

We will see that the task of designing a source–channel feedback generation rule for packet combining-based ARQ can be mapped to a classical sequential decision problem [2]. Consequently, it will be possible to use an algorithm based on dynamic programming to find the optimal feedback generation rule and reproduction rule that minimizes a Lagrangian sum of rate and distortion. We will also see that the distortion metric plays an important role, not only in the source reproduction, but also in the feedback generation. Because the optimal solution is computationally complex, we will discuss simpler alternatives for feedback generation.

As said, we start by considering a transmission protocol which is most generally described as Type I hybrid ARQ with packet combining (*e.g.*[1]) at the receiver. As earlier, consider the transmission of a k-dimensional random source vector X taking values in $\mathcal{X} \subset \mathcal{R}^k$, over a memoryless noisy channel with discrete input alphabet $\mathcal{I}$, possibly continuous valued output alphabet $\mathcal{Y}$ and known transition probabilities. The source vector is quantized by a fixed, predesigned k-dimensional vector quantizer (VQ) with M cells.

Each VQ cell is assigned an L-dimensional channel codeword (or *"packet"*) by a fixed, predesigned channel coding scheme. As the encoder is taken to be passive, let $S : \mathcal{X} \to \mathcal{I}^L$, denote the (fixed) map for the codeword assignment. Note that the map includes quantization, index assignment and channel coding, if any. Therefore, for a realization x of random vector X, $S(x)$ denotes the codeword to be transmitted over the channel. $S(x)$ takes M possible values denoted by S_i, $i = 1, 2, \ldots, M$, in $\mathcal{I}^L$. The transmission proceeds as follows. Codeword $S(X)$ is transmitted and then a feedback of ACK/NACK is requested. On receiving NACK, a *copy of* $S(X)$ is retransmitted. This is continued until an ACK is received. At the end of the nth retransmission, the receiver uses all the available noisy copies Y_i of $S(X)$ to generate the ACK/NACK feedback. As the channel is assumed to be memoryless, Y_i for $i = 1, 2, \ldots$, are statistically independent given the codeword $S(X)$.

As the encoding rule is fixed, in what follows we shall drop the symbol $\mathbf{S}$ from the expressions of distortion $D(\mathbf{S}, \phi, \mathbf{c})$ and rate $R(\mathbf{S}, \phi)$ in (6.2) and (6.5).

6.2.1 Decoder Design Problem

For the general system described in Section 6.1, the quantizer, the assigned channel codewords, and the decoder structure determine the average rate and distortion. For a fixed quantizer and channel codeword assignment, the general design problem is the minimization of a Lagrangian sum of the expected distortion $D(\phi, \mathbf{c})$ and average rate $R(\phi)$

with respect to ϕ and $\mathbf{c}$. Recalling the definition of the function $\psi^n(\vec{Y}_1^n)$ describing the NACK/ACK sequence in (6.1), mathematically, for a nonnegative Lagrangian penalty λ, the problem can be written as:

$$\min_{\phi,\mathbf{c}} E\left[\sum_{n=1}^{\infty} \left(d(X, c^n(\vec{Y}_1^n)) + \lambda n L \right) \psi^n(\vec{Y}_1^n) \right].$$ (6.6)

Let π_i^0 be the probability that the source vector lies in the ith VQ cell, i.e. $\pi_i^0 = \Pr(S(X) = S_i), i = 1, 2, \ldots, M$. Let $\vec{\pi}^0 \triangleq \{\pi_i^0, i = 1, 2, \ldots, M\}$. Also let s_i denote the centroid of the ith VQ cell, i.e. $s_i = E[X|S(X) = S_i]$. Then, under general conditions, we can write $D(\phi, \mathbf{c})$ as:

$$D(\phi, \mathbf{c}) = \underbrace{\sum_{i=1}^{M} \pi_i^0 E\left[\| X - s_i \|^2 \, \Big| \, S(X) = S_i \right]}_{D_s} +$$

$$\underbrace{\sum_{i=1}^{M} \pi_i^0 E\left[\sum_{n=1}^{\infty} \| s_i - c^n(\vec{Y}_1^n) \|^2 \psi^n(\vec{Y}_1^n) \, \Big| \, S(X) = S_i \right]}_{D_c(\phi,\mathbf{c})},$$ (6.7)

where D_s, the distortion due to the VQ, is a term independent of ϕ and $\mathbf{c}$. Therefore the design problem reduces to the following:

$$\min_{\phi,\mathbf{c}} J(\phi, \mathbf{c}, \vec{\pi}^0, \lambda) \text{ where } J(\phi, \mathbf{c}, \vec{\pi}^0, \lambda) \triangleq D_c(\phi, \mathbf{c}) + \lambda R(\phi).$$ (6.8)

For reasons soon to become clear, we have explicitly shown the dependence of the objective function on the prior probability vector π^0.

An examination of the expression for the objective function $J(\phi, \mathbf{c}, \lambda)$ reveals that $J(\phi, \mathbf{c}, \vec{\pi}^0, \lambda)$ is the Bayesian risk in a classical sequential decision problem [2]. The corresponding terminology is as follows. The collection of VQ cell indices $\{i = 1, 2, \ldots, M\}$ is the *parameter space*. π_i^0 is the *a priori* probability of parameter i used for computation of the Bayesian risk. Y_i's are the observation random variables which are conditionally independent and identically distributed, given the parameters. The set of reproduction vectors $C \subset \mathcal{X}$ is the *action space*. The feedback generation rule ϕ represents the *stopping rule*. A NACK feedback corresponds to a request for another observation. The reproduction vector map $c^n : \mathcal{Y}^{nN} \to C$ is the *terminal decision rule*. The *loss function*, or penalty for taking an action $c \in C$ when the parameter is i, is given by the squared error $\| s_i - c \|^2$. The increase in rate at a given step, λN, is the *cost of the incremental observation*.

Given the aforementioned interpretation of the joint source–channel decoding problem as a classical definition of sequential decision problem with Bayesian risk, the optimal joint source–channel decoder is the solution to the sequential decision problem given by (6.8). The solution provides a feedback generation rule which explicitly considers the trade-off between distortion and rate and makes use of the available source statistics.

Notice that there is flexibility in choosing the reproduction vectors, i.e. the elements of reproduction codebook C. If they are chosen as the centroids of the source encoder maps, i.e. if $C = \{s_i, i = 1, 2, \ldots, M\}$, then the problem is an M-*ary sequential detection problem with Bayes penalties* $C_{i,j} = \| s_i - s_j \|^2$. This problem has been studied in the context of signal detection (*e.g.* [3]). The nonsequential analog in the context of JSCC has also been studied (*e.g.* [4]). A finite but densely populated codebook can also be used for reproduction. Such table-lookup codebooks for reproduction vectors are considered in [5] for the nonsequential case. It can be seen that any maximum a posteriori estimate of the source will lie in the convex closure of the centroids s_i of the source encoder cells. Therefore, most generally, the set of reproduction vectors, the action space, should be the set of convex combinations of the centroids s_i. When discussing performance results later on, we will use as the reproduction codebook C, the collection of all convex combinations of source encoder centroids s_i. This set includes the minimum mean squared error (MMSE) estimate of the source encoder centroids.

Having seen how the design problem is one of a sequential decision, it is important to see next the characteristics associated with the optimal sequential design. Let $\pi_i^n(\vec{y}_1^n)$ denote the posterior probability of codeword S_i given the observations $\vec{y}_1^n$. That is, $\pi_i^n(\vec{y}_1^n) \overset{\Delta}{=} \Pr(S(X) = S_i|\vec{y}_1^n)$ for $i = 1, 2, \ldots, M$. Let $\vec{\pi}^n(\vec{y}_1^n) \overset{\Delta}{=} \{\pi_i^n(\vec{y}_1^n), i = 1, 2, \ldots, M\}$. Let $f(y_n|S_i)$ denote transition probabilities for the codewords computed from the transition probabilities for the channel. Then for a given observation vector $\vec{y}_1^n$, the following relationship exists between $\vec{\pi}^n(\vec{y}_1^n)$, $\vec{\pi}^{n-1}(\vec{y}_1^{n-1})$ and y_n:

$$\pi_i^n(\vec{y}_1^n) = \frac{\pi_i^{n-1}(\vec{y}_1^{n-1})f(y_n|S_i)}{\sum_{j=1}^M \pi_j^{n-1}(\vec{y}_1^{n-1})f(y_n|S_j)}. \tag{6.9}$$

Let this function, which is independent of time index n, be denoted by $H(\vec{\pi}, y)$. Then $\vec{\pi}^n(\vec{y}_1^n) = H(\vec{\pi}^{n-1}(\vec{y}_1^{n-1}), y_n)$.

Let Γ denote the simplex of all probability distributions over transmit codewords S_i, i.e. $\Gamma \overset{\Delta}{=} \{a_1, a_2, \ldots, a_M : 0 \le a_i \le 1, \sum_{i=1}^M a_i = 1\}$. All posterior probability distributions $\vec{\pi}^n$ belong to Γ. The function $\rho : \Gamma \to [0, \infty)$ can be defined as:

$$\rho(\vec{\pi}) \overset{\Delta}{=} \min_{c \in C} \sum_{i=1}^M \| s_i - c \|^2 \pi_i. \tag{6.10}$$

Then, the following main result follows from the theory of sequential decisions.

Proposition 6.1 *For every $\lambda \ge 0$, there exists a unique cost-to-go function $V(\cdot, \lambda) : \Gamma \to \mathcal{R}$ which satisfies the following dynamic programming equation for all $\vec{\pi} \in \Gamma$:*

$$V(\vec{\pi}, \lambda) = \min \left(\lambda L + E[V(H(\vec{\pi}, Y), \lambda)|\vec{\pi}], \rho(\vec{\pi}) \right). \tag{6.11}$$

This result is an implicit relation describing the cost-to-go function, which is defined over all probability distributions in Γ. This is a minimum of two terms. The second is the distortion component of the reproduction, and the first involves an expectation of the function on the left-hand side, making the equation implicit. In fact, the expectation term is important enough to introduce a second notation for it. Let us denote by $A(\vec{\pi}, \lambda)$, the expected cost-to-go given a codeword posterior probability distribution as follows:

$$A(\vec{\pi}, \lambda) \overset{\Delta}{=} E[V(H(\vec{\pi}, Y), \lambda)|\vec{\pi}] = \sum_{i=1}^M \pi_i E[V(H(\vec{\pi}, Y), \lambda)|S_i]. \tag{6.12}$$

With this definition, we can characterize an optimal pair of feedback generation and the reproduction rules in the form of the following proposition.

Proposition 6.2 *Consider the feedback generation rule ϕ^* and the reproduction rule c^*, given as:*

- *Feedback generation rule:*
  - *$\phi^{*n}(\vec{y}_1^n) = 1$, i.e. send ACK if $\rho(\pi_i^n(\vec{y}_1^n)) \leq \lambda L + A(\vec{\pi}^n(\vec{y}_1^n), \lambda)$.*
  - *$\phi^{*n}(\vec{y}_1^n) = 0$, i.e. send NACK otherwise.*
- *Reproduction rule: Whenever $\phi^{*n}(\vec{y}_1^n) = 1$, set $c^{*n}(\vec{y}_1^n) = \arg\min_{c \in C} \sum_{i=1}^{M} \| s_i - c \|^2 \pi_i^n(\vec{y}_1^n)$.*

Then ϕ^* and c^* are optimal, that is, they solve problem (6.8).

Note that the optimal reproduction rule and optimal feedback generation rule, for each λ, are time-invariant functions of $\vec{\pi}$. They, therefore, follow the decoder structures as shown in Figures 6.5 and 6.6. The rule is optimal in the framework but is complex as it involves (i) computing the posterior probability vector $\pi_i^n(\vec{y}_1^n)$, via Bayes (ii) computing the distortion terms $\rho(\pi_i^n(\vec{y}_1^n))$ from (6.10), and (iii) knowing the shape of and computing the value of the expected cost-to-go function $A(\vec{\pi}^n(\vec{y}_1^n), \lambda)$. Computation (i) is moderately complex depending on the size of the codebook, and computation (ii) is relatively easy but computation (iii) is hard unless approximations are used. The outline of proofs for Propositions 6.1 and 6.2 is shown in the following sequence of facts:

1. For any feedback generation rule ϕ, the optimal reproduction rule depends on $\vec{y}_1^n$ through the posterior probabilities $\vec{\pi}^n(\vec{y}_1^n)$. The optimal reproduction rule is given by $c^{*n}(\vec{y}_1^n) = \arg\min_{c \in C} \sum_{i=1}^{M} \| s_i - c \|^2 \pi_i^n(\vec{y}_1^n)$

2. For any $\vec{\pi} \in \Gamma$, let $V_0(\vec{\pi}, \lambda) \overset{\Delta}{=} \rho(\vec{\pi})$ and

$$V_T(\vec{\pi}, \lambda) \overset{\Delta}{=} \inf_{\phi, \phi^T \vec{y}_1^T = 1 \forall \vec{y}_1^T} J(\phi, c^*, \vec{\pi}, \lambda) \text{ for } T = 1, 2, \dots \quad (6.13)$$

$V_T(\vec{\pi}, \lambda)$ is the minimum Bayesian risk over all feedback generation rules which are forced to send ACK at step T, when the prior probability is some $\vec{\pi} \in \Gamma$. Then, the following decomposition holds for a memoryless channel:

$$V_T(\vec{\pi}, \lambda) = \min\left(\lambda L + E[V_{T-1}(H(\vec{\pi}, Y), \lambda)|\vec{\pi}], \rho(\vec{\pi})\right)$$

$$= \min\left(\lambda L + \sum_{i=1}^{M} \pi_i \int_{y^L} f_{Y|S_i}(y) V_{T-1}(H(\vec{\pi}, y), \lambda) dy, \rho(\vec{\pi})\right). \quad (6.14)$$

3. $V_T(\vec{\pi}, \lambda) \geq V_{T+1}(\vec{\pi}, \lambda) \geq V_{T+2}(\vec{\pi}, \lambda) \geq \dots$ Therefore, $V_T(\vec{\pi}, \lambda)$, as $T \to \infty$ converges pointwise to a function that can be shown to be $V(\vec{\pi}, \lambda)$ satisfying (6.11).

The structure of the optimal decoder obtained in Proposition 6.2 is shown in Figure 6.7. At the same time, it is worth noticing that the aforementioned optimal general solution is exceedingly complex. This complexity can be attributed to the following processing as shown in Figure 6.7.

Complexity of state estimation: The optimal decoder, i.e. the optimal feedback generation rule as well as the optimal reproduction rule, is computed from the state which is the posterior probability distribution over transmit codewords. The state space is the M-dimensional

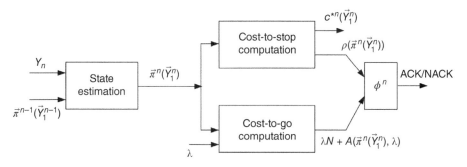

Figure 6.7 Feedback generation with state estimation.

probability simplex, where M is the number of possible input vectors. For even moderately long size of the vectors, and moderate source coding rate, M can be prohibitively large. As state estimation has to be done at all steps during the transmission, it is a big contributor to implementation complexity.

Complexity of design and implementation of optimal feedback generation rule: The feedback generation rule ϕ^* compares the conditional expected channel distortion given current observations $\rho(\pi_i^n(\vec{y}_1^n))$ with the cost of sending a NACK, that is $\lambda L + A(\vec{\pi}^n(\vec{y}_1^n), \lambda)$. This requires the knowledge of the functions $V(\vec{\pi}, \lambda)$ and $A(\vec{\pi}, \lambda)$ *for all* posterior probability distributions $\vec{\pi} \in \Gamma$. It turns out that the determination of these functions is highly challenging. The general solution in Proposition 6.2 has only been characterized in a very few cases such as binary sequential hypothesis testing [2], and only approximate methods have been developed for the case of M-ary detection, i.e. the case of $C = \{s_i, i = 1, 2, \ldots, M\}$ and 1-0 penalty (*e.g.* [3]). M-SPRT uses expressions similar to Wald's approximations to approximate $V(\pi, \lambda)$ and $A(\pi, \lambda)$ [2, 3].

The high complexity associated with the optimal solution motivates suboptimal schemes which consider the distortion metric explicitly. Three suboptimal approaches are worthy of note:

1. Distortion-based feedback generation rule (which we will call "scheme DIST")
2. Finite horizon optimal rules (which we will call "scheme FINHZN")
3. Finite lookahead rules (which we will call "scheme FINLKHD").

Scheme DIST: Distortion-based Feedback Generation Rule

Notice that ϕ^* in Proposition 6.2 compares the conditional expected channel distortion given current observations, given by $\rho(\pi_i^n(\vec{y}_1^n))$, to $\lambda L + A(\vec{\pi}^n(\vec{y}_1^n), \lambda)$, which varies with $\vec{\pi}$ and λ. The first term in the summation, λL, clearly increases with λ. Even the second term, the function $A(\vec{\pi}, \lambda)$ turns out to be a monotonically increasing function of λ, for every prior $\vec{\pi} \in \Gamma$.

Proposition 6.3 *For $\lambda_1 \geq \lambda_2$, $A(\vec{\pi}, \lambda_1) \geq A(\vec{\pi}, \lambda_2)$, for all $\vec{\pi} \in \Gamma$.*

An outline of the proof of Proposition 6.3 is sketched below.

Proof Outline of Proposition 6.3: The proof relies on using induction to establish monotonicity of finite horizon cost-to-go with respect to λ, followed by a limiting operation. Let $V_T(\vec{\pi}, \lambda)$ be the cost-to-go for the finite horizon problem with horizon T, as defined in (6.13) for $T = 0, 1, 2, \ldots$. As in (6.12), define $A_T(\vec{\pi}, \lambda) \triangleq E[V_{T-1}(H(\vec{\pi}, Y), \lambda)|\vec{\pi}]$ for $T = 1, 2, \ldots$. The cost-to-go $V_0(\vec{\pi}, \lambda)$ is independent of λ, and hence $A_1(\vec{\pi}, \lambda)$ is monotonically increasing with λ. Assume $A_T(\vec{\pi}, \lambda)$ is a monotonically increasing function of λ. Then as $V_T(\vec{\pi}, \lambda)$ is a minimum of two monotonically increasing functions, it is monotonically increasing. Consequently, $A_{T+1}(\vec{\pi}, \lambda)$, which is an expectation over monotonically increasing functions, is also monotonically increasing. It can be shown that $A_T(\vec{\pi}, \lambda)$ converges to $A(\pi, \lambda_1)$, and hence $A(\pi, \lambda_1)$ is monotonically increasing.

Now we establish how the parameter λ determines the operating point of rate vs. distortion trade-off and motivate **Scheme DIST**. Consider the behavior of ϕ^* and $\mathbf{c}^*$ for different values of the Lagrangian rate penalty λ. It is easy to see that the reproduction rule $\mathbf{c}^*$ remains unchanged. On the other hand, increasing λ results in greater rate penalty and hence a smaller rate. This implies that the decision to send NACK will be taken more infrequently as λ increases. Hence, the behavior of $A(\vec{\pi}, \lambda)$ is similar to distortion, as larger rate penalty λ leads to larger distortion. Consequently, the complexity associated with having to know the functions $V(\vec{\pi}, \lambda)$ and $A(\vec{\pi}, \lambda)$ for all posterior probability distributions $\vec{\pi} \in \Gamma$ in the optimal solution can be addressed by using the distortion itself as a suboptimal mechanism to determine the feedback.

Therefore, in the "DIST" suboptimal scheme, the feedback generation rule is determined using the distortion. To get the first suboptimal feedback generation rule $\hat{\phi}$, the function $\lambda L + A(\vec{\pi}, \lambda)$, which varies with $\vec{\pi}$, is replaced with a function $\delta(\lambda)$ which is independent of $\vec{\pi}$. Hence, the feedback generation rule $\hat{\phi}$ in the "DIST" suboptimal scheme is as follows:

- Feedback generation rule for scheme DIST:
  - $\hat{\phi}^n(\vec{y}_1^n) = 1$ i.e. send ACK if $\rho(\pi_i^n(\vec{y}_1^n)) \leq \delta$.
  - $\hat{\phi}^n(\vec{y}_1^n) = 0$, i.e. send NACK otherwise.
- Reproduction rule: The reproduction rule is the same as the optimal, Whenever $\hat{\phi}^n(\vec{y}_1^n) = 1$, set $\hat{c}^n(\vec{y}_1^n) = c^{*n}(\vec{y}_1^n) = \arg\min_{c \in C} \sum_{i=1}^M \| s_i - c \|^2 \pi_i^n(\vec{y}_1^n)$.

Varying δ from small to large values captures the rate-distortion trade-off/throughput-reliability trade-off in ARQ with packet combining. Note that, like the optimal rules, the scheme "DIST" also results in time-invariant feedback generation rules. Large values of δ result in high throughput and small values of δ result in low distortion. It turns out that for sequential detection of 1-bit equiprobable quantized symmetrical sources, $A(\vec{\pi}, \lambda)$ is indeed independent of $\vec{\pi}$, and hence for this special case, the proposed scheme coincides with the optimal solution.

Scheme FINHZN: Finite Horizon Optimal Rules

A *T-horizon optimal* feedback generation rule is obtained by minimizing $J(\phi, \mathbf{c}, \lambda)$ over only those feedback generations rules for which $\phi^T(\vec{y}_1^T) = 1$ for all $\vec{y}_1^T$ for some fixed integer T. That is, such a feedback generation rule is the solution of the optimization problem in (6.13). These are straightforward to design as the feedback generation maps are computed explicitly instead of being governed by an implicit formula. These result in time-varying feedback generation rules but have the advantage of bounded delay and bounded memory requirements.

Scheme FINLKHD: Finite Lookahead Rules

A class of time-invariant suboptimal rules, called *T-step lookahead rules*, is obtained by executing, at each step, the *T*-horizon optimal feedback rule designed for the next *T* steps. For large enough *T*, such a rule can be expected to approximate the optimal feedback generation rule.

In what follows, to simplify the naming of these suboptimal schemes and aid in the clarity of the presentation, we will collectively refer to the schemes "DIST," "FINHZN," and "FINLKHD" together as *distortion-aware feedback generation rules* or simply *distortion-aware* schemes. To summarize, the features of these distortion-aware schemes are the following:

1. The distortion metric plays a significant part in the feedback generation.
2. Channel statistics and source statistics are used, both for the reproduction rule as well as for feedback generation.
3. Independent of the source statistics, the sensitivity of the bits to channel errors, measured from their contribution to distortion, may still be different for different bits. The distortion-aware schemes therefore ascribe possibly unequal importance to the transmit symbols.
4. There is a direct way of controlling the trade-off between quality and rate.

These four features of the distortion-aware schemes, the fallouts of the analysis of the optimal solutions of the Lagrangian formulation, are also the features which distinguish these suboptimal joint source–channel approaches aforementioned from conventional tandem protocol designs. The natural question to ensue, then, is how do the performance of these schemes compare against conventional approaches? The conventional approach to generating ACK/NACK feedback has been through the use of error detection at the receiver. A NACK is generated if there are detectable but uncorrectable errors in the received sequence of channel symbols. The detection is accomplished by adding redundancy and using error detection codes such as CRC. In order to compare the feedback schemes that have been described earlier with conventional approaches, we briefly pause the presentation to describe some schemes with conventional approaches that will be used for comparison.

Scheme CRC-baseline: Baseline CRC-based system: Figure 6.8 describes the decoder for a baseline packet combining system based on CRC. The main features of the baseline system are (i) maximum likelihood (ML) estimation of transmit bits, (ii) check of integrity of the bits by error detection, and (iii) reproduction of the source by inverse quantization.

Scheme CRC-MMSE: CRC-based system with pseudo-MMSE decoding: The distortion-aware schemes expect to improve upon the baseline CRC-based system by use of (i) a different feedback generation rule and (ii) reproduction by MMSE estimation of the source as opposed to inverse quantization. To be able to discern the gains due to these two separate factors, we can conceive another CRC-based system which uses CRC for feedback generation but uses MMSE estimation of the source for reproduction. In order to keep the reproduction rule identical to the distortion-aware schemes, for MMSE estimation we must use only the information bits, i.e. the bits in the received symbols, excluding the CRC

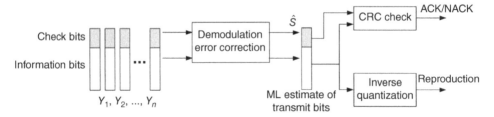

Figure 6.8 Receiver for baseline CRC-based system.

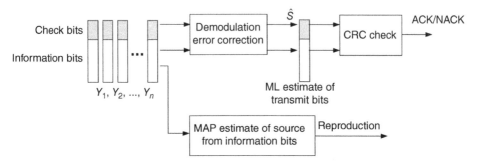

Figure 6.9 Receiver for CRC-based system with pseudo-MMSE decoding.

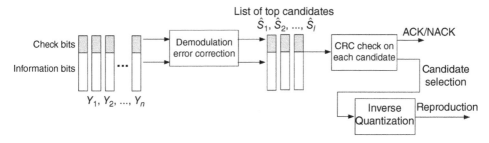

Figure 6.10 Receiver for CRC-based system with list decoding.

bits. As CRC bits are ignored for reproduction, we name this system CRC-Based system with Pseudo-MMSE decoding. This system is shown in Figure 6.9.

Scheme CRC-list: CRC-based system with list decoding: Some control over the throughput-reliability trade-off can be obtained in a CRC-based system with the help of list decoding. The CRC-based system with list decoding is shown in Figure 6.10. In list decoding, instead of generating a single ML estimate of the transmit bits, a finite list of the most likely candidate estimates is generated. If any of the candidates satisfies the CRC, an ACK is generated and that candidate is used for reproduction by inverse quantization. If no candidate satisfies the CRC, a NACK is generated. Clearly, by varying the size of the list, throughput can be traded for reliability.

Also, we can conceive a **CRC-based system with list and pseudo-MMSE decoding (scheme CRC-List-MMSE)** where list decoding is used for feedback generation, but MMSE decoding from the information bits alone is used for reproduction.

Recall from the four features of the distortion-aware schemes that they attempt to minimize distortion and make use of source statistics. Because of this, and in addition to comparison against the CRC-based systems, which represent the conventional error detection-based techniques, it is of interest to compare the performance gain/loss of the distortion-aware schemes vs. *optimized* techniques designed to minimize bit error rate (BER) for a given throughput. For such a comparison, we can conceive *zero-redundancy BER-based* feedback generation rules, which are obtained as suboptimal solutions (analogous to the FINHZN and FINLKHD schemes) to a modification of the sequential decision problem (6.8) where the action space C is the collection of source encoder indices or codewords $\{0,1\}^L$, and the loss function is bit-wise Hamming distance (number of bits that two sequences differ from each other, see Section 3.1.1) between the true parameter (transmitted source encoder index) and the reproduction. Thus in this case, the objective is to minimize, for different values of Lagrange multiplier λ,

$$E\left[\sum_{n=1}^{\infty} \left(Ham(S(X), c^n(Y_1^n)) + \lambda n L \right) \psi^n(Y_1^n) \right], \tag{6.15}$$

where $Ham : \{0,1\}^L \times \{0,1\}^L \to \mathcal{N}_+$ is the Hamming distance between two binary vectors.

After pausing to introduce conventional CRC-based schemes and zero-redundancy BER-based schemes for performance comparison purposes, we are now able to continue with the analysis of the differences between these schemes and the distortion-aware techniques JSCC systems with feedback. To do this, we consider transmission of synthetic random sources (memoryless unit-variance Gaussian sources) quantized by tree-structured vector quantizers (TSVQs) over a memoryless noisy channel. We assume that the channel input is binary, such as the one obtained by BPSK modulation. Therefore, we will be referring to channel input symbols as bits and the codewords will be sequences of bits with length L. The channel is a binary input, ternary output discrete memoryless channel obtained by quantizing the output of BPSK transmission over an additive white Gaussian noise (AWGN) channel into three regions, $(-\infty, -t_0], (-t_0, t_0)$, and $[t_0, \infty)$. For each signal-to-noise ratio (SNR) of the AWGN channel, the threshold t_0 corresponds to the value that maximizes the information theoretic capacity of the resulting discrete channel. This channel is useful for the analysis at hand because it captures the features of both hard decoding and soft decoding. Also, for the design described, which requires numerical computation of expectations, it helps that the set of all possible channel outputs be finite. The schematic of quantization of the AWGN channel and the corresponding discrete channel is depicted in Figure 6.11 with numerical values of the transition probabilities as shown in the Figure, exemplified for different AWGN SNRs in Table 6.1

Figures 6.12–6.15 compare the distortion-aware schemes with the CRC-based schemes when transmitting a unit variance Gaussian source quantized by a 16-level tree-structured scalar quantizer [6]. The 16 levels of the quantizer were mapped into $L = 4$ bits using natural binary indexing. These 4-bit codewords were transmitted across the discrete memoryless noisy channels using schemes DIST, FINHZN with $T = 3$ and $T = 4$, one-step FINLKHD, as well as the CRC-based Schemes.

Figure 6.12 and 6.13 show the results for the channels which are equivalent to AWGN channels with SNR 0 and 3 dB, respectively. End-to-end total SNR is the mean squared error per sample expressed in dB. The points on the curve DIST are obtained by simulation

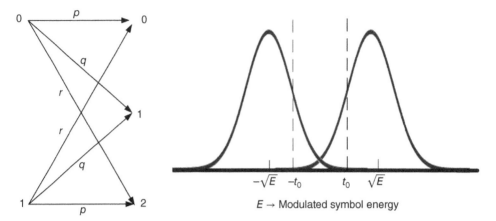

Figure 6.11 Discrete two-input three-output channel is obtained as BPSK over quantizing AWGN channel.

Table 6.1 Transition probabilities of the derived discrete channel for different AWGN SNRs

SNR dB →	−1	0	1	2	3	4	5
p	0.7716	0.819	0.864	0.9049	0.9387	0.9644	0.9819
q	0.1905	0.1530	0.1164	0.0828	0.05414	0.03181	0.01636
r	0.03782	0.02768	0.01908	0.01220	0.007121	0.0037	0.001675

as the distortion threshold δ is varied from large values to small values. Similarly, the points for one-step FINLKHD were obtained by simulation as the Lagrange multiplier λ was varied from large values to small values. The simulations are based on 1 200 000 samples of a unit variance Gaussian source. For each sample, the channel was used nearly 20 times. The points for FINHZN $T = 3$ and $T = 4$ were obtained by numerical calculation. They are operational rate-distortion performance curves obtained by pruning a depth-T pruned tree-structured vector quantizers (PTSVQs) [7] with $3^L = 81$ children per node. The relationship between PTSVQ and the design problem is explained later in this chapter, in Section 6.3. Three simple CRC-based transmitters, with 1-bit, 2-bit, and 3-bit CRC's applied to each 4-bit packet, are also shown for comparison in the figures. The decoders are CRC-List and CRC-List-MMSE for different list sizes. The CRC-List scheme with list size 1 is the baseline CRC-based system. CRC-List-MMSE with list size of 1 corresponds to the CRC-MMSE scheme. Results were obtained for list sizes of $1, 2, 4, \ldots, 2^n$ for a transmitter which uses an n-bit CRC. The numbers next to points for CRC-List represent the list size used.

The plots also show results for fixed horizon schemes which are in fact schemes with *repetition coding and no feedback*. In these schemes, the codeword is repeated a fixed number of times. The decoder performs an MMSE estimation of the source from the received copies. The performance at the highest transmission rate achieved by a T-horizon FINHZN scheme is equal to that of a fixed horizon scheme transmitting T copies.

A feature that is immediately noticeable in the results is the high flexibility offered by the distortion-aware schemes. The Lagrangian approach yields a continuum of operating points for each of the distortion-aware schemes. The CRC-based schemes, on the other hand, provide limited flexibility, operating at discrete set of points. Secondly, though the distortion-aware schemes are suboptimal, they consistently outperform the CRC-based schemes for a wide range of transmission rates. For the noisier channel, namely the one corresponding to an AWGN of 0 dB SNR, the gains of DIST and 1-step FINLKHD are nearly 2 dB at almost all transmission rates. The gains of distortion-aware schemes for the AWGN channel of 3 dB SNR are lower, but they still outperform all CRC-based schemes except one. The CRC-List scheme with code rate 4/7, which adds a 3-bit CRC for every four information bits, with list size 2 outperforms the distortion-aware schemes, but this is in the context that the distortion-aware schemes in the plots have no redundancy added. Another interesting observation is that for high-redundancy CRC schemes, such as CRC-List with code rate 4/7, CRC-List outperforms CRC-List-MMSE. This observation can be explained from the fact that the extra diversity provided by redundant bits more than compensates for the suboptimality of ML decoding over MMSE decoding. This is not the case for high code rate (i.e. low redundancy) CRC schemes (Figure 6.14). The high code rate CRC-based schemes used rates 4/5, 8/10, and 16/19, which are 1-bit CRC added for every 4 information bits, 2-bit CRC added for every 8 information bits, and 3-bit CRC added for every 16 information bits, respectively. For these code rates, CRC-List-MMSE generally performs better than CRC-List. Also, a fourth noticeable feature is that the schemes DIST and 1-step FINLKHD perform nearly identically.

The source coder used in Figures 6.12–6.15 is a 4-bit TSVQ with average distortion D_s equal to 0.0097 which is nearly 20 dB. The source distortion becomes the dominating factor for higher transmission rates. Figure 6.15 plots only the channel-induced distortion D_c, expressed in dB, for these schemes for the channel with SNR equal to 0 dB. As channel distortion can be driven arbitrarily close to zero, the channel-induced SNR for schemes DIST and 1-step FINLKHD does not saturate, unlike the curves in Figure 6.12. Moreover, the curves for DIST and 1-step FINLKHD are nearly linear, implying that the distortion drops exponentially with transmission rate. Also, they are at a sharper slope than the fixed horizon schemes. This shows that the gain in SNR of DIST and 1-step FINLKHD over schemes not using feedback increases with transmission rate. In addition, note that the FINHZN schemes are efficient at low transmission rates, but their performance curves saturate as the rate approaches the corresponding fixed horizon schemes.

As discussed earlier (see (6.15)), it is of interest to learn of the impact on performance given by the distortion metric in the feedback generation rule by comparing against a scheme that has for loss function the bit-wise Hamming distance. For this, we consider performance comparison against zero-redundancy BER-based schemes.

Figures 6.16 and 6.17 depict for various schemes, the average rate vs. total SNR performance as λ is varied from small to large, for channels obtained from AWGN channels with SNR 0 and 3 dB, respectively. In all the curves, including the zero-redundancy BER-based schemes, *the reproduction rule is chosen to be the MMSE estimate of the source.* The rate-distortion performance of the distortion-aware schemes DIST, FINHZN with $T = 3$, and 1-step FINLKHD is compared against zero redundancy packet combining feedback generation rules (i) BER-based FINHZN with $T = 3$ and (ii) BER-based 1-step FINLKHD.

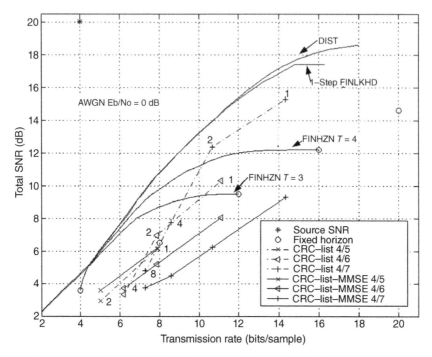

Figure 6.12 Performance (total SNR vs. transmission rate) of various schemes of scalar IID Gaussian source quantized with 4-bit TSVQ over noisy channel (equivalent to AWGN SNR = 0 dB).

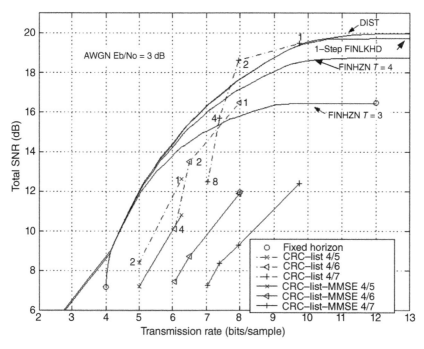

Figure 6.13 Performance (total SNR vs. transmission rate) of various schemes of scalar IID Gaussian source quantized with 4-bit TSVQ over noisy channel (equivalent to AWGN SNR = 3 dB).

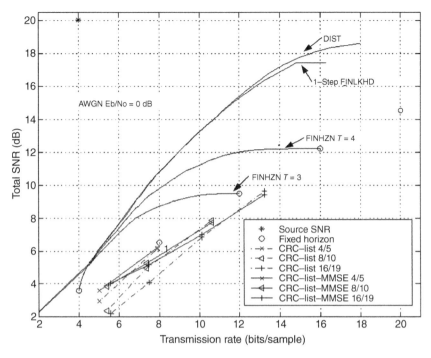

Figure 6.14 Performance (total SNR vs. transmission rate) of high-rate CRC-based schemes, IID Gaussian source, dim = 1, TSVQ 4 bit/sample, equivalent to AWGN SNR = 0 dB.

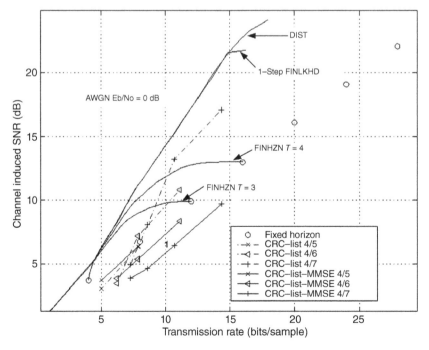

Figure 6.15 Channel Distortion for Various Schemes, IID Gaussian source, dim = 1, TSVQ 4 bit/sample, equivalent to AWGN SNR = 0 dB.

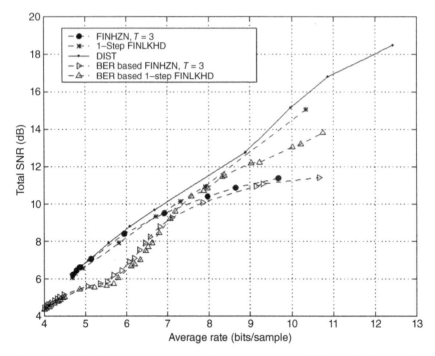

Figure 6.16 Performance comparison with zero-redundancy BER-based schemes. Gaussian source, TSSQ with 4 bits/sample. AWGN channel SNR = 0 dB.

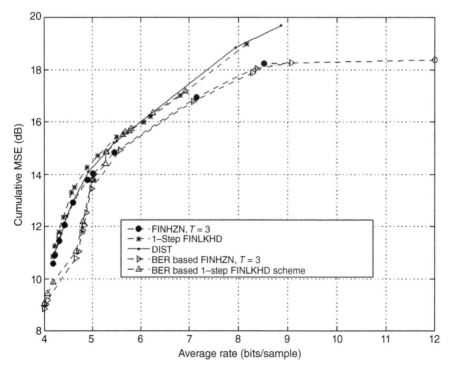

Figure 6.17 Performance comparison with zero-redundancy BER-based schemes. Gaussian Source, TSSQ with 4 bits/sample. AWGN channel SNR = 3 dB.

Although the BER-based zero-redundancy schemes behave like the conventional CRC schemes, that is, they treat all the source encoder bits equally, *they make use of source statistics for ACK/NACK generation*. The only difference between the BER-based schemes and distortion-aware schemes is the distortion metric. From the figures, it is evident that the channel distortion/rate performance of the distortion-aware schemes is almost always superior to the BER-based schemes. But the most interesting feature is that, at high transmission rates per source sample, the BER-based zero-redundancy schemes seem to catch up with the corresponding distortion-aware schemes. The distortion-aware schemes show high gains in the regime where transmission rate per source sample is low. The highest performance improvement is about 2 dB in both cases. Another advantage of the curves for the distortion-aware schemes is their high positive slope in the low rate region, compared to the zero-redundancy BER-based schemes. This has implications in applications with progressive transmission of a source message, where a rapid improvement in source quality as a function of bit rate is desirable.

6.3 Pruned Tree-Structured Quantization in Noise and Feedback

We now study another approach to JSCC with feedback that is based on the progressive transmission of an embedded source encoder (recall the overview of embedded coding seen in Section 2.4). Specifically, we consider optimal decoding schemes where a progressively transmitted embedded source coder suffers channel noise and each step of the progressive transmission is driven by feedback from the receiver to the transmitter. In this framework, the transmitter is active, that is, on receiving a NACK it does not retransmit the codeword transmitted earlier but instead transmits new information. We restrict our attention to the finite horizon case, where the transmission is not allowed to continue beyond a fixed number of steps, say, T. As done earlier, we focus on the receiver side by describing the structure of the optimal decoder and feedback generation mechanism.

In the absence of channel coding, the progressive coding can be done using a tree-structured quantizer. The tree-structured quantizer is capable of coding in several stages; each stage provides a refinement of the previous stage. If some form of variable length coding is available, then an effective way of obtaining a collection of quantizers from a single tree-structured quantizer is by pruning [7]. The resulting collection of quantizers is called PTSVQs. There is an elegant theory associated with pruning. Pruned tree-structured quantizers first appeared in the context of decision trees where Breiman et al. [8] presented an algorithm for pruning. It was later generalized to other contexts, such as tree-structured quantization, regression trees, quantization of noisy sources and variable-order Markov modeling [6, 7]. We will see soon that there is a close link between PTSVQ and transmission using an embedded source-coder over a channel with ACK/NACK feedback. This concept taken in a generalized form allows to carry out joint source–channel PTSVQ, or PTSVQ in the presence of noise and feedback. In addition to establishing the close link, we will explain that there is a *feedback-threshold* function which reveals the simple structure behind the optimal feedback generation rules for all Lagrangian penalties.

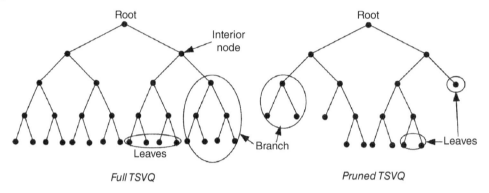

Figure 6.18 TSVQ and Pruned TSVQ

6.3.1 Pruned Tree-Structured Vector Quantizers

A T-stage TSVQ is a collection of T VQs, one associated with each stage, such that every VQ cell of the ith stage is obtained by partitioning some cell, its "parent," at the $i-1$th stage, for $i = 1,2, \ldots, T$. The quantizer at the 0th stage consists of one cell. The parent–child relationship between cells gives a full balanced tree of VQ cells. A cell in a PTSVQ is called a *leaf* if it has no children, else, it is called an *interior node*. Without loss of generality, we shall assume that the tree is binary, i.e. each cell is either a leaf or has exactly two children. Figure 6.18 illustrates a full TSVQ constructed from binary partitioning of parent cells, as just described.

Let the collection of cells in a TSVQ denoted by Z_0, be denoted by $\hat{Z}_0$. A PTSVQ, Z', is obtained from a full TSVQ by selecting a subset $\hat{Z} \subset \hat{Z}_0$ of the cells, with the property that a cell $t \in \hat{Z}$ if and only if its parent cell $parent(t) \in \hat{Z}$. We say that $Z \preceq Z_0$. The relation $\preceq$ is naturally extended as a partial order for comparing two pruned TSVQs – or just pruned trees. For two pruned trees Z and Z', $Z' \preceq Z$ if $\hat{Z}' \subset \hat{Z}$. Figure 6.18 illustrate a pruned tree obtained from a full tree.

The encoding and decoding of a PTSVQ is analogous to that of a full TSVQ. A source vector is quantized in stages, till a leaf cell that contains the vector is found. The path from the root node to the leaf is used for encoding the vector, and a representative vector, the "centroid" of the leaf cell, is used for reproduction.

For a given source with known statistics, a rate and an average distortion can be associated with every PTSVQ. The rate is measured as either (i) the expected length of the path from the root to the leaf or (ii) the expected entropy of such a path. We shall assume the former definition of the rate. Let the distortion-rate pair for a PTSVQ $Z \preceq Z_0$ be denoted by $(D(Z), R(Z))$. Then the collection of optimal rate-distortion pairs, namely those on the lower convex hull of the set $\{(D(Z), R(Z)) : Z \preceq Z_0\}$, has the following interesting property [7].

Theorem 6.1 *PTSVQ property* The collection of points on the lower convex hull of the set $\{(D(Z), R(Z)) : Z \preceq Z_0\}$ can be obtained by repeatedly pruning a single tree. In other words, there is a sequence of PTSVQs, $\ldots Z_k \preceq Z_{k-1} \preceq \ldots \preceq Z_2 \preceq Z_1 \preceq Z_0$, which traces the convex hull.

This elegant result also leads to the generalized algorithm due to Breiman, Friedman, Olshen, and Stone for obtaining the points on the convex hull [6–8].

We will now proceed to dive deeper into the progressive transmission of a TSVQ-encoded source in the presence of channel noise and ACK/NACK feedback and obtain a PTSVQ-like property for the optimal decoding schemes. In that sense, the following sections present a generalization of PTSVQ.

6.3.2 Progressive Transmission with ACK/NACK Feedback of TSVQ-Encoded Sources

Earlier in this chapter, we considered how to carry out joint source–channel decoding when the transmitter does retransmission of the codeword. There, the feedback generation map selection could be used to control the throughput-reliability or rate-distortion trade-off. Here we consider a slightly more general case in which, on receipt of a NACK, the transmitter proceeds with the transmission of new information. Clearly, this contains, as a special case, the case of retransmission of the same codeword. Still the idea remains to obtain the optimal decoding schemes for this case.

The study of the more general case of retransmitting of new information as a response to a NACK reception widens the interpretation of ACK/NACK feedback. Conventionally, ACK/NACK feedback was used for indicating if the transmit codewords were decoded with acceptable reliability or not. This conventional interpretation turns out to be narrow in light of the possibility of the transmitter transmitting new information when receiving a NACK feedback. In this case it is necessary to consider that the NACK feedback serves a dual purpose. Firstly, the NACK feedback is used to indicate that the previous transmission was corrupted beyond recovery by the channel noise. In addition, the NACK feedback also serves the purpose of controlling the rate-distortion performance of the joint source–channel coder. This is because a NACK may be sent even when the previous transmission was noiseless, but it is favorable for rate-distortion trade-off that further information about the same source vector be sent. In other words, the NACK feedback is used as a permission to continue transmission of new or old information about the same source vector to refine its reconstruction. This new interpretation essentially says that NACK feedback can be used for rate control. We shall see that the decoder structure, in fact, has a property like that of the PTSVQ; namely, the optimal decoding schemes at different rate-distortion trade-offs are embedded.

However, we note that this is still not the most general transmission scheme conceivable because the transmitter is still not active. In the present scenario, the transmitter transmits a fixed sequence of codewords for a given source vector and stops when an ACK is received.

To study the case of transmission of new information when receiving a NACK feedback, as earlier, consider the transmission of a k-dimensional random source vector X taking values in $\mathcal{X} \subset \mathcal{R}^k$, over a noisy channel with discrete input alphabet $\mathcal{I}$ and possibly continuous valued output alphabet $\mathcal{Y}$. X is quantized by a TSVQ with depth T which generates a channel codeword $S_n(X) \in \mathcal{I}^L$ for each stage $n = 1, 2, \ldots, T$. We assume that the TSVQ and the codeword allocation are predesigned and fixed. If codeword $S_k(X)$ is transmitted, a noisy version of the codeword $Y_n \in \mathcal{Y}^L$ is received. We need not assume that the channel is memoryless. We shall just assume that the statistics of the source, i.e. the distribution of

X and that of the channel, i.e. the joint distributions of $Y_1, Y_2, \ldots, Y_T$ are known for each value of X. For simplicity, we shall assume that conditional probability densities of the kind $f(y_{i_1} | y_{i_2}, y_{i_3}, \ldots y_{i_k}, x)$ can be computed for all values of $y_{i_k} \in \mathcal{Y}$ and $x \in \mathcal{X}$.

The transmission proceeds as follows. First, the codeword $S_1(X)$ is transmitted and then a feedback of ACK/NACK is requested. On receiving NACK, which is taken as a **"permission to continue transmission,"** $S_2(X)$ is transmitted. This way, codewords $S_1(X), S_2(X), \ldots, S_n(X), \ldots$ are transmitted one by one until either an ACK is received or $S_T(X)$ has been transmitted.

The *feedback generation rule* ϕ at the receiver is specified by a sequence of *feedback generations maps* $\phi^n : \mathcal{Y}^{nL} \rightarrow \{0,1\}, n = 1,2, \ldots$. At the nth step, let the received realizations of the noisy copies be $y_1, y_2, \ldots, y_n$ for $y_i \in \mathcal{Y}^L$. Then an ACK is transmitted if $\phi^n(y_1, y_2, \ldots, y_n) = 1$. A NACK is transmitted if $\phi^n(y_1, y_2, \ldots, y_n) = 0$. The *reproduction rule* $\mathbf{c}$ at the receiver is specified by a sequence of reproduction maps $c^n : \mathcal{Y}^{nL} \rightarrow C \subset \mathcal{X}$. C is the *reproduction codebook*. If an ACK is generated at the nth step, i.e. if $\phi^n(y_1, y_2, \ldots y_n) = 1$, then the source is reconstructed as $c^n(y_1, y_2, \ldots, y_n)$. It is not necessary for C to be discrete. Again, let $\vec{y}_1^n$ and $\vec{Y}_1^n$ be the shorthand for denoting the sequences $y_1, y_2, \ldots, y_n$ and random vectors $Y_1, Y_2, \ldots, Y_n$, respectively.

In an ACK/NACK-based transmission, no further transmission is sent after ACK is received. Therefore, we are safe in requiring a consistency condition on the feedback generation maps ϕ^n, that if a the map declares ACK for a received vector of length n, it will declare ACK for all received vectors of length $n + 1$ and higher where the former vector is a prefix. This is expressed as an implication:

$$\phi^n(\vec{y}_1^n) = 1 \implies \phi^{n+1}(\vec{y}_1^{n+1}) = 1 \; \forall \vec{y}_1^{n+1} \in \mathcal{Y}^{nN}, n = 1,2, \ldots, T-1. \tag{6.16}$$

Recall from (6.1) that $\psi^n(\vec{y}_1^n) = \prod_{i=1}^{n-1}(1 - \phi^i(\vec{y}_1^i))\phi^n(\vec{y}_1^n)$. Then $\psi^n(\vec{y}_1^n) = 1$ for all those sequences $\vec{y}_1^n$ which generate an ACK only at the nth step and not earlier. In this section, we consider only a finite-stage TSVQ; hence, we require that $\phi^T(\vec{Y}_1^T) = 1$ always. This implies $E\left[\sum_{n=1}^T \psi^n(\vec{Y}_1^n)\right] = 1$.

The average rate per source vector, that is the expected number of channel symbols put on the channel before stopping (i.e. before an ACK is received), is given by:

$$R(\phi) = E\left[\sum_{n=1}^T nL\psi^n(\vec{Y}_1^n)\right]. \tag{6.17}$$

Let $d(\cdot, \cdot)$ denote the squared error distortion measure. Then for given ϕ and $\mathbf{c}$, the expected distortion is computed as:

$$D(\phi, \mathbf{c}) = E\left[\sum_{n=1}^T d(X, c^n(\vec{Y}_1^n))\psi^n(\vec{Y}_1^n)\right]. \tag{6.18}$$

Note that, although specifying the collection of maps $\phi^n, n = 1,2, \ldots, T$ is not the same as specifying the collection $\psi^n, n = 1,2, \ldots, T$, the performance measures $D(\phi, \mathbf{c})$ and $R(\phi)$ depend only on $\psi^n, n = 1,2, \ldots, T$. For a nonnegative multiplier $\lambda \geq 0$, define

$$J(\phi, \mathbf{c}, \lambda) \stackrel{\Delta}{=} D(\phi, \mathbf{c}) + \lambda R(\phi) \tag{6.19}$$

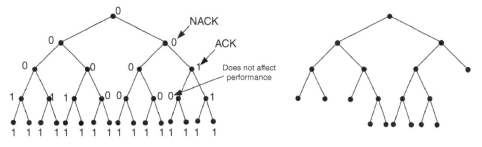

Feedback generation rule for full TSVQ   *Equivalent pruned TSVQ*

Figure 6.19   Feedback generation rule over a full TSVQ and equivalent pruned TSVQ.

Then the problem of decoder design can be expressed as:

$$\min_{\phi, \mathbf{c}} J(\phi, \mathbf{c}, \lambda) = \min_{\phi, \mathbf{c}} E\left[\sum_{n=1}^{T} \left(d(X, c^n(\vec{Y}_1^n)) + \lambda nL\right) \psi^n(\vec{Y}_1^n)\right], \qquad (6.20)$$

This problem is a Bayesian sequential decision problem. We shall refer to the Lagrangian sum of distortion and transmission rate as "Bayesian risk" or simply "risk."

It is straightforward to see that there is close relationship between PTSVQ and transmission of a TSVQ over a noiseless channel with ACK/NACK feedback. In fact, over a discrete output noiseless channel, there is a one-to-one relationship between all possible pruned instances of a tree, and all possible feedback generation rules (specified in terms of ψ^n's, as opposed to ϕ's). Any pruning of a full tree can be represented in terms of some feedback generation rule ϕ and vice versa. Over a noiseless channel, a sequence of received codewords, $\vec{y}_1^n$, (which is the same as the sequence of transmit codewords) is equivalent to a path from the root node to a node at depth n. If $\phi^n(\vec{y}_1^n) = 0$, a NACK is transmitted, then the node corresponding to $\vec{y}_1^n$ is an interior node. Alternatively, if $\psi^n(\vec{y}_1^n) = 1$ then $\vec{y}_1^n$ corresponds to a leaf in the pruned tree. Figure 6.19 illustrates a binary TSVQ transmitted over a noiseless binary channel, with the values of some feedback generation rule ϕ and the equivalent PTSVQ.

As before, the goal of the decoder design is to find the optimal reproduction rule $\mathbf{c}$ and the optimal feedback generation rule ϕ. This optimal decoder design is obtained as the solution to the sequential design problem given by (6.20). The solution to (6.20) can be obtained by a dynamic programming argument [2, 9]. To see how this is the case, we first state the optimal reproduction rule.

Theorem 6.2   *Optimal Reproduction Rule*   Let $c^{*n}(\vec{Y}_1^n)$ be a Bayes estimate of the source based on a fixed number of received codewords $\vec{Y}_1^n$. Then for every feedback generation rule ϕ and λ, $J(\phi, \mathbf{c}, \lambda)$ is minimized with respect to $\mathbf{c}$ by the functions $c^{*n}(\vec{Y}_1^n)$ for $n = 0, 1, 2, \ldots, T$. Here,

$$c^{*0} = \arg\min_{c \in C} E\left[d(X, c)\right], \text{ and}$$

$$c^{*n}(\vec{y}_1^n) = \arg\min_{c \in C} E\left[d(X, c) \mid \vec{Y}_1^n = \vec{y}_1^n\right]. \qquad (6.21)$$

The proof is straightforward and can be found in [2, 9].

It can be seen that *the optimal reproduction rule turns out to be independent of the feedback generation rule ϕ and of λ*. This implies that one can always use the same reproduction rule for all methods of generating feedback and for all penalties λ on the rate. We shall assume in the remainder of this section that $\mathbf{c}^*$ is the reproduction rule.

To obtain the feedback rule ϕ^* which minimizes $J(\phi, \mathbf{c}^*, \lambda)$ for a fixed λ, define

$$\rho_n(\vec{Y}_1^n) \overset{\Delta}{=} E\left[d(X, c^{*n}(\vec{Y}_1^n))|\vec{Y}_1^n\right] \text{ and}$$

$$U_n(\vec{Y}_1^n, \lambda) \overset{\Delta}{=} \rho_n(\vec{Y}_1^n) + \lambda nL, \tag{6.22}$$

where $U_n(\vec{Y}_1^n, \lambda)$ is the conditional risk of stopping, i.e. sending ACK, at the nth step, having received $\vec{Y}_1^n$.

As we have assumed that at each step, the possible number of codewords transmitted is finite and also that the total number of steps in transmission is finite. Under these conditions, it is straightforward to show that $\rho_n(\vec{Y}_1^n)$ and consequently $J(\phi, \mathbf{c}^*, \lambda)$ is bounded for each λ.

Next, we use the technique of induction as used in dynamic programming to design the feedback generation rule. Let us define a feedback generation rule ϕ^* as follows.

Suppose $T-1$ noisy codewords $\vec{Y}_1^{T-1}$ have already been received. If an ACK is to be sent at that point, then the conditional risk is $U_{T-1}(\vec{Y}_1^{T-1})$. On the other hand, if a NACK is sent, then another noisy copy will have to be received, and in that case the conditional risk is $E[U_T(\vec{Y}_1^T)|\vec{Y}_1^{T-1}]$. Thus at the $T-1^{st}$ step, for a point $\vec{y}_1^{T-1} \in \mathcal{Y}^{L(T-1)}$ the risk is minimized if we define $\phi^{*T-1}(\vec{y}_1^{T-1})$ as:

$$\phi^{*T-1}(\vec{y}_1^{T-1}) \overset{\Delta}{=} \begin{cases} 1 \text{ if } U_{T-1}(\vec{y}_1^{T-1}, \lambda) \le E[U_T(\vec{Y}_1^T, \lambda)|\vec{y}_1^{T-1}] \\ 0 \text{ otherwise.} \end{cases} \tag{6.23}$$

Let $G_{T-1}(\vec{y}_1^{T-1}, \lambda)$ denote the minimum posterior risk, given the received codewords $\vec{y}_1^{T-1}$ and penalty λ at the $T-1^{st}$ step, Clearly $G_{T-1}(\vec{y}_1^{T-1}, \lambda)$, is given by:

$$G_{T-1}(\vec{y}_1^{T-1}, \lambda) \overset{\Delta}{=} \min[U_{T-1}(\vec{y}_1^{T-1}, \lambda), E[U_T(\vec{Y}_1^T, \lambda)|\vec{y}_1^{T-1}]]. \tag{6.24}$$

This cost is obtained by using the feedback generation map $\phi^{*T-1}(\vec{y}_1^{T-1})$. Similarly, the "minimum" posterior risk $G_n(\vec{y}_1^n)$ at any step n, can be defined as follows:

$$G_T(\vec{y}_1^T, \lambda) \overset{\Delta}{=} U_T(\vec{y}_1^T, \lambda)$$

$$G_{n-1}(\vec{y}_1^{n-1}, \lambda) \overset{\Delta}{=} \min[U_{n-1}(\vec{y}_1^{n-1}, \lambda), E[G_n(\vec{Y}_1^n, \lambda)|\vec{y}_1^{n-1}]] \text{ for } n = 2,3,\dots,T,$$

$$G_0 \overset{\Delta}{=} \min[U_0, E[G_1(\vec{Y}_1^1, \lambda)]]. \tag{6.25}$$

Hence, the feedback generation maps $\phi^{*n}(\vec{y}_1^n)$ can be defined inductively as:

$$\phi^{*T}(\vec{y}_1^T) \overset{\Delta}{=} 1 \text{ and}$$

$$\phi^{*n-1}(\vec{y}_1^{n-1}) \overset{\Delta}{=} \begin{cases} 1 \text{ if } U_{n-1}(\vec{y}_1^{n-1}, \lambda) \le E[G_n(\vec{Y}_1^n, \lambda)|\vec{y}_1^{n-1}] \\ 0 \text{ otherwise, for } n = 1,2,\dots T. \end{cases} \tag{6.26}$$

This collections of maps $\phi^{*n}(\vec{y}_1^n)$ for $n = 0,1,\dots T$, are an optimal feedback generation rule. Note that G_n and hence ϕ^{*n} vary with λ. These observations lead to the following result [2, 9] for the optimal feedback generation rule.

Theorem 6.3 *Optimal Feedback Generation Rule* The collection of maps $\phi^* \stackrel{\Delta}{=} \{\phi^{*n}, n = 0,1, \dots T\}$ is the optimal feedback generation rule for a given λ, i.e. for any other feedback generation rule ϕ, the following holds:

$$J(\phi^*, \mathbf{c}^*, \lambda) \leq J(\phi, \mathbf{c}^*, \lambda). \tag{6.27}$$

The aforementioned results define the optimal feedback generation rules and optimal reproduction rules for the JSCC problem with ACK/NACK feedback using standard techniques from sequential analysis. We will see next that the optimal feedback generation rules and the optimal reproduction rules have a property analogous to PTSVQ. For this, we would like to examine the property in Theorem 6.1 in terms of the interpretation of PTSVQ as a sequential decisions with ACK/NACK feedback. Consider the noiseless channel case and recall the relationship between a PTSVQ and feedback generation rule that we discussed earlier and exemplified in Figure 6.19. Let ϕ_1 and ϕ_2 be the corresponding feedback generation rules for two pruned trees Z_1 and Z_2, respectively. It can be verified easily that $Z_1 \preccurlyeq Z_2$ if and only if $\phi_1^n \geq \phi_2^n$ for $n = 0,1, 2, \dots, T$. Therefore, the PTSVQ property stated in Theorem 6.1 can be restated in terms of the feedback generation rules and lead to the following theorem as a generalization to noisy channels:

Theorem 6.4 Let $\lambda_1 \geq \lambda_2 \geq 0$. Let ϕ_1^* and ϕ_2^* denote the optimal feedback generation rules for the problem (6.20), as given by Eq. (6.26) corresponding to λ_1 and λ_2, respectively. Then $\phi_1^{*n}(\vec{y}_1^n) \geq \phi_2^{*n}(\vec{y}_1^n)$ for all $\vec{y}_1^n \in \mathcal{Y}^{nL}$ for $n = 0,1, \dots T$. In other words, $\phi_1^{*n}(\vec{y}_1^n) = 0 \implies \phi_2^{*n}(\vec{y}_1^n) = 0$ and $\phi_2^{*n}(\vec{y}_1^n) = 1 \implies \phi_1^{*n}(\vec{y}_1^n) = 1$.

In order to prove Theorem 6.4, we need to show first a small result. First, define maps $W_n(\vec{y}_1^n, \lambda) \stackrel{\Delta}{=} G_n(\vec{y}_1^n, \lambda) - L\lambda n$ for $n = 0,1, 2, \dots, T$. It can be easily verified by induction that (6.26) can be reformulated as:

$$\phi^{*T}(\vec{y}_1^T) = 1 \text{ and}$$

$$\phi^{*n-1}(\vec{y}_1^n) = \begin{cases} 1 \text{ if } \rho_{n-1}(\vec{y}_1^{n-1}) \leq \lambda L + E[W_n(\vec{Y}_1^n, \lambda)|\vec{y}_1^{n-1}] \\ 0 \text{ otherwise, for } n = 1,2, \dots T. \end{cases} \tag{6.28}$$

Also,

$$W_T(\vec{y}_1^T, \lambda) = \rho_T(\vec{y}_1^T)$$

$$W_{n-1}(\vec{y}_1^{n-1}, \lambda) = \min\left[\rho_{n-1}(\vec{y}_1^{n-1}, \lambda), \lambda L + E[W_n(\vec{Y}_1^n, \lambda)|\vec{y}_1^{n-1}]\right] \text{ for } n = 2,3, \dots, T,$$

$$W_0 = \min[\rho_0, \lambda L + E[W_1(\vec{Y}_1^1, \lambda)]]. \tag{6.29}$$

Second, to prove Theorem 6.4, we also need the following lemma. The lemma proposes monotonicity and continuity for an intermediate calculation.

Lemma 6.1 The functions $W_n(\vec{y}_1^n, \lambda)$ for given received codewords $\vec{y}_1^n$ is a continuous and monotonically increasing function of λ, for $n = 0,1, \dots, T$.

Proof of Lemma 6.1: By induction. Clearly, $W_T(\vec{y}_1^T, \lambda)$ is independent of λ and hence is a monotonically increasing and continuous function of λ for all $\vec{y}_1^n$. Assume that $W_m(\vec{y}_1^m, \lambda)$ is a monotonically increasing continuous function of λ, for $m = n + 1, \dots T$. Let $\lambda_1 \geq \lambda_2 \geq 0$. Then, we have, for monotonicity,

$$W_n(\vec{y}_1^n, \lambda_1) = min[\rho_n(\vec{y}_1^n), \lambda_1 L + E[W_{n+1}(\vec{Y}_1^{n+1}, \lambda_1)|\vec{y}_1^n]]$$

$$\geq min[\rho_n(\vec{y}_1^n), \lambda_2 L + E[W_{n+1}(\vec{Y}_1^{n+1}, \lambda_2)|\vec{y}_1^n]]$$

$$= W_n(\vec{y}_1^n, \lambda_2). \tag{6.30}$$

Analogously, $W_n(\vec{y}_1^n, \lambda_1)$ is the minimum of two continuous functions.

We can now describe the proof of Theorem 6.4 as follows.

Proof of Theorem 6.4: From (6.28) and Lemma 6.1, if $\lambda_1 \geq \lambda_2 \geq 0$, then $\rho_n(\vec{y}_1^n) \leq \lambda_2 L + E[W_{n+1}(\vec{Y}_1^{n+1} n, \lambda_2)|\vec{y}_1^n] \leq \lambda_1 L + E[W_{n+1}(\vec{Y}_1^{n+1} n, \lambda_1)|\vec{y}_1^n]$. This means that $\phi_2^{*n}(\vec{y}_1^n) = 1 \implies \phi_1^{*n}(\vec{y}_1^n) = 1$, which proves the theorem.

The main implication of Theorem 6.4 is that it shows that optimal feedback generation rules for different λ are embedded. This property is very useful for progressive transmission. Consider the transmission of a single-source vector using the transmission scheme discussed here. Suppose the transmission starts with the decoder using an optimal feedback generation rule for a certain λ. Theorem 6.4 implies that, at any step in the transmission, the decoder can switch to an optimal feedback generation rule for a lower λ, i.e. a higher rate, *without losing optimality*. It never happens that, for a set of received codewords, the feedback generation rule designed for a higher rate sends an ACK and one designed for lower rate sends a NACK.

The Feedback-Threshold Function

Moreover, Theorem 6.4 opens up new paths to further characterize the collection of optimal feedback generation rules. This section of the chapter is more technical mathematically but leads to a deeper understanding of the collective behavior of the feedback generation rules for the entire convex hull of the operating points. For the remainder of this chapter, let ϕ_λ^* and ϕ_λ^{*n} denote the optimal feedback generation rule and the optimal feedback generation maps, respectively, for the Lagrangian rate penalty λ. For any sequence of received codewords $\vec{y}_1^n \in \mathcal{Y}^{nL}$, consider the set $B(\vec{y}_1^n) \overset{\Delta}{=} \{\lambda : \phi_\lambda^{*n}(\vec{y}_1^n) = 1\}$. From (6.28), this set is the same as $\{\lambda : \rho_n(\vec{y}_1^n) \leq \lambda L + E[W_{n+1}(\vec{Y}_1^{n+1}, \lambda)|\vec{y}_1^n]\}$. As the function $\lambda L + E[W_{n+1}(\vec{Y}_1^{n+1}, \lambda)|\vec{y}_1^n]$ is continuous as a function of λ, and $B(\vec{y}_1^n)$ is the inverse image of $[\rho_n(\vec{y}_1^n), \infty)$ under that function, $B(\vec{y}_1^n)$ is a closed set. From Theorem 6.4, $B(\vec{y}_1^n)$ is of the form $[\lambda_0, \infty)$, for some number $\lambda_0 \geq 0$, which depends on $\vec{y}_1^n$. The function $\Lambda^{*n}(\vec{y}_1^n) : \mathcal{Y}^{nL} \to [0, \infty)$ can be defined as:

$$\Lambda^{*n}(\vec{y}_1^n) \overset{\Delta}{=} inf\{\lambda \in B(\vec{y}_1^n)\}. \tag{6.31}$$

Then clearly, the random variable $\Lambda^{*n}(\vec{Y}_1^n)$, has the property that $\phi_\lambda^{*n}(\vec{Y}_1^n) = 1$ if and only if $\Lambda^{*n}(\vec{Y}_1^n) \leq \lambda$ for all $\lambda \geq 0$. Hence, we have the following interesting result.

Theorem 6.5 The optimal feedback generation rules ϕ_λ^* satisfy

$$\phi_\lambda^{*n}(\vec{Y}_1^n) = u(\lambda - \Lambda^{*n}(\vec{Y}_1^n)) \text{ (a.e.) for } n = 0,1, \dots T, \text{ and } \lambda \geq 0, \tag{6.32}$$

where $u(\cdot)$ is the unit step function, i.e. $u(\lambda) = 1$ if $\lambda \geq 0$ and 0 otherwise.

The proof follows the outline arguments discussed prior to the theorem statement. Note that (6.32) is true only for the optimal feedback generation rules obtained from (6.26) (or from (6.28)).

We shall refer to the functions $\Lambda^{*n}(\vec{y}_1^n)$ as the ***feedback-threshold functions or maps***. The result is interesting and useful because it "reduces" the task of designing a different feedback generation rule for every λ to constructing a single collection of maps Λ^{*n} from which all optimal feedback generation rules can be obtained.

To characterize $\Lambda^{*n}(\vec{Y}_1^n)$ further, consider the following definitions. Define, for any feedback generation map ϕ with $\phi^T(\vec{Y}_1^T) = 1$, the function $D^T(\phi, \vec{Y}_1^T) \overset{\Delta}{=} \rho^T(\vec{Y}_1^T)$. And define recursively

$$\Delta D^n(\phi, \vec{Y}_1^n) \overset{\Delta}{=} \rho^n(\vec{Y}_1^n) - E\left[D^{n+1}(\phi, \vec{Y}_1^{n+1}) | \vec{Y}_1^n\right] \text{ for } n = 0,1,\dots T-1, \tag{6.33}$$

$$D^n(\phi, \vec{Y}_1^n) \overset{\Delta}{=} \rho^n(\vec{Y}_1^n) - (1 - \phi^n(\vec{Y}_1^n))\Delta D^n(\phi, \vec{Y}_1^n) \text{ for } n = 0,1,\dots T-1. \tag{6.34}$$

Analogously define

$$R^T(\phi, \vec{Y}_1^T) \overset{\Delta}{=} TL \tag{6.35}$$

$$\Delta R^n(\phi, \vec{Y}_1^n) \overset{\Delta}{=} E\left[R^{n+1}(\phi, \vec{Y}_1^{n+1}) | \vec{Y}_1^n\right] - nL \text{ for } n = 0,1,\dots T-1, \tag{6.36}$$

$$R^n(\phi, \vec{Y}_1^n) \overset{\Delta}{=} nL + (1 - \phi^n(\vec{Y}_1^n))\Delta R^n(\phi, \vec{Y}_1^n) \text{ for } n = 0,1,\dots T-1. \tag{6.37}$$

Notice that $D^n(\phi, \vec{Y}_1^n)$ and $R^n(\phi, \vec{Y}_1^n)$ depend only on $\phi^i(\vec{Y}_1^i), i = n, n+1, \dots, T$. Also $\Delta D^n(\phi, \vec{Y}_1^n)$ and $\Delta R^n(\phi, \vec{Y}_1^n)$ do not depend on the value of $\phi^n(\vec{Y}_1^n)$ but only on $\phi^i(\vec{Y}_1^i), i = n+1, \dots, T$. It is straightforward to verify that $D^n(\phi, \vec{Y}_1^n)$ equals $D(\phi, \mathbf{c}^*)$ for $n = 0$, where $D(\phi, \mathbf{c})$ is defined in (6.18) and $\mathbf{c}^*$ is defined in (6.21). Similarly, $R^n(\phi, \vec{Y}_1^n)$ equals $R(\phi)$ in (6.17). Also $\Delta R^n(\phi, \vec{Y}_1^n) \geq L > 0$ for any n. Equations (6.34) and (6.37) isolate the dependence of $D^n(\phi, \vec{Y}_1^n)$ and $R^n(\phi, \vec{Y}_1^n)$ on the function $\phi^n(\vec{Y}_1^n)$. Also, by definition, $(1 - \phi^n(\vec{Y}_1^n)) \geq 0$. Therefore, the following lemma about separation of minimizations holds.

Lemma 6.2 For any Lagrange multiplier $\lambda \geq 0$, and for $n = 0,1,\dots T-1$,

$$\min_{\phi^i(\vec{y}_1^i), i=n,n+1,\dots,T} D^n(\phi, \vec{y}_1^n) + \lambda R^n(\phi, \vec{y}_1^n) =$$

$$\rho^n(\vec{y}_1^n) + \lambda nL + \min_{\phi^n(\vec{y}_1^n)} \left((1 - \phi^n(\vec{y}_1^n)) \left(\min_{\phi^i(\vec{y}_1^i), i=n+1,\dots,T} \{-\Delta D^n(\phi, \vec{y}_1^n) + \lambda \Delta R^n(\phi, \vec{y}_1^n)\} \right) \right)$$

Consequently, to minimize $D^n(\phi, \vec{y}_1^n) + \lambda R^n(\phi, \vec{y}_1^n)$, we must set $\phi^n(\vec{y}_1^n) = 0$ if and only if $\min_{\phi^i(\vec{y}_1^i), i=n+1,\dots,T} \left(-\Delta D^n(\phi, \vec{y}_1^n) + \lambda \Delta R^n(\phi, \vec{y}_1^n)\right)$ is negative.

With these results, we are now equipped to consider the following important theorem that characterizes the feedback-threshold function $\Lambda^{*n}(\vec{y}_1^n)$.

Theorem 6.6

$$\Lambda^{*n}(\vec{y}_1^n) = \sup_{\phi^i(\vec{y}_1^i), i=n,n+1,\dots,T} \frac{\Delta D^n(\phi,\vec{y}_1^n)}{\Delta R^n(\phi,\vec{y}_1^n)}. \tag{6.38}$$

Proof: First, note from the recursive definitions in (6.33) and (6.36) that if the functions $\phi^i(\vec{y}_1^i), i = n+1, \dots, T$ minimize $D^{n+1}(\phi,\vec{y}_1^{n+1}) + \lambda R^{n+1}(\phi,\vec{y}_1^{n+1})$ for all $\vec{y}_1^{n+1}$, then they minimize $-\Delta D^n(\phi,\vec{y}_1^n) + \lambda \Delta R^n(\phi,\vec{y}_1^n)$. Second, define

$$\hat{\lambda}^n(\vec{y}_1^n) \overset{\Delta}{=} \sup_{\phi^i(\vec{y}_1^i), i=n,n+1,\dots,T} \frac{\Delta D^n(\phi,\vec{y}_1^n)}{\Delta R^n(\phi,\vec{y}_1^n)}.$$

We shall show that $\Lambda^{*n}(\vec{y}_1^n) = \hat{\lambda}^n(\vec{y}_1^n)$, i.e. it is optimal to set $\phi_\lambda^{*n}(\vec{y}_1^n) = 1$ if $\lambda \geq \hat{\lambda}^n(\vec{y}_1^n)$ and to set $\phi_\lambda^{*n}(\vec{y}_1^n) = 0$ if $\lambda < \hat{\lambda}^n(\vec{y}_1^n)$. To see this, let us consider the three cases $\lambda > \hat{\lambda}^n(\vec{y}_1^n)$, $\lambda < \hat{\lambda}^n(\vec{y}_1^n)$, and $\lambda = \hat{\lambda}^n(\vec{y}_1^n)$, one at a time.

Case 1: For any $\lambda > \hat{\lambda}^n(\vec{y}_1^n)$ and any ϕ, we have,

$$-\Delta D^n(\phi,\vec{y}_1^n) + \lambda \Delta R^n(\phi,\vec{y}_1^n) = \left(-\frac{\Delta D^n(\phi,\vec{y}_1^n)}{\Delta R^n(\phi,\vec{y}_1^n)} + \lambda\right) \Delta R^n(\phi,\vec{y}_1^n)$$

$$> 0 \quad \text{as } \Delta R^n(\phi,\vec{y}_1^n) > 0.$$

Therefore, by Lemma 6.2, $\phi_\lambda^{*n}(\vec{y}_1^n) = 1$.

Case 2: Similarly, if $\lambda < \hat{\lambda}^n(\vec{y}_1^n)$ then, by definition of supremum, there is a feedback generation rule ϕ' such that $\lambda < \frac{\Delta D^n(\phi',\vec{y}_1^n)}{\Delta R^n(\phi',\vec{y}_1^n)} \leq \hat{\lambda}^n(\vec{y}_1^n)$. Consequently, $-\Delta D^n(\phi',\vec{y}_1^n) + \lambda \Delta R^n(\phi',\vec{y}_1^n) < 0$. By Lemma 6.2, we must set $\phi_\lambda^{*n}(\vec{y}_1^n) = 0$.

Case 3: If $\lambda = \hat{\lambda}^n(\vec{y}_1^n)$, for any ϕ, $-\Delta D^n(\phi,\vec{y}_1^n) + \lambda \Delta R^n(\phi,\vec{y}_1^n) \geq 0$. Therefore, we can safely set $\phi_\lambda^{*n}(\vec{y}_1^n) = 1$ without any penalty. Therefore, we can set $\phi_\lambda^{*n}(\vec{y}_1^n) = u(\lambda - \hat{\lambda}(\vec{y}_1^n))$. Hence, Theorem 6.6 holds.

Case 3 leads to a more explicit characterization of the feedback-threshold functions $\Lambda^{*n}(\vec{Y}_1^n)$ as seen next.

Lemma 6.3 Consider the design of the optimal feedback generation rule for $\hat{\lambda}^n(\vec{y}_1^n)$, i.e. the solution to the minimization problem:

$$\min_{\phi^i(\vec{y}_1^i), i=n+1,\dots,T} \left\{-\Delta D^n(\phi,\vec{y}_1^n) + \hat{\lambda}(\vec{y}_1^n)\Delta R^n(\phi,\vec{y}_1^n)\right\}. \tag{6.39}$$

Then $\phi_{\hat{\lambda}^n(\vec{y}_1^n)}^*$ is a solution to the aforementioned minimization if and only if

$$-\Delta D^n(\phi_{\hat{\lambda}^n(\vec{y}_1^n)}^*,\vec{y}_1^n) + \hat{\lambda}(\vec{y}_1^n)\Delta R^n(\phi_{\hat{\lambda}^n(\vec{y}_1^n)}^*,\vec{y}_1^n) = 0.$$

Proof: Establishing sufficiency is straightforward as, for any feedback generation rule, and hence for $\phi_{\hat{\lambda}^n(\vec{y}_1^n)}^*$, by definition of $\hat{\lambda}^n(\vec{y}_1^n)$ we must have

$$\frac{\Delta D^n(\phi_{\hat{\lambda}(\vec{y}_1^n)}^*,\vec{y}_1^n)}{\Delta R^n(\phi_{\hat{\lambda}(\vec{y}_1^n)}^*,\vec{y}_1^n)} \leq \hat{\lambda}(\vec{y}_1^n)$$

$$\implies -\Delta D^n(\phi_{\hat{\lambda}^n(\vec{y}_1^n)}^*,\vec{y}_1^n) + \hat{\lambda}(\vec{y}_1^n)\Delta R(\phi_{\hat{\lambda}^n(\vec{y}_1^n)}^*,\vec{y}_1^n) \geq 0. \tag{6.40}$$

We show the necessity as follows. By definition of the minimum,

$$-\Delta D^n(\phi^*_{\hat{\lambda}^n(\vec{y}_1^n)},\vec{y}_1^n) + \hat{\lambda}(\vec{y}_1^n)\Delta R(\phi^*_{\hat{\lambda}^n(\vec{y}_1^n)},\vec{y}_1^n)$$

$$\leq -\Delta D^n(\phi,\vec{y}_1^n) + \hat{\lambda}^n(\vec{y}_1^n)\Delta R^n(\phi,\vec{y}_1^n) \text{ for all } \phi$$

$$\implies 0 \leq -\Delta D^n(\phi^*_{\hat{\lambda}^n(\vec{y}_1^n)},\vec{y}_1^n) + \hat{\lambda}^n(\vec{y}_1^n)\Delta R(\phi^*_{\hat{\lambda}^n(\vec{y}_1^n)},\vec{y}_1^n)$$

$$\leq \Delta R^n(\phi,\vec{y}_1^n)\left(-\frac{\Delta D^n(\phi,\vec{y}_1^n)}{\Delta R^n(\phi,\vec{y}_1^n)} + \hat{\lambda}^n(\vec{y}_1^n)\right)$$

As $\Delta R^n(\phi,\vec{y}_1^n)$ is bounded above by the quantity TL, and by definition of supremum, the second term on the right-hand side can be made arbitrarily small, we must have the left-hand side equal to zero. Therefore, Lemma 6.3 is established. Finally, by establishing Lemma 6.3, it is now possible to assert the uniqueness of $\Lambda^{*n}(\vec{Y}_1^n)$.

Theorem 6.7 For some nonnegative λ, if

$$\min_{\phi^i(\vec{y}_1^i),i=n+1,\dots,T} \left\{-\Delta D^n(\phi,\vec{y}_1^n) + \lambda\Delta R^n(\phi,\vec{y}_1^n)\right\} = 0 \tag{6.41}$$

then $\lambda = \hat{\lambda}^n(\vec{y}_1^n)$. Therefore, (6.41) is a necessary and sufficient condition for computation of $\hat{\lambda}^n(\vec{y}_1^n)$ and hence that of $\Lambda^{*n}(\vec{y}_1^n)$.

Proof: We have already seen in Lemma 6.3 that $\hat{\lambda}^n(\vec{y}_1^n)$ satisfies (6.41). It is straightforward to check that, for $a, b \geq 0$ and $c, d > 0$, if λ_1 and λ_2 are such that $-a + \lambda_1 b = 0 \leq -c + \lambda_1 d$ and $-c + \lambda_2 d = 0 \leq -a + \lambda_2 b$, then $\lambda_1 = \lambda_2$. Therefore, no other λ can satisfy (6.41), which proves Theorem 6.7.

Moreover, the following two results explore further the structure of Λ^{*n}. We do so by proving the existence of a feedback-threshold function. This proof is slightly more involved and mathematically technical than rest of the treatment in this book so as to retain the validity of result in a general setting.

Lemma 6.4 For any observation vector $\vec{Y}_1^{n+1}$, $\Lambda^{*n}(\vec{Y}_1^n) \geq \Lambda^{*n+1}(\vec{Y}_1^{n+1})$ a.e.

Proof: Suppose that the Lemma does not hold true. In this case, then, there is a λ such that $\Lambda^{*n}(\vec{y}_1^n)\langle\lambda \leq \Lambda^{*n+1}(\vec{y}_1^{n+1})$. Then from (6.32), $\phi^{*n}_\lambda(\vec{y}_1^n) = 1$ but $\phi^{*n+1}_\lambda(\vec{y}_1^{n+1}) = 0$. This is a contradiction. A more complete proof involves showing that if $P[\Lambda^{*n}(\vec{Y}_1^n) < \Lambda^{*n+1}(\vec{Y}_1^{n+1})] > 0$ then there is a λ such that the event $\{\Lambda^{*n}(\vec{Y}_1^n) < \lambda\} \cap \{\lambda \leq \Lambda^{*n+1}(\vec{Y}_1^{n+1})\}$ has a nonzero probability. This is seen as follows. It can be readily shown that the event $\Lambda^{*n}(\vec{Y}_1^n) < \Lambda^{*n+1}(\vec{Y}_1^{n+1})$ is equal to $\cup_{\lambda, \text{ rational}} \{\{\Lambda^{*n}(\vec{Y}_1^n) < \lambda\} \cap \{\lambda \leq \Lambda^{*n+1}(\vec{Y}_1^{n+1})\}\}$, as rational numbers are dense on the real line. Secondly as rational numbers are countable, we have the union bound:

$$0\langle P[\Lambda^{*n}(\vec{Y}_1^n) < \Lambda^{*n+1}(\vec{Y}_1^{n+1})] \leq \sum_{\lambda, \text{ rational}} P[\{\Lambda^{*n}(\vec{Y}_1^n) < \lambda\} \cap \{\lambda \leq \Lambda^{*n+1}(\vec{Y}_1^{n+1})\}].$$

A countable sum of nonnegative numbers is nonzero if and only if there is at least one nonzero term in the summation. Therefore, there is a λ such that $P[\{\Lambda^{*n}(\vec{Y}_1^n)\langle\lambda\} \cap \{\lambda \leq \Lambda^{*n+1}(\vec{Y}_1^{n+1})\}] > 0$. For such a λ, by (6.32), the event $\{\phi^{*n}_\lambda(\vec{Y}_1^n) = 1\} \cap \{\phi^{*n+1}_\lambda(\vec{Y}_1^{n+1}) = 0\}$ has a nonzero probability. This contradicts the consistency condition given by (6.16).

Finally, we characterize the function Λ^{*n} more precisely.

Lemma 6.5 If $\lambda \geq ess\ sup_{\vec{y}_1^n} \Lambda^{*n}(\vec{y}_1^n)$ then $W_n(\vec{Y}_1^n, \lambda) = \rho_n(\vec{Y}_1^n)$.

where *ess sup* indicates the essential supremum, i.e. supremum almost everywhere.
Therefore we obtain,

Theorem 6.8

$$\Lambda^{*n}(\vec{y}_1^n) = \max \{\rho_n(\vec{y}_1^n) - E[\rho_{n+1}(\vec{Y}_1^{n+1}|\vec{y}_1^n)], ess\ sup_{\vec{y}_1^{n+1}} \Lambda^{*n}(\vec{y}_1^{n+1})\}. \tag{6.42}$$

Theorem 6.8 allows us to describe a single collection of "feedback-threshold" functions $\Lambda^{*n}(\vec{y}_1^n)$, that can be used, computed, or approximated to obtain any optimal rate-distortion point in the framework. This technical looking result therefore completes the program in this book on developing understanding of source–channel decoding with distortion-aware ACK/NACK feedback for the fairly general case of progressive transmission of a tree-structured vector-quantized source. The next set of sections show some practical methods to achieve further participation of the receiver in the source–channel coding with feedback, primarily with the view of the common theme of progressive transmission.

6.3.3 Progressive Transmission and Receiver-Driven Rate Control

The embeddedness of the optimal policies and the existence of $\Lambda^*(Y_1^n)$ is a very useful property that can come in handy in a variety of application scenarios. Note that, by extending the definition of a NACK feedback to mean a permission to continue transmission, ACK/NACK can be actively used for receiver-driven rate control. As the optimal policies are embedded, the progressive transmission, say that of an image, can be accomplished without losing optimality at the terminal and at intermediate transmission budgets. The quality of the received image can be successively improved as new bits are received (by means of successive transmissions of NACK feedback). The optimal feedback generation rules reveal a very simple structure in the form of Theorem 6.5. If the feedback-threshold function $\Lambda^*(Y_1^n)$ is known or if it can be approximated, then the ACK/NACK generation for a range of operating points can be accomplished at once. Secondly, the receiver can switch from operating at a low average transmission rate to a higher average transmission rate, in the middle of a transmission, without losing the optimality of the rate-distortion trade-off. This rate control technique can be potentially useful in the following situations.

Delay-limited reconstruction: In interactive applications such as videoconferencing, a quick reconstruction at low transmission budget for the foreground, and slow but detailed and error-free reconstruction of the background might be used, provided such a separation is available. Controlling the ACK/NACK of the appropriate packets may allow a trade-off between reconstruction speed and quality. The last part of this chapter explores this situation in detail.

Bandwidth/data rate-limited reconstruction: While receiving statistically multiplexed streams of variable rate at a receiver, the transmission rates of one or more of the streams can be controlled using appropriate ACK/NACK feedback.

Computation-limited and buffer-size limited reconstruction: Similarly, for a multitasking environment such as a server at a base station, the CPU usage and memory allocated to an incoming stream over a noisy channel can be variable. Based on the current processing capability, some amount of control can be exercised by appropriate operating point selection in the feedback generation rules.

Tolerance-limited reconstruction: In digital encoding of video, the intra-coded frames generate a lot more data than predictive or inter-coded frames. The predictive frames can be thought of as incremental information. In a low noise environment, fewer intra-coded frames can be transmitted, while in a noisy environment they need to be more frequent. The switching between the two for best rate-distortion performance can be accomplished by the use of ACK/NACK feedback.

6.4 Delay-Constrained JSCC Using Incremental Redundancy with Feedback

Real-time transmission of multimedia sources requires adhering to stringent delay constraints. This makes traditional FEC (with no feedback) be in principle better suited for error control because in this scenario, unlike for the case of data transmission, the use of feedback-based schemes (such as Hybrid ARQ) is limited by the delay constraint. However, as expressed earlier, the use of feedback allows for more efficient error control techniques. We now study a JSCC scheme for real-time traffic that, nevertheless, uses feedback-based error control. This scheme for real-time wireless traffic, first introduced in [10], uses incremental redundancy error control with feedback while simultaneously meeting a hard delay constraint. In this scheme, the delay constraint is met by optimally changing the source coding rate at appropriate times. The analysis of the system dynamics follows the introduction of an entity called the *codetype*. The codetype allows the modeling of the system as a Markov chain from which it is possible to derive an optimized design for the source and channel rate allocations using dynamic programming. In addition, by designing a constant bit rate and fixed delay system that transmits with fixed-size frames, it becomes possible to avoid some of the drawbacks of ARQ-based systems, namely delay accumulation, delay jitter, and complex buffer management. The technique that will be discussed in this section uses feedback to achieve a lower end-to-end distortion for the same channel conditions than a system with pure FEC. Equivalently, for the same level of distortion, the technique presents a gain in the channel SNR with respect to one without feedback. This gain could be valuable in multiuser networks where the SNR gain translates into increased capacity for handling multiuser interference, hence supporting a larger number of simultaneous calls. Consequently, this case will be studied as a potential application of the system with feedback. We will also see that the resulting system has the added advantage that it is more robust under channel mismatch conditions.

6.4.1 System Description

For the system we will study next, the main components of the mobile terminal are shown in Figure 6.20. The source encoder is required to be capable of operating at several different

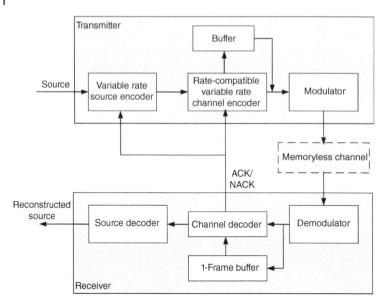

Figure 6.20 Block diagram of the mobile terminal for scheme implementing delay-constrained JSCC using incremental redundancy with feedback.

rates. The source encoder operates on a fixed block of source samples and produces, for each particular rate, a fixed number of bits, referred to as *source blocks*. The source blocks are encoded, into *codewords*, by a concatenation of an outer error detection code and an inner error correction code selected from a *rate-compatible* family of codes (such as those seen earlier in Section 3.4).

Transmission over the wireless channel is synchronous: *Fixed-size frames* – generated one frame per source block – are transmitted over the channel. A frame may contain codewords and partial codewords corresponding to one or more source blocks. The transmission takes place using ACK/NACK feedback according to the following protocol.

Consider three consecutively transmitted frames, referred in time as previous, current, and next frames, and the corresponding source blocks as previous, current, and next source blocks. If the feedback received at the transmitter for the previous frame was an ACK, then the current frame would consist entirely of the bits for the codeword of the current source block. This frame is transmitted over the channel and is decoded at the receiver to recover the source block. The success or failure of decoding is determined by the error detection code. This result is sent back to the encoder by one bit (ACK/NACK) through a feedback channel which is assumed to be error-free and instantaneous. If an ACK is received, the encoder proceeds with the transmission of the next frame in the same manner it did for the current frame. If a NACK is received (indicating failed decoding of the current source block), the encoder selects a lower rate channel code for the current source block. Due to the rate compatibility of the channel code family, only some extra parity bits need to be sent to make the lower rate codeword for the current source block. These bits, the *incremental redundancy*, are sent as a part of the next frame along with the codeword for the next source block. As the frame size is fixed, the source encoding rate and the channel

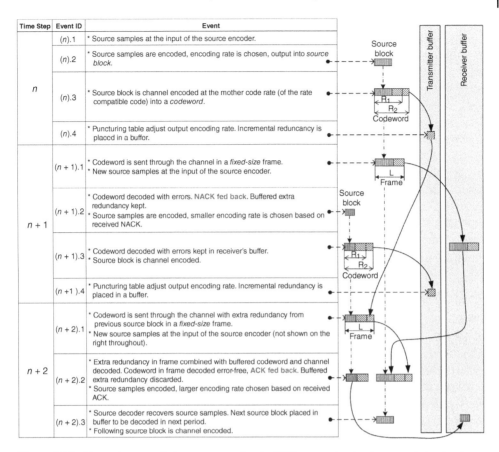

Time Step	Event ID	Event
n	(n).1	* Source samples at the input of the source encoder.
	(n).2	* Source samples are encoded, encoding rate is chosen, output into *source block*.
	(n).3	* Source block is channel encoded at the mother code rate (of the rate compatible code) into a *codeword*.
	(n).4	* Puncturing table adjust output encoding rate. Incremental reduncancy is placed in a buffor.
n + 1	(n + 1).1	* Codeword is sent through the channel in a *fixed-size* frame. * New source samples at the input of the source encoder.
	(n + 1).2	* Codeword decoded with errors. NACK fed back. Buffered extra redundancy kept. * Source samples are encoded, smaller encoding rate is chosen based on received NACK.
	(n + 1).3	* Codeword decoded with errors kept in receiver's buffer. * Source block is channel encoded.
	(n +1).4	* Puncturing table adjust output encoding rate. Incremental redundancy is placed in a buffer.
n + 2	(n + 2).1	* Codeword is sent through the channel with extra redundancy from previous source block in a *fixed-size* frame. * New source samples at the input of the source encoder (not shown on the right throughout).
	(n + 2).2	* Extra redundancy in frame combined with buffered codeword and channel decoded. Codeword in frame decoded error-free, ACK fed back. Buffered extra redundancy discarded. * Source samples encoded, larger encoding rate chosen based on received ACK.
	(n + 2).3	* Source decoder recovers source samples. Next source block placed in buffer to be decoded in next period. * Following source block is channel encoded.

Figure 6.21 Example transmission sequence for the JSCC protocol using incremental redundancy with feedback. For clarity of the illustration, the channel code is depicted as being systematic, although this is not a requirement for the JSCC system (the channel code does need to be rate-compatible when punctured).

code for the next source block are selected so as to accommodate the incremental redundancy for the current source block. On receiving the next frame, in addition to decoding the next codeword, the decoder combines the incremental redundancy for the current source block with its previously received bits and decodes it to the best of its capability. At this point, the current decoded source block is accepted without any further error detection (error detection may still be done but not for the purpose of deciding the ACK/NACK feedback but, rather, to indicate the error status to the subsequent source decoder). This way the system is restricted to at most one transmission of incremental redundancy. Hence, the maximum delay experienced in the decoding of any source block is one frame. The decoder processes the next frame by the same protocol while keeping track of the possibly altered coding rate allocations. Figure 6.21 illustrates this procedure for the case when a block of source samples that are encoded at a transmission time step n result in a NACK feedback following its first decoding attempt. This NACK feedback triggers the transmission of incremental redundancy within the following frame, which forces a smaller source encoding rate for the block of source samples that are encoded at a transmission time step $n + 1$. Upon combining the incremental redundancy with the already previously received redundancy, the resulting lower rate channel code results in a successful decoding and the transmission of an ACK feedback.

The key feature of this protocol is that it alters the source coding rate of the next source block to accommodate for the recovery of errors in the current source block (as is the case illustrated in Figure 6.21). Hence, even if the decoding of the next source block is successful, it may result in reconstruction of the corresponding source samples with a higher average distortion than those in the current frame. In return, the incremental redundancy bits help decode the current frame correctly. In this system, unlike a retransmission-based system, the delays for the source blocks do not accumulate. Also, because of the hard constraint on the delay, the decoder can operate with a fixed delay, i.e. delay jitter is eliminated. Further, the buffering mechanisms at the receiver and the transmitter require small memory and are synchronous because of the use of fixed frame sizes (in spite of possibly large variations in the sizes of the incremental redundancies).

6.4.2 Optimal Source and Channel Rate Allocations Design

As noted, a frame may be composed of either one codeword or one codeword and incremental redundancy for the codeword of the previous source block. The different compositions of the frames is described by introducing the *codetype* of a codeword. The codetype of a codeword is "1" if all its bits occupy an entire frame. The codetype of a codeword is "m" if it makes a frame by being combined with incremental redundancy (extra bits) for (the previous) codeword of codetype "$m - 1$." Let us assume that there is a maximum number N of codetypes. Let m_n and B_n, respectively, denote the codetype of the codeword used in the nth frame, and the feedback received after the nth frame. Then the evolution of the codetype can be described as:

$$m_n = \begin{cases} 1 & \text{if } B_{n-1} = ACK, \\ m_{n-1} + 1 & \text{if } B_{n-1} = NACK \text{ and } m_{n-1} < N, \\ N & \text{if } B_{n-1} = NACK \text{ and } m_{n-1} = N. \end{cases} \tag{6.43}$$

Consider a codeword of codetype m. Let $r_s(m)$ denote the number of source bits and $r_c(m, 0)$ and $r_c(m, 1)$ denote the number of channel code parity bits for the first attempt and the incremental redundancy bits, respectively. The channel decoder decodes $r_s(m)$ source bits, from a received word of length $r_s(m) + r_c(m, 0)$ bits in the first attempt and of length $r_s(m) + r_c(m, 0) + r_c(m, 1)$ in the second attempt. Note that although we are describing the channel coding set up as though the channel code is systematic, we do this for clarity of the explanation and it is not a requirement for the design itself.

Next, consider transmission over a memoryless channel. Denote by $FER(r_s, l)$ the minimum frame error rate (FER) (probability of uncorrectable post decoding error) for a channel code from the code family which encodes at least r_s source bits into a codeword of length of at the most l bits. Defining $FER(r_s, l)$ this way extends the definition of FER (usually defined for elements of channel code family) to hold for arbitrary numbers r_s, l. Then the probability of a frame error after the first decoding attempt for codetype m, i.e. probability of a NACK feedback, is $f(m, 0) \overset{\Delta}{=} FER(r_s(m), r_s(m) + r_c(m, 0))$. Let $f(m, 1)$ denote the probability of residual frame error after the second decoding attempt. $f(m, 1)$ can be closely approximated by $FER(r_s(m), r_s(m) + r_c(m, 0) + r_c(m, 1))$.

Next, assume that the source coder generates independent source blocks, and the end-to-end quality can be closely measured by the average of distortions experienced by individual frames. Hence, we can develop the expression for the expected distortion faced by a frame, encoded with r_s source bits, r_{c0} parity bits in the first transmission and r_{c1} parity bits in the incremental redundancy transmission as follows:

$$D(r_s, r_{c0}, r_{c1}) \overset{\Delta}{=} (1 - FER(r_s, r_s + r_{c0} + r_{c1}))D_s(r_s) \tag{6.44}$$
$$+ FER(r_s, r_s + r_{c0} + r_{c1})D_c(r_s, r_{c0} + r_{c1}).$$

Here, $D_s(r_s)$ is the distortion due to the source encoder and $D_c(r_s, r_{c0} + r_{c1})$ is the average end-to-end distortion conditioned on the event of frame error. While $D_s(r_s)$ is only a function of the source rate r_s, D_c depends on $r_s, r_{c0} + r_{c1}$ and channel SNR, E_s/N_0. Note that the probability of NACK after first transmission, $(FER(r_s, r_s + r_{c0}))$, does not directly appear in the expression of average distortion. This is so because only the residual error event contributes to the channel-induced distortion, and also that the probability of the residual error event is well approximated by the probability of failure of the codeword corresponding to two transmissions, i.e. $FER(r_s, r_s + r_{c0} + r_{c1})$, irrespective of the probability of NACK after first transmission.

Consider the evolution of the codetypes as described by (6.43). Since each codetype depends only on the previous codetype, for transmission over a memoryless channel, we can model the relation between codetypes with a Markov chain, as shown in Figure 6.22, where each codetype is one state of the chain. Let π_m denote the steady-state probability of codetype m. Given the Markov chain model the stationary distributions for each codetype, π_m, are

$$\pi_1 = \left[1 + \sum_{i=1}^{N-2}\left(\prod_{j=1}^{i} f(j, 0)\right) + \frac{\left(\prod_{j=1}^{N-1} f(j, 0)\right)}{1 - f(N, 0)} \right]^{-1}, \tag{6.45}$$

$$\pi_m = \pi_1 \prod_{j=1}^{m-1} f(j, 0) \quad \text{for } m = 2, \ldots, N-1, \tag{6.46}$$

$$\pi_N = \frac{\pi_1}{1 - f(N, 0)} \prod_{j=1}^{N-1} f(j, 0). \tag{6.47}$$

We assume that the source encoder generates independent source blocks, and the end-to-end quality can be measured by the average of distortions suffered by individual frames. Hence, from (6.44), the average distortion $D(m)$ for a source block encoded with codetype m is

$$D(m) = [1 - f(m, 1)]D_s(r_s(m)) + f(m, 1)D_c(r_s(m), r_c(m, 0), r_c(m, 1)), \tag{6.48}$$

and the average distortion is then,

$$\overline{D} = \sum_{m=1}^{N} \pi_m D(m). \tag{6.49}$$

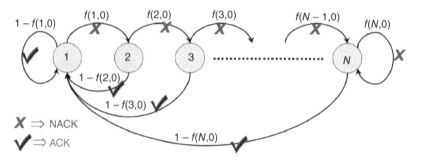

Figure 6.22 Markov chain of codetypes.

Alternatively, the average distortion of source blocks not suffering from post-decoding channel-induced errors, i.e. error-free source blocks, is given by:

$$\overline{D}_{errorfree} = \frac{\sum_{m=1}^{N} \pi_m [1 - f(m, 1)] D_s(r_s(m))}{\sum_{m=1}^{N} \pi_m [1 - f(m, 1)]}.$$

For real-time sources, it is common practice to constrain the FER to values small enough such that the more damaging contribution due to channel errors to the end-to-end distortion is kept at a negligible value. If at design time we follow this idea and we impose a *maximum residual FER constraint*, i.e. we require that the final FER is such that $f(m, 1) < \epsilon$ for all m for some small number $\epsilon > 0$, then we can approximate $\overline{D}_{errorfree}$ by setting $D(m) = D_s(r_s(m))$ into (6.49), as:

$$\overline{D}_{errorfree} = \sum_{m=1}^{N} \pi_m D_s(r_s(m)). \tag{6.50}$$

Note that (6.50) is of the same form as (6.49).

The design problem is that of optimally assigning incremental redundancy in a two-step transmission framework (i.e. sending incremental redundancy at most once per source block) with fixed length frames, so that the average distortion, in (6.49) or alternatively, the error-free distortion subject to FER constraints (6.50), be minimized. In other words, the design of the system involves assignment of source and channel code rates, i.e. selection of $r_s(m)$, $r_c(m, 0)$, and $r_c(m, 1)$ for each codetype m, so that the average distortion in (6.49), or alternatively, the error-free distortion subject to FER constraints, is minimized. As we will see next, an optimal solution for this design problem can be found through a dynamic programming approach that leverages the underlaying Markov chain of codetypes. We will get to this point by following a number of steps:

Step 1 – Distortion as a ratio of costs: Let $\mathcal{B}$ denote the set of all possible valid code assignments. Each element is defined by specifying for each of the N codetypes the triplet source block size, number of channel code parity bits in the first transmission attempt, and number of incremental redundancy bits. Note that $\mathcal{B}$ is a finite set, defined by the combination of available operating modes in both the source and channel encoders constrained by the fixed frame size condition and, optionally, a FER goal. Let a code assignment be denoted by $\mu \in \mathcal{B}$. Let us denote by $f_\mu(j, 0)$ and $f_\mu(j, 1)$ the probability of post-decoding frame errors after the first and second transmission, respectively, of the jth code used in assignment μ.

Note that, since only one transmission of incremental redundancy is allowed, $f_\mu(j, 1)$ can also be interpreted as the residual FER. Similarly, let $r_{s_\mu}(j)$, $r_{c_\mu}(j, 0)$, and $r_{c_\mu}(j, 1)$ denote the number of source bits, the number of channel bits in the first step, and number of channel bits in the second step, of codetype j of assignment μ. Let $D_\mu(j)$ denote the resulting distortion defined as in (6.48), for the jth codetype of assignment μ. Then examining (6.49), we can rewrite $\overline{D}$ as:

$$\overline{D} = \frac{D_\mu(1) + \sum_{i=1}^{N-2}\left(\prod_{j=1}^{i} f_\mu(j, 0)\right)D_\mu(i) + \frac{\left(\prod_{j=1}^{N-1} f_\mu(j,0)\right)}{1-f_\mu(N,0)}D_\mu(N)}{1 + \sum_{i=1}^{N-2}\left(\prod_{j=1}^{i} f_\mu(j, 0)\right) + \frac{\left(\prod_{j=1}^{N-1} f_\mu(j,0)\right)}{1-f_\mu(N,0)}} \tag{6.51}$$

This expression is the ratio of two costs evaluated on the Markov chain. The numerator and the denominator are the average distortion cost and the average number of steps starting at codetype 1 and before returning to codetype 1.

Step 2 – Optimization problem: Let $\delta(\mu)$ and $Q(\mu)$ denote the numerator term and denominator term of $\overline{D}$ in (6.51) when code assignment μ is used. Hence, the optimization problem can be stated as:

$$\min_{\mu \in B} \frac{\delta(\mu)}{Q(\mu)} \tag{6.52}$$

where all code assignments $\mu \in B$ satisfy

$$r_{s_\mu}(m) + r_{c_\mu}(m, 0) + r_{c_\mu}(m-1, 1) \le L \tag{6.53}$$

$$r_{s_\mu}(1) + r_{c_\mu}(1, 0) \le L \tag{6.54}$$

$$r_{s_\mu}(N) + r_{c_\mu}(N, 0) + r_{c_\mu}(N, 1) \le L \tag{6.55}$$

and, optionally, the residual FER constraint:

$$f_\mu(m, 1) < \epsilon. \tag{6.56}$$

Step 3 – Minimum as a zero of the Lagrangian: Let λ be a real number. Define for a code assignment μ, the Lagrangian $\mathcal{L}_\mu(\lambda) \triangleq \delta(\mu) - \lambda Q(\mu)$ and consider the optimization problem:

$$\min_{\mu \in B} \mathcal{L}_\mu(\lambda) = \min_{\mu \in B} \delta(\mu) - \lambda Q(\mu). \tag{6.57}$$

Let $\mu^*(\lambda)$ denote the assignment achieving the minimum in the problem (6.57). Then, we have the following claim.

Claim: If for some λ^*, $\min_{\mu \in B}\mathcal{L}_\mu(\lambda^*) = 0$, then $\lambda^* = \min_{\mu \in B}\frac{\delta(\mu)}{Q(\mu)}$ and $\mu^*(\lambda^*)$ is the solution to $\min_{\mu \in B}\frac{\delta(\mu)}{Q(\mu)}$.

Proof: Note that for all $\mu \in B$, $Q(\mu) > 0$. If $\min_{\mu \in B}\frac{\delta(\mu)}{Q(\mu)} = \lambda^*$ then $\frac{\delta(\mu)}{Q(\mu)} \ge \lambda^*$ for all μ and hence $\mathcal{L}_\mu(\lambda^*) = \delta(\mu) - \lambda^* Q(\mu) \ge 0$. Also if $\mu^* = \arg\min_{\mu \in B}\frac{\delta(\mu)}{Q(\mu)}$ then $\frac{\delta(\mu^*)}{Q(\mu^*)} = \lambda^*$, hence $\mathcal{L}_{\mu^*}(\lambda^*) = 0$.

Therefore, the optimization problem "reduces" to finding a λ and corresponding $\mu^*(\lambda)$ for which $\mathcal{L}_{\mu^*(\lambda)}(\lambda) = 0$. This problem is easier because $\delta(\mu)$ and $Q(\mu)$ are of the same form and hence $\delta(\mu) - \lambda Q(\mu)$ can be written as:

$$\delta(\mu) - \lambda Q(\mu) = \left[D_\mu(1) - \lambda\right] + \sum_{i=1}^{N-2} \left(\prod_{j=1}^{i} f_\mu(j,0)\right) \left[D_\mu(i) - \lambda\right]$$

$$+ \frac{\left(\displaystyle\prod_{j=1}^{N-1} f_\mu(j,0)\right)}{1 - f_\mu(N,0)} \left[D_\mu(N) - \lambda\right] \tag{6.58}$$

Step 4 – Optimization by dynamic programming: Solving the problem in (6.57) for a fixed λ can be done through dynamic programming as described next. Consider (6.58) and define

$$J_\mu(N,\lambda) = \frac{(D_\mu(N) - \lambda)}{1 - f_\mu(N,0)} \quad \text{and}$$

$$J_\mu(m,\lambda) = (D_\mu(m) - \lambda) + f_\mu(m,0)J_\mu(m+1,\lambda), \quad \text{for } m = 1,2,\ldots,N-1, \tag{6.59}$$

then $\delta(\mu) - \lambda Q(\mu) = J_\mu(1,\lambda)$. Note that

1. $J_\mu(m,\lambda)$ depends only on the assignments of the codes for codetypes $m, m+1, \ldots, N$ and not on the earlier codewords.
2. On the other hand, the constraints in (6.53) relates codetype m to codetype $m-1$ only through the parameter $r_{c\mu}(m-1,1)$, as $r_{s\mu}(m) + r_{c\mu}(m,0) = L - r_{c\mu}(m-1,1)$.

Hence, the following algorithm can be devised. Let $r_{smin} + r_{cmin} < l \le L$ where $r_{smin} + r_{cmin}$ is the smallest possible length of a codeword in the available channel code family. Then define

$$J_N^*(l,\lambda) = \min_{r_s,r_{c0}} \frac{(D(r_s,r_{c0},r_{c1}) - \lambda)}{1 - FER(r_s,l)}$$

$$\text{subject to } r_s + r_{c0} = l, \ FER(r_s,L) < \epsilon. \tag{6.60}$$

$D(r_s,r_{c0},r_{c1})$ is as given in the definition (6.44), unless the approximation in (6.50) is in effect which results in $D(r_s,r_{c0},r_{c1}) \approx D_s(r_s)$. $J_N^*(l,\lambda)$ denotes the smallest value of $J_\mu(N,\lambda)$ satisfying the constraint on the final FER.

Similarly, for $m = 1,2,\ldots,N-1$, define recursively

$$J_m^*(l,\lambda) = \min_{r_s,r_{c0},r_{c1}} (D(r_s,r_{c0},r_{c1}) - \lambda) + FER(r_s,l)J_{m+1}^*(L - r_{c1},\lambda)$$

$$\text{subject to } r_s + r_{c0} = l, \ FER(r_s,r_s + r_{c0} + r_{c1}) < \epsilon. \tag{6.61}$$

This is the Bellman equation for optimality. The value $J_1^*(L,\lambda)$ is the value of the minimum in problem (6.57) and the minimizing parameters in the aforementioned optimization give code for mth codetype in $\mu^*(\lambda)$.

Step 5 – Policy iteration algorithm for optimal λ: The aforementioned dynamic programming algorithm (6.61) can obtain for any λ the optimal code assignment from $\mathcal{B}$. But we are interested in the specific value of λ which yields $\mathcal{L}_{\mu^*(\lambda)}(\lambda) = 0$. The optimal value of λ can be obtained by policy iteration algorithm [11]. The policy iteration algorithm updates, at each step, the N codes and a value of λ. The algorithm guarantees that in a finite number of steps, the assignment converges to the optimal. The resulting policy iteration algorithm for optimal λ is shown as Algorithm 3.

6.4.3 Performance

We will now study how the use of the delay-constrained JSCC scheme with feedback-based transmission of incremental redundancy affects the performance of a communication system. To see this, we will consider the communication of speech over a wireless code-division multiple access (CDMA) system. In CDMA, all users transmit simultaneously over the same frequency band. Since perfect separation between calls cannot be achieved under practical conditions, each user interferes all the others to some extend. Clearly, this interference affects the quality of service (QoS) of the communication. In CDMA, this QoS could be measured in terms of the FER when considering delay-sensitive real-time sources such as speech (this is because in a general case of transmitting real-time traffic there is a limit to the number of times that is possible to attempt the retransmission of failed frames). Consequently, the number of users that may be supported simultaneously in a CDMA network is determined by the fact that as more users are admitted, the interference grows up to the point when it is not possible to accept more calls and achieve the promised QoS.

Consider, then, the uplink of a single-cell, chip-sampled direct-sequence CDMA system using BPSK modulation and matched filter receivers, over an AWGN channel with variance σ^2. In this case, using a Gaussian approximation, the effect of multiuser interference can be represented by an equivalent signal-to-interference plus noise ratio (SINR). A target SINR is set for each user with the goal of maintaining a desired QoS level. We assume that the system is able to estimate in each frame an "equivalent cross-correlation" γ that represents the combined effect of all interferences [12]. Then, under ideal power control, power assignments and the required SINR values by each of the M users satisfy

$$\beta_i = \frac{P_i}{\sigma^2 + \gamma^2 \sum_{\substack{j=1 \\ j \neq i}}^{M} P_j}, \qquad i = 1, 2, \ldots, M, \tag{6.62}$$

where P_i is the ith user's received power necessary to obtain the target SINR β_i. In this application, we consider that the required SINR is set so as to maintain certain end-to-end distortion. Before continuing, we note in passing that the M equations (6.62) are sufficiently general that with slight modifications they can also be used for a general multiuser wireless

---

Algorithm 6.1 Policy iteration algorithm for optimal λ

1: Set $\lambda_0 = a$ (arbitrary constant).
2: Obtain $\mu^1 = \mu^*(\lambda_0)$ by (6.60) and (6.61).
3: $k = 1$.
4: Set $\lambda_k = \frac{\delta(\mu^k)}{Q(\mu^k)}$.
5: Set $\mu^{k+1} = \mu^*(\lambda_k)$ by solving (6.60) and (6.61).
6: If $\mu^{k+1} \neq \mu^k$ Set $k = k + 1$ and go to Step 4. Else, stop. The converged assignment is the optimal assignment and the final λ is the optimal λ.

---

communications system representing, for example, inter-cell interference in a cellular system. However, for clarity of presentation, we focus the forthcoming presentation on the CDMA case.

Next, the system of M equations (6.62) in the unknowns P_i can be solved to obtain

$$P_i = \frac{\frac{\sigma^2/\gamma^2}{1+1/(\gamma^2\beta_i)}}{1 - \sum_{j=1}^{M} \frac{1}{1+1/(\gamma^2\beta_j)}}. \tag{6.63}$$

Note that this result establishes a limit on the number M of simultaneous active transmissions because the denominator in (6.63) could potentially become negative, which would make for an unfeasible power allocation solution. What this result indicates is that the user capacity in the system is determined by the fact that as more users are admitted into the network, the interference grows up to the point where it is not possible to accept more calls and deliver the promised QoS. Specifically, as more users are admitted into the system,

$$\sum_{j=1}^{M} \frac{1}{1 + \frac{1}{\gamma^2\beta_j}}$$

grows up to a point where it is too close to 1. At this point, referred to as the *congestion point*, power assignment becomes excessive. Therefore, a practical limit on the user capacity is one determined by the condition:

$$\sum_{j=1}^{M} \frac{1}{1 + \frac{1}{\gamma^2\beta_j}} = 1 - \epsilon, \tag{6.64}$$

where ϵ is a small positive number. Also, it can be shown that in order to minimize the average distortion, all users should be assigned the same target SINR. Therefore, the maximum number of users that can be simultaneously supported in the network, the user capacity, for a target SINR β is

$$M = (1 - \epsilon)\left[1 + \frac{1}{\gamma^2\beta}\right]. \tag{6.65}$$

As said earlier, we will consider the case when the real-time traffic is formed by conversational speech. The source encoder is the GSM advance multi-rate (AMR) narrowband speech encoder [13]. This encoder operates with 20 ms frames, 5 ms look-ahead, and at eight possible encoding rates: 12.2, 10.2, 7.95, 7.4, 6.7, 5.9, 5.15, and 4.75 kbps. Eighteen different speech sequences, from male and female speakers, extracted from the NIST speech corpus [14], were used as the input sequences generated by the source. We also used the GSM AMR encoder's error concealment option because it is possible that there could be uncorrected errors after channel decoding. For the channel encoder, we used a memory 4, puncturing period 8, rate-compatible punctured convolutional (RCPC) code [15] decoded with a soft Viterbi decoder. The frame was chosen to be $L = 500$ bits long.

To measure end-to-end distortion, we chose a log-spectral distortion calculated by numerical approximation of the function [16]:

$$SD(\hat{A}(f), A(f)) = \sqrt{\int |W_B(f)|^2 \left|10\log\frac{|\hat{A}(f)|^2}{|A(f)|^2}\right|^2}, \tag{6.66}$$

where $A(f)$ and $\hat{A}(f)$ are the FFT-approximated spectra of the original and the reconstructed speech frames, respectively, and W_B is a subjective sensitivity weighting function defined by:

$$W_B(f) = \frac{1}{25 + 75(1 + 1.4(f/1000)^2)^{0.69}}. \tag{6.67}$$

This distortion is measured on a frame-by-frame basis and then averaged over all frames. Contrary to the common practice when using spectral distortion measures, in the present case, we include in the average calculation all frames, even those that are outliers (i.e. frames which contribute to excessive distortion). This is because to measure end-to-end distortion, it is necessary to consider also frames affected by channel errors which are reconstructed through error concealment. For this same reason, the spectrums are calculated by approximating their FFT and not through its linear predictive coding (LPC) parameters representation, which produce a smoothed version of the real spectrum that cannot accurately represent the effects of the artifacts introduced by channel-induced errors. Finally, performance in terms of distortion will be measured in a normalized format, as the ratio of the spectral distortions with that of the speech sequence encoded at the highest rate (12.2 kbps) in the absence of channel noise. This is so the performance results are independent of source coder parameters and normalization constants.

Figure 6.23 shows the measured normalized distortion as a function of the channel SNR, for the designs with and without a residual FER constraint. While the case without a residual FER constraint represents optimum source and channel allocation minimizing the end-to-end distortion, it is also important and convenient to consider the design under residual FER constraint because this corresponds to the case when the source encoding contributes most to the end-to-end distortion and the effects of the more annoying channel-induced errors are kept at an imperceptible level. It is because of this that the target residual FER was set at 10^{-2}. For both cases with and without a residual FER constraint, we show the results for incremental redundancy with two codetypes ($N = 2$) and a pure FEC system (no feedback) which corresponds to one codetype ($N = 1$). Note that even in the case for the pure FEC system, the allocation is optimized for each channel SNR value. We can see that the feedback-based system for the case of no residual FER constraint presents an SNR gain of 0.85 dB in most of the cases with extremes values of 1 dB (15% increase in distortion) and 0.5 dB (8.5% increase in distortion) approximately. When using the residual FER constraint, the SNR gain fluctuates between 1.25 dB (5% increase in distortion) and 2.5 dB (15% increase in distortion) approximately. It is worthwhile to note that it is to expect that the larger the number N of codetypes, the better performance would be. In this sense, the choice of $N = 2$ represents the most modest of the performance improvement that could be achieved with the use of feedback-based incremental redundancy transmission. However, the choice of number of codytypes is usually dictated by the real-time constraint of the traffic. For example, in the present case, the choice of $N = 2$ would introduce an extra frame duration of delay (20 ms). This may not seem too much (is 10% of the maximum 200 ms delay that could be considered as acceptable for conversational speech) but is a delay that will be added to the already existing delays (20 ms for source encoding delay, plus medium access control, potential transcoding, and other queuing at a later point in the network, etc.), and so it becomes a very reasonable decision.

We just have seen how for a given channel, the feedback-based system achieves consistently lower distortion compared to schemes not using feedback (alternatively, the

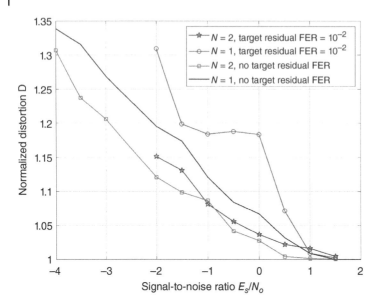

Figure 6.23 Normalized distortion vs. channel SNR for AWGN channels with and without residual FER constraints.

feedback-based system obtains useful gains in channel SNR needed to achieve the same level of distortion). A significant additional advantage of a feedback-based scheme is its inherent adaptability to variation in channel conditions. Hence, a feedback-based system is expected to achieve a smoother degradation in performance as the channel worsens in comparison to the one for which the parameters were designed for. Figure 6.24 shows the channel mismatch performance of transmission schemes with and without feedback ($N = 2$ and $N = 1$, respectively) both designed for optimal performance for a fixed channel SNR (0 and -1 dB in the figure). Notice that not only the scheme with feedback outperforms the scheme with no feedback at the design SNRs, and the scheme with feedback is considerably robust to channel mismatch. Its performance degrades much slower than the scheme without feedback and hence the performance gap between the two schemes widens.

In addition, ideally, one would like the system parameters, namely the source encoding rate, the transmission rate, and the error protection schemes, and the protocol to be completely tuned to the channel. But finiteness of the number of available choices for rate, channel codes, and source-coding rates dictates that there can only be a finite number of rate allocations. Hence, the performance of any fixed rate nonadaptive system may depend crucially on the frame length, as the minimum achievable distortion may become a non-convex function of the frame length. It is interesting to note that the JSCC feedback-based system we are discussing does not suffer from this drawback. The minimum achievable distortion for a given channel varies smoothly as a function of the frame length, despite having the same finite family of source and channel codes for design. Figure 6.25 illustrates this observation. It shows the minimum achievable average distortion for an AWGN channel with SNR 0 dB, as a function of the frame length in bits. The scheme with no feedback has a

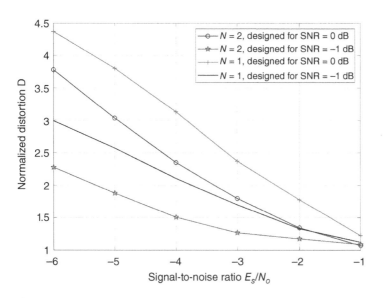

Figure 6.24 Comparison of robustness under channel mismatch conditions.

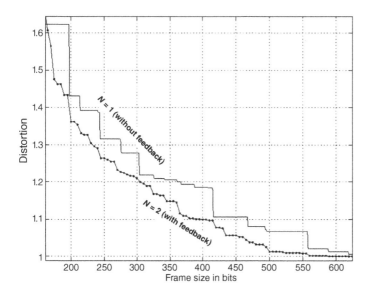

Figure 6.25 Smoothness of performance with respect to data rate/ packet size variation: channel SNR 0 dB.

performance curve which has a staircase-like structure. Hence, it suffers from discontinuities as two frame lengths close by can yield widely different performances. On the other hand, the performance of the proposed scheme show that for any two frame lengths that are close by, not only the scheme with feedback achieves considerably lower distortions but also the variation of minimum distortion is small. Hence, the designer can select frame lengths according to other conveniences and need not worry about any particular selection being unfortunate on the given channel.

We finally turn our attention to an application case that shows how the performance improvement in the delay-constrained JSCC scheme with feedback-based transmission of incremental redundancy can be leveraged to increase the network load that can be supported in a CDMA system. As shown in (6.65), the call admission control problem in CDMA can be framed as one where users are admitted up to a maximum number M, determined by the SINR required to guarantee some end-to-end distortion level. On the one side this model better represents the problem of call admission control from the network operator's viewpoint, i.e. calls requesting service when there are a certain maximum number of calls already being served are rejected in order to maintain the quality for the existing calls. On the other side, in CDMA it is possible to accept more calls beyond the set limit. This is why researchers have studied the Earlang capacity of CDMA systems following a $M/M/\infty$ model. Here, the numbers of calls that can be supported is calculated based on the goal that the outage probability remains below some threshold, where outage occurs when the system exceeds some operational parameter (typically related to interference in CDMA) [17, 18].

In [17], the outage condition was defined in terms of the relation between the amount of interference density and background noise level. Noticing again the equivalence between (6.63) and (6.64) with (4) and (6) in Ref. [18] and reminding that that in order to minimize the average distortion all users should be assigned the same target SINR β, it is straightforward to show that the outage condition derived from the feasibility constraint (6.64) can be written as:

$$\sum_{i=1}^{N} \omega_i > (1 - \epsilon)\left(1 + \frac{1}{\gamma^2 \beta}\right)$$ (6.68)

where ω_i is a binary random variables that is equal to 1 when user i is in a talk spurt. Let $\rho = P(\omega_i = 1)$. Typically $\rho = 0.4$. If the outage probability is defined as:

$$P_{out} = P\left[\sum_{i=1}^{N} \omega_i > (1 - \epsilon)\left(1 + \frac{1}{\gamma^2 \beta}\right)\right],$$ (6.69)

the Erlang capacity is the maximum offered load such that the outage probability is kept below some target, typically 1% or 2%. Let the random variable $Z = \sum_{i=1}^{N} \omega_i$. Defining

$$K_0 = (1 - \epsilon)\left(1 + \frac{1}{\gamma^2 \beta}\right),$$ (6.70)

and using the characteristic function it can be shown that, [17],

$$P_{out} = e^{-\rho v/\mu} \sum_{\lfloor K_0 \rfloor}^{\infty} \frac{(\rho v/\mu)^k}{k!},$$ (6.71)

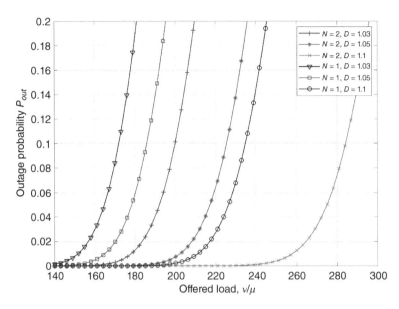

Figure 6.26 Outage probability in a CDMA system with no FER constraint.

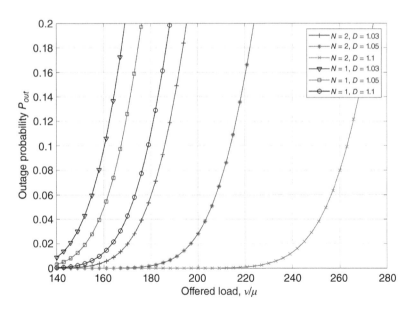

Figure 6.27 Outage probability in a CDMA system with residual FER constraint.

where v is the arrival rate for new calls, μ is the ending rate for existing calls, and v/μ is called the offered load for a network. In this case, we can see again that if the same level of end-to-end distortion can be achieved at a reduced target SINR, then K_0 will increase, which will also decrease the outage probability. Equivalently, for the same outage probability and as lower target SINR are required, it is possible to support larger offered loads. Following up on this idea, it is interesting to compared how systems with and without feedback perform in terms of the outage probability as a function of network offered load. This comparison can be seen in Figures 6.26 and 6.27 for designs with no and with residual FER constraint, respectively. Since the outage probability depends on the target distortion through the target SINR in K_0 (6.70), we show performance for three different target distortion values 1.03, 1.05, and 1.1. Clearly, given an outage probability, in all cases the system with feedback can support a larger offered load than the system with no feedback. In general, the limit on the offered load is defined in terms of some small target outage probability, typically 1% or 2%. If the outage probability goal is 2% then, for the design with no FER constraint, the increase in supported offered load is 17%, 22%, and 22%, for target distortion 1.03, 1.05, and 1.1, respectively. For the same outage probability goal, in the case of design with FER constraint, the increase in supported offered load is 17%, 29%, and 49% for target distortion 1.03, 1.05, and 1.1, respectively.

References

1 Wicker, S. (1995). *Error Control Systems for Digital Communication and Storage*. Prentice Hall.

2 Ferguson, T.S. (1967). *Mathematical Statistics: A Decision Theoretic Approach*. New York: Academic Press.

3 Baum, C.W. and Veeravalli, V.V. (1994). A sequential procedure for multihypothesis testing. *IEEE Transactions on Information Theory* 40 (6): 1994–2007.

4 Farvardin, N. (1990). A study of vector quantization for noisy channels. *IEEE Transactions on Information Theory* 36 (4): 799–809.

5 Farvardin, N. and Vaishampayan, V. (1991). On the performance and complexity of channel optimized vector quantizers. *IEEE Transactions on Information Theory* 37 (1): 155–160.

6 Gersho, A. and Gray, R.M. (1991). *Vector Quantization and Signal Compression*. Kluwer Academic.

7 Chou, P.A., Lookabaugh, T.A., and Gray, R.M. (1989). Optimal pruning with applications to tree-structured source coding and modeling. *IEEE Transactions on Information Theory* 35 (2): 299–315.

8 Breiman, L., Friedman, J.H., Olshen, R.A., and Stone, C.J. (1984). *Classification and Regression Trees*. Belmont, CA: Wadsworth International Group.

9 Ghosh, M., Mukhopadhyay, N., and Sen, P.K. (1997). *Sequential Estimation*. Wiley.

10 Kwasinski, A., Chande, V., and Farvardin, N. (2003). Delay-constrained joint source-channel coding using incremental redundancy with feedback. *Proceedings 2003 IEEE Information Theory Workshop*, pp. 283–286.

11 Bertsekas, D. (1995). *Dynamic Programming and Optimal Control*. Athena Scientific.

12 Kwasinski, A. and Farvardin, N. (2003). Resource allocation for CDMA networks based on real-time source rate adaptation. *IEEE International Conference on Communications, 2003. ICC '03*, Volume 5 (11–15 2003), pp. 3307–3311.

13 ETSI/GSM (1998). Digital cellular telecommunications system (Phase 2+); Adaptive Multi-Rate (AMR) speech transcoding (GSM 06.90 version 7.2.1). *Document ETSI EN 301 704 V7.2.1 (2000-04)*.

14 DARPA TIMIT (1990). Acoustic-phonetic continuous speech corpus CD-ROM. *Document NISTIR 4930, NIST Speech Disk 1-1.1*.

15 Hagenauer, J. (1988). Rate compatible punctured convolutional (RCPC) codes and their applications. *IEEE Transactions on Communications* 36 (4): 389–399.

16 Bernard, A., Liu, X., Wesel, R.D., and Alwan, A. (2002). Speech transmission using rate-compatible trellis codes and embedded source coding. *IEEE Transactions on Communications* 50 (2): 309–320.

17 Viterbi, A.J. (1995). *CDMA, Principles of Spread Spectrum Communications, Addison-Wesley Wireless Communications Series*. Addison-Wesley.

18 Sampath, A., Mandayam, N.B., and Holtzman, J.M. (1997). Erlang capacity of a power controlled integrated voice and data CDMA system. *IEEE 47th Vehicular Technology Conference*, Volume 3, pp. 1557–1561.

7

Quantizers Designed for Noisy Channels

In this chapter we will study a joint source–channel coding (JSCC) technique that evolved from the design of source encoders. In Chapter 2, we reviewed source coding techniques that represent samples from a source using a number of bits for each sample. This process often involves compressing the source so that its representation matches the number of bits to what is feasible on the communication channel. A concern is the distortion introduced in the source representation when limiting the number of bits per sample. While source encoding is usually motivated by channel capacity, the study of basic source coders does not consider the effects of channel errors. This view is changed in this chapter, as the effects of errors and distortion introduced by the channel are included as an integral part of the design. The source coder design will not only consider the representation of the source but also how this representation interacts with the channel to result in an overall distortion.

There are two main approaches for designing quantizers for noisy channels. The first approach, *channel-optimized quantization*, includes the effects of channel errors in the distortion measure. The second approach involves designing a standard quantizer, but then considering channel effects when deciding on the coding of the resulting indexes. Both approaches will be found in this chapter, and many designs combine both.

7.1 Channel-Optimized Quantizers

Channel-optimized quantizers are designed by including both the distortion from the source representation and the distortion from channel errors in an overall distortion model used to optimize the quantizer design. This area was pioneered in [1], where the following key ideas were developed. Figure 7.1 shows a simplified block diagram with the main components in the communication system for which a channel-optimized quantizer will be used. This diagram is slightly more general than the problem discussed in this section, as the encoder function $f(\cdot)$ is not considered in designing a channel-optimized quantizer, but it will be addressed later in this chapter. For the design of a channel-optimized quantizer, consider a continuous random variable X that characterizes the samples from a continuous memoryless source. A scalar quantizer maps each sample x from the source into a quantization level y chosen from the set $\{q_j\}_{j=1}^M$. This quantizer will be designed considering the effects of channel errors. The quantizer does not need to be uniform, so it is necessary to define the thresholds that mark the range of values from X that map into a

Joint Source-Channel Coding, First Edition. Andres Kwasinski and Vinay Chande.

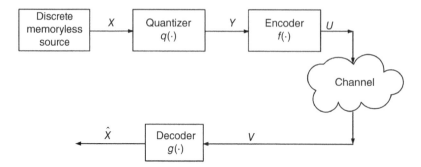

Figure 7.1 Block diagram of the system to study scalar quantizer design for a noisy channel.

quantization levels q_j. These quantization levels are the reconstruction levels assigned to the decoder output, $\hat{X}$. Let $\{T_j\}_{j=1}^{M+1}$ be the set of *quantization thresholds* which are defined such that $y = q_j$ when $T_j < x \leq T_{j+1}$ (where typically $T_1 = -\infty$ and $T_{M+1} = \infty$). Then the stochastic properties of the system from the quantizer assignment Y to the decoder output $\hat{X}$ are characterized by the channel transition matrix $\mathbf{P}$ with elements:

$$\mathbf{P}_{jk} = \mathbf{P}[\hat{X} = q_k | Y = q_j].$$

These probabilities encompass the effects of all system components between the quantizer and its decoder. This includes the communication channel but may include other components too, such as a channel encoder and decoder for error control. In this case, the transition matrix $\mathbf{P}$ would need to include the effects from error correction, but the formulation itself need not be changed.

The system performance metric used to optimize the design is the mean squared error (MSE):

$$\epsilon^2 = E[(X - \hat{X})^2], \tag{7.1}$$

which can be expanded into a quantization error component, $\epsilon_q^2 = E[(X - Y)^2]$ and a channel error component, $\epsilon_c^2 = E[(Y - \hat{X})^2]$, [2],

$$\epsilon^2 = \epsilon_q^2 + \epsilon_c^2 + \epsilon_m. \tag{7.2}$$

Note that ϵ^2 is just the end-to-end distortion, and we already discussed in Section 4.1 its split into different components. In particular, (7.2) is the same as (4.3), only with a more compact notation where $\epsilon_m = 2E[(X - Y)(Y - \hat{X})]$. This expression is instructive in exposing the error components, but to find an expression in terms of the quantizer parameters and the channel transition matrix, it is better to revert to the MSE definition (7.1):

$$\epsilon^2 = E[(X - \hat{X})^2]$$

$$= E[X^2] + E[\hat{X}^2] - 2E[X\hat{X}]$$

$$= \int_{-\infty}^{\infty} x^2 f_X(x)dx + \sum_{k=1}^{M} q_k^2 \sum_{j=1}^{M} P[\hat{X} = q_k | Y = q_j] \int_{T_j}^{T_{j+1}} f_X(x)dx +$$

$$- 2\sum_{k=1}^{M} q_k \sum_{j=1}^{M} P[\hat{X} = q_k | Y = q_j] \int_{T_j}^{T_{j+1}} x f_X(x)dx, \tag{7.3}$$

where $f_X(x)$ is the probability density function (pdf) for the source sample random variable X. It is interesting to consider how (7.3) would change when designing a quantizer without channel considerations. This can be done by re-deriving $\epsilon^2 = E[(X - \hat{X})^2]$ considering only source quantization effects, or, using the ongoing derivation, eliminating channel error effects by setting **P** to be the identity matrix: $P_{jk} = P[\hat{X} = q_k | Y = q_j] = 1$ if $j = k$ and $P_{jk} = 0$ if $j \neq k$. In this case, we get the MSE $\tilde{\epsilon}^2$:

$$\tilde{\epsilon}^2 = \int_{-\infty}^{\infty} x^2 f_X(x)dx + \sum_{j=1}^{M} q_j^2 \int_{T_j}^{T_{j+1}} f_X(x)dx - 2\sum_{j=1}^{M} q_j \int_{T_j}^{T_{j+1}} x f_X(x)dx. \tag{7.4}$$

Comparing (7.3) and (7.4), the idea behind channel-optimized quantizer design becomes clear: design an optimal quantizer by including in the optimization metric (MSE in this case) the effects of both source quantization and channel errors. Then ϵ^2 in (7.3) depends on the source characteristics through the pdf $f_X(x)$, on the quantization design choices through the quantization levels q_j, the thresholds T_j and the number of quantization levels M, and also on the channel effects through the probabilities $P[\hat{X} = q_k | Y = q_j]$. Since (7.3) and (7.4) state two different optimization metrics, the resulting quantizers (levels q_j and thresholds T_j) would be different. If we design a quantizer ignoring the channel effects, we can then subtract (7.4) from (7.3) when using the same q_j and T_j and find the MSE difference that results from ignoring the channel effects:

$$\sum_{k=1}^{M} q_k^2 \sum_{\substack{j=1 \\ j \neq k}}^{M} P[\hat{X} = q_k | Y = q_j] \int_{T_j}^{T_{j+1}} f_X(x)dx +$$

$$-2\sum_{k=1}^{M} q_k \sum_{\substack{j=1 \\ j \neq k}}^{M} P[\hat{X} = q_k | Y = q_j] \int_{T_j}^{T_{j+1}} x f_X(x)dx. \tag{7.5}$$

As expected, this result shows that the MSE penalty from not considering channel errors increases with larger channel error probabilities (i.e. larger components $P[\hat{X} = q_k | Y = q_j]$, $j \neq k$).

Resuming the design of the channel-optimized quantizer, the task is to minimize ϵ^2 with respect to the variables that are the set of levels q_k and the thresholds T_j in (7.3). This can be accomplished by solving the system of equations:

$$\frac{\partial \epsilon^2}{\partial q_k} = 0, \quad k = 1, 2, \ldots, M,$$

$$\frac{\partial \epsilon^2}{\partial T_j} = 0, \quad j = 2, 3, \ldots, M.$$

This system of equations results in the following expressions to calculate q_k and T_j:

$$q_k = \frac{\sum_{j=1}^{M} P[\hat{X} = q_k | Y = q_j] \int_{T_j}^{T_{j+1}} x f_X(x)dx}{\sum_{j=1}^{M} P[\hat{X} = q_k | Y = q_j] \int_{T_j}^{T_{j+1}} f_X(x)dx}, \quad k = 1, 2, \ldots, M, \tag{7.6}$$

$$T_j = \frac{1}{2} \frac{\sum_{k=1}^{M} q_k^2 (P_{jk} - P_{(j-1)k})}{\sum_{k=1}^{M} q_k (P_{jk} - P_{(j-1)k})}, \quad j = 2, 3, \ldots, M. \tag{7.7}$$

The levels q_k depend on the values T_j (but not on other levels $q_j, j \neq k$). Similarly, the values T_j depend on the q_k (but not on other values $T_k, k \neq j$). Therefore, to calculate the values q_k and T_j, there are two possible approaches. One is to solve the system of equations given by (7.6) and (7.7) numerically, and the other is reminiscent of the iterative Lloyd algorithm used to design scalar quantizers (see Algorithm 2.1 in Chapter 2). In the latter case, during each iteration the values q_k are used to calculate updated values for T_j, which in turn are used to update the values q_k. The iterations continue until the updated results remain close enough to the previous values. The iterative solution may converge to a local rather than global optimum. This outcome may be managed by careful selection of initial conditions or by techniques for avoiding local optima (such as simulated annealing). One of the implementation difficulties that arises in the iterative algorithm with a noisy channel is that there is no guarantee that the algorithm will converge. The condition for convergence requires that at each iteration the MSE is reduced from the previous iteration. Mathematically, this means that

$$\frac{\partial^2 \epsilon^2}{\partial T_j^2} > 0,$$

$$\Rightarrow 2 f_X(T_j)(E[\hat{X}|T_j] - E[\hat{X}|T_{j-1}]) > 0,$$

and, thus, the conditions $E[\hat{X}|T_j] > E[\hat{X}|T_{j-1}]$ are sufficient for convergence. Also, when solving (7.6) and (7.7), it is necessary to check that the result satisfies the *realizability condition*:

$$T_{j-1} < T_j, \quad j = 2, 3, \ldots, M + 1, \tag{7.8}$$

which should be considered as a constraint in any solution algorithm.

When solving (7.6) and (7.7), it is important to consider properties and symmetries that may exist in the channel matrix, as they may help simplify the solution and ensure the convergence and feasibility of the result.

To compare this quantizer design with a non-channel-optimized one, we can again make $P_{jk} = P[\hat{X} = q_k | Y = q_j] = 1$ if $j = k$ and $P_{jk} = P[\hat{X} = q_k | Y = q_j] = 0$ if $j \neq k$, in which case (7.6) and (7.7) become

$$q_k = \frac{\int_{T_j}^{T_{j+1}} x f_X(x) dx}{\int_{T_j}^{T_{j+1}} f_X(x) dx}, \quad k = 1, 2, \ldots, M, \tag{7.9}$$

$$T_j = \frac{1}{2}(q_j + q_{j-1}). \quad j = 2, 3, \ldots, M. \tag{7.10}$$

These results are the same as those reviewed in Chapter 2 resulting from the optimal quantizer design, when ignoring channel effects (see (2.3) and (2.4)). Consequently, the observations mentioned at that time can now be directly extended to the case of a channel-optimized quatizer design. Specifically, the solution (7.6) could be seen as the quantization levels being still the "center of mass," or centroid, of each quantization interval, for a "mass" distributed according to a mass density given by the pdf of the source, but now the centroid positions are affected by another interacting "field" given by the channel effect probabilities.

7.2 Scalar Quantizer Design

In the previous section, we discussed the basic concept of channel-matched quantizer design as presented in [1]. This work was followed by refinements, including a more in-depth study of conditions necessary for an optimum design and for the convergence of the usually iterative design algorithm. Also, it was observed that when transmitting over a noisy channel, it was not always the case that having higher quantization resolution resulted in less overall distortion. This was contrary to the usual behavior for non-channel-optimized channel quantizers in isolation, where more quantization levels resulted in less reconstruction distortion. In the design of quantizers for noisy channels, one may then ask how many quantization levels are best. It is also important to integrate in the design the best way of assigning codewords to quantization interval indexes so as to further reduce the effect of channel errors. We will study all these design problems. In this section, we will consider scalar quantizer design, saving vector quantizers for the following sections.

We start by following the work in [3]. Now the design will include all blocks in the diagram of Figure 7.1, including the encoder function $f(\cdot)$. The process starts with the output of a discrete-time memoryless source. Each sample at the output of the source follows the statistics of a random variable X, with probability density function $f_X(x)$, zero mean (always possible to achieve with simple analog circuitry in the source sampler), and variance $E[X^2] = \sigma_X^2 < \infty$. The source samples are input into a scalar quantizer, which maps each sample $x \in \mathbb{R}$ into a quantization level $y = q(x)$ chosen from M possible values $\{q_j\}_{j=1}^M$, each $q_j \in \mathbb{R}$. The mapping is defined according to a set of thresholds $\{T_j\}_{j=1}^{M+1}$ (where, as before, typically $T_1 = -\infty$ and $T_{M+1} = \infty$) that partition the support of X into M regions $A_j = (T_j, T_{j+1}], j = 1, 2, \ldots, M$, such that $y = q(x) = q_j$ when $T_j < x \le T_{j+1}$ or, equivalently when $x \in A_j$.

The quantizer output is subsequently input to a channel encoder. Its operation is determined by a mapping function $f(\cdot)$ from the M quantization levels q_j onto codewords formed from r bits, where $2^r \ge M$. Following all the processing of the random source samples, the codewords at the output of the channel encoder are also a random variable, denoted U. The codewords from the channel encoder are transmitted to the receiver through a channel modeled as a binary symmetric channel (BSC) with crossover probability p_e. The channel output is a random variable denoted V. The combined effect of the BSC on the r-bit transmitted codewords can be characterized by a channel transition matrix $\mathbf{P}$ with elements:

$$P_{m,k} = P[V = v_k | U = u_m], \quad m, k = 1, 2, \ldots, M.$$

These probabilities depend on the BSC crossover probability p_e through the relation:

$$P_{m,k} = p_e^{d_H}(1 - p_e)^{(r-d_H)}, \tag{7.11}$$

where d_H is the Hamming distance between the realizations u_m and v_k of the random variables U and V, respectively. The transmitted codeword u_m results from mapping a quantization level q_j to $u_m = f(q_j)$, where q_j is related to a quantization interval $A_j = (T_j, T_{j+1}]$. With this, and since in general $2^r \geq M$, we have that the subscript m is in fact a function of j. This observation will be used shortly in the notation. For simplicity of notation, we also introduce $N = 2^r$.

At the receiver, the sample v_j from the channel is input to a decoder where it is mapped to a reconstruction level of the transmitted source given by the relation $\hat{X} = g(V)$. Here, the possible reconstruction levels are given by the set $\{\gamma_j\}_{j=1}^N$ (it is not required for N to equal M). Measuring distortion through MSE, the end-to-end distortion for this system is given by:

$$\epsilon^2 = E[(X - \hat{X})^2] = \sum_{k=1}^{N} \sum_{j=1}^{M} P_{m(j),k} \int_{T_j}^{T_{j+1}} (x - \hat{x}_k)^2 f_X(x) dx. \tag{7.12}$$

Similar to the case in the previous section, designing to minimize the MSE will mean finding the optimum set of quantization thresholds and reconstruction levels and, new to this section, will also mean finding the optimum encoding function f. The introduction of the encoding function into the system design creates a number of differences from the channel-matched quantizer studied previously. One is that the number of reconstruction levels need not be the same as the number of quantization levels. Another difference, of course, is how to incorporate the design of the channel encoder into the search for the optimum quantization levels and thresholds.

In the previous section, we mentioned that the effects from the error correction mechanism can be incorporated in the design by including them in the matrix transition probability **P**. For example, a model of a BSC may use as crossover probability the post-channel decoding bit error rate (BER). Yet a true JSCC scheme would be designed differently. We will next explain this JSCC design approach.

The minimization of the distortion in (7.12) involves finding optimum mappings $q(\cdot), f(\cdot)$, and $g(\cdot)$. A joint source–channel design implies that instead of designing separate mappings $q(\cdot)$ and $f(\cdot)$, the design will seek an optimum joint mapping given by the function $\Psi(x) = f[q(x)] = u$. Following an overall approach analogous to the one described in the previous section, the joint source–channel encoding function $\Psi(x)$ is designed assuming that the decoder function $g(v)$ is fixed, and then the decoder function is designed assuming that the joint source–channel encoding function is fixed. This approach to the optimization of communication system components at the transmitter and receiver was originally discussed in [4].

Assuming that the decoder function $g(v)$ is fixed, the design of the optimum encoder follows a path similar to the one described previously. Yet we briefly summarize to introduce a formulation relevant to the broader problem. Finding the optimum encoder means finding the optimum partitions for the support of X that minimizes the MSE ϵ^2. These partitions are defined by the condition:

$$A_i(g) = \left\{ x : E[(x - \hat{X})^2 | U = u_i] \leq E[(x - \hat{X})^2 | U = u_j], \forall j \neq i \right\}. \tag{7.13}$$

In the notation used to name the partition, the dependence on the choice for the decoder function is now explicit. To see where this definition comes from, consider first writing the

MSE using the law of total expectation:

$$\epsilon^2 = E[(X - \hat{X})^2]$$
$$= \int_{-\infty}^{\infty} E[(X - \hat{X})^2 | X = x] \, f_X(x) dx.$$

Since $f_X(x)$ is nonnegative, minimizing the MSE requires minimizing just the expectation $E[(X - \hat{X})^2 | X = x]$. Since X and U are related through a functional mapping, conditioning the expectation on $X = x$ is equivalent to conditioning it on $U = u$ and, thus, $E[(X - \hat{X})^2 | X = x] = E[(X - \hat{X})^2 | U = u]$, which leads to the definition (7.13).

Now, the expression $E[(X - \hat{X})^2 | U = u]$ can be expanded as: $E[X^2 | U = u] + E[\hat{X}^2 | U = u] - 2E[X\hat{X} | U = u]$. Since X does not depend on U and $E[X^2] = \sigma_X^2$, the expression further becomes $\sigma_X^2 + E[\hat{X}^2 | U = u] - 2E[X\hat{X} | U = u]$. Now the partition definition (7.13) can be written as:

$$A_i(g) = \left\{ x : E[\hat{X}^2 | U = u_i] - 2xE[\hat{X} | U = u_i] \le E[\hat{X}^2 | U = u_j] - 2xE[\hat{X} | U = u_j], \; \forall j \ne i \right\}.$$

It follows that the partitions can be defined as the intersection of segments on the real line, $A_{ij}(g)$:

$$A_i(g) = \bigcap_{\substack{j=1 \\ j \ne i}}^{N} A_{ij}(g), \tag{7.14}$$

where

$$A_{ij}(g) = \left\{ x : 2x \left(E[\hat{X} | U = u_j] - E[\hat{X} | U = u_i] \right) \le E[\hat{X}^2 | U = u_j] - E[\hat{X}^2 | U = u_i] \right\}. \tag{7.15}$$

Furthermore, by defining

$$\Upsilon_{(2)ij} = E[\hat{X}^2 | U = u_j] - E[\hat{X}^2 | U = u_i],$$
$$\Upsilon_{(1)ij} = 2E[\hat{X} | U = u_j] - 2E[\hat{X} | U = u_i],$$

the segments $A_{ij}(g)$ can be written using an affine inequality on x as:

$$A_{ij}(g) = \left\{ x : \Upsilon_{(1)ij} x - \Upsilon_{(2)ij} \le 0 \right\}. \tag{7.16}$$

From here, the thresholds demarcating the limits of the segments can be written as:

$$T_{ij} = \frac{\Upsilon_{(2)ij}}{\Upsilon_{(1)ij}},$$

for $\Upsilon_{(1)ij} \ne 0$, and the segments themselves are as follows:

$$A_{ij}(g) = \begin{cases} (-\infty, T_{ij}], & \Upsilon_{(1)ij} > 0, \\ [T_{ij}, \infty), & \Upsilon_{(1)ij} < 0, \\ (-\infty, \infty), & \Upsilon_{(1)ij} = 0, \Upsilon_{(2)ij} \le 0, \\ \emptyset, & \Upsilon_{(1)ij} = 0, \Upsilon_{(2)ij} < 0. \end{cases}$$

From here, (7.14) results in

$$
A_i(g) = \begin{cases} \emptyset, & \Upsilon_{(1)ij} = 0, \Upsilon_{(2)ij} < 0, \text{for some } j, \\ \mathbb{R}, & \Upsilon_{(1)ij} = 0, \Upsilon_{(2)ij} \leq 0, \text{for all } j, \\ (T_i^l, T_i^u], & \text{otherwise}, \end{cases}
$$

where

$$
T_i^u = \min_{j:\Upsilon_{(1)ij}>0} \{T_{ij}\}, \tag{7.17}
$$

and

$$
T_i^l = \max_{j:\Upsilon_{(1)ij}<0} \{T_{ij}\}. \tag{7.18}
$$

In summary, the definition for the optimum design of the encoding function, given a decoder function $g(\cdot)$ is

$$
\Psi(x) = u_i, \text{for } x \in A_i(g), \text{and } i = 1, 2, \ldots, N. \tag{7.19}
$$

It is important to note here than when the channel is very noisy, a lower distortion may be achieved through a more robust quantizer that uses a small number of quantization intervals. In these cases, the best design could be achieved using a smaller number of intervals than the number of available codewords N. In terms of the design equations shown earlier, this situation appears as a case where $\Upsilon_{(1)ij} \neq 0$ and $T_i^u < T_i^l$. Of course, implied in the aforementioned formulation is a realizability constraint that $T_i^u > T_i^l$ (see the definitions (7.17), as a limit on $\Upsilon_{(1)ij} > 0$ approaching from above, and (7.18), as a limit on $\Upsilon_{(1)ij} < 0$ approaching from below). Therefore, in the cases where $\Upsilon_{(1)ij} \neq 0$ and $T_i^u < T_i^l$ (that violates the realizability constraint), the design decision shall be not to map any value of x into a u_i. In other words, for an i such that $\Upsilon_{(1)ij} \neq 0$ and $T_i^u < T_i^l$, $A_i(g) = \emptyset$.

We now turn our attention to the design of the decoding function $g(\cdot)$. The process is analogous to the design of the encoder, only now we assume that that encoding function $\Psi(x)$ is fixed and the goal is to find the best mapping $\hat{X} = g(V)$. The optimum decoding mapping equals the conditional expectation of the input given the output of the channel:

$$
g(v_i) = E[X|V = v_i], \tag{7.20}
$$

for $i = 1, 2, \ldots, N$.

Having established the conditions for the optimal encoder and decoder, we now describe algorithms to find them. Algorithm 7.1 is given in [3] to determine the optimal encoder for a given decoder.

Note that the algorithm denotes the codewords as u_i' and not u_i. This notation refers to the same codewords but organized in a different order. Recall that the codewords u_i are formed from the mapping function $f(\cdot)$ (now embedded in the joint source–channel encoding function $\Psi(\cdot)$) applied to the M quantization levels to form r-bit codewords, where $N = 2^r \geq M$. The aforementioned algorithm is based on returning the codewords reordered in such a way that

$$
E[\hat{X}|U = u_1'] \leq E[\hat{X}|U = u_2'] \leq \cdots \leq E[\hat{X}|U = u_M'], \tag{7.21}
$$

where the prime used to denote the codeword u_i' refers to the new sorting order. This reordering of the codewords is key for developing the encoder because it simplifies the

---

Algorithm 7.1 Algorithm to determine the optimal scalar encoder for a given decoder

---

1: Set $i = 1$.

2: Compute T_i^u. Let $j(i)$ be such that $T_i^u = T_{ij(i)}$. Eliminate allcodewords $u'_{i+1}, u'_{i+2}, \ldots,$ $u'_{j(i)-1}$. Set $T_{j(i)}^l = T_i^u$.

3: Set $i = j(i)$.

4: **if** $i < N$ **then**

5:    GO TO Step 2.

6: **else**

7:    Set $T_i^u = \infty$

8:    STOP.

9: **end if**

---

algorithm to the level of simply partitioning the real line following the same sequential order in (7.21). Formally, it was shown in [3] that for a given function $g(\cdot)$, when the codewords are sorted following (7.21), for $i < j$ and $T_k^u > T_k^l$, $k = i, j$, the segment $A_i(g)$ lies to the left of the segment $A_j(g)$. The proof for this property also proves the converse implication, where obtaining an optimal encoding mapping for which the segment $A_i(g)$ lies to the left of the segment $A_j(g)$ implies that $E[\hat{X}|U = u'_i] \le E[\hat{X}|U = u'_j]$. This property provides the means for assigning codewords to the best partitions derived from the encoding function (7.19). The aforementioned algorithm further relies on a theorem that links together the condition for realizability, the ordered partitioning of the real line on the support for X, and the codeword assignment. This theorem, studied in [3], states that when the codewords are ordered to follow (7.21), given the well-defined ith partition (this is $T_i^u > T_i^l$), and supposing that $T_i^u = T_{ij}$ for some $j = i+1, i+2, \ldots, N$, then the following three statements are true:

- If $j > i + 1$, all codewords $u'_{i+1}, u'_{i+2}, \ldots, u'_{j-1}$ are not well defined, that is, $T_k^u < T_k^l$ for $k = i+1, i+2, \ldots, j-1$.
- The jth codeword is well defined.
- $T_i^u = T_j^l$.

These relations are illustrated in Figure 7.2a. Figure 7.2b summarizes the aforementioned algorithm for encoder design. In the illustration, it can be seen how the algorithm's iterations successively find the codewords, sorted as in (7.21), by first determining the upper threshold for the corresponding partitioning segment, then eliminating not well-defined codewords, and finally initializing the lower threshold for the segment associated with the following codeword.

Figure 7.3 depicts a flowchart that combines the presentation in this section into the design of the encoder and decoder for a range $0 \le p_e \le p_{e_{max}}$ of the BSC's crossover probability. The algorithm calculates the optimum encoder and decoder for the range of p_e values, computing the solution as p_e is decreased from $p_{e_{max}}$ in constant steps $\Delta p_e > 0$, and also obtains the resulting MSE curve $\epsilon^2(p_e)$. Because the solution may correspond to local minima, the algorithm performs successive design passes, alternating the calculations when p_e increases from 0 and decreases from $p_{e_{max}}$. The design procedure ends when curves $\epsilon^2(p_e)$ resulting from two successive passes differ by a small enough difference, in which case it is considered that the solution is at least a good local optimum solution.

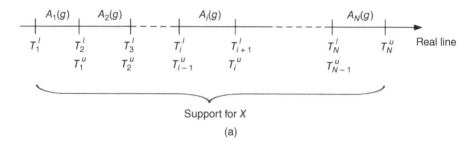

Support for X

(a)

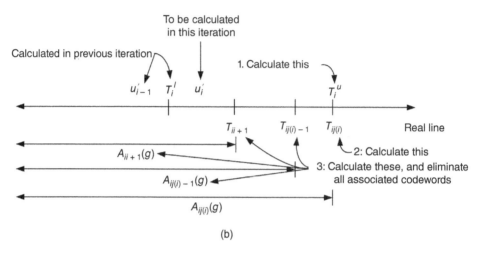

(b)

Figure 7.2 (a) Sorted segments for the encoding partition and their upper and lower limits. (b) Geometry of the encoder design.

The performance of this joint source–channel encoder/decoder design for a noisy channel was evaluated in [3] by measuring the attainable end-to-end signal-to-noise (SNR) ratio (measured in decibels). The input source (X) statistics were assumed to follow a zero-mean unit-variance Gaussian, Laplacian, or uniform distribution. The channel was a BSC with possible crossover probabilities, p_e, equal to 0.005, 0.01, 0.05, or 0.1. The SNR performance with this noisy channel design was compared against that obtained with a system following the optimum performance theoretically attainable (OPTA) and against a system designed to minimize the error introduced in the quantizer but ignoring the effects of the channel.

Recall that the OPTA system is one where source coding follows the limits given by information theory's rate-distortion and channel capacity theorems. In the present system, the channel capacity is given by $C = 1 + p_e\log_2 p_e + (1 - p_e)\log_2(1 - p_e)$ bits per channel use. Consequently, the minimum attainable distortion is $D_m = D(rC)$ where r is the number of channel uses per symbol and $D(\cdot)$ is the distortion-rate function, which depends on the statistics for the source.

The system that minimizes the error introduced in the quantizer but ignores channel effects uses a Lloyd-Max quantizer. This design calculates the quantization values using (2.3). Following quantization, the output is encoded using the natural binary code or the

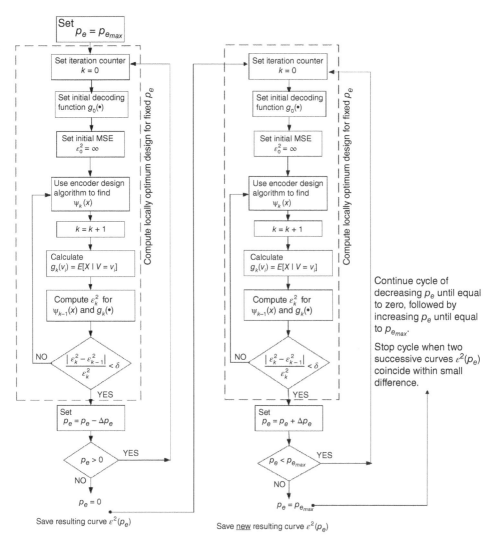

Figure 7.3 Encoder and decoder design for a range $0 \le p_e \le p_{e_{max}}$ of the crossover probability for the binary symmetric channel.

folded binary code, which are shown in Table 7.1 along with the Gray code. Both the natural binary code and the folded binary code were also used in the joint source–channel encoder/decoder design for a noisy channel.

The comparison results are shown in Tables 7.2–7.4 for the Gaussian-distributed, Laplacian-distributed, and uniformly distributed sources, respectively. In the tables, DNCNBC and DNCFBC refer to the joint source–channel encoder/decoder design for a noisy channel using the natural and folded binary codes, while LMNBC and LMFBC refer to the Lloyd-Max design optimizing the quantizer only and using the natural and folded binary codes. OPTA refers to the OPTA result.

Table 7.1 Four-bit natural binary code (NBC), folded binary code (FBC), and Gray code (GC).

NBC	FBC	GC
0000	0000	0000
0001	0001	0001
0010	0010	0011
0011	0011	0010
0100	0100	0110
0101	0101	0111
0110	0110	0101
0111	0111	0100
1000	1111	1100
1001	1110	1101
1010	1101	1111
1011	1100	1110
1100	1011	1010
1101	1010	1011
1110	1001	1001
1111	1000	1000

The results show that the joint source–channel encoder/decoder design for a noisy channel has higher SNR than the design that ignores the channel. The performance advantage is more evident when the channel is worse. The magnitude of performance improvement depends on the statistics of the source. Also, since the design for a noisy channel is a joint source–channel design, there is little difference between using the natural binary code or the folded binary code. This is not always the case for the design that ignores the channel when optimizing the quantizer alone. The OPTA system shows the best results, which can only be achieved under ideal asymptotic settings.

7.3 Vector Quantizer Design

Vector quantization (VQ) is a popular technique for the representation of a source because in general it can achieve smaller bit rates than scalar quantization, for the same level of distortion. However, this more compact representation with less redundancy tends to be more sensitive to channel impairments. On an end-to-end comparison between vector and scalar quantizers, while the vector quantizer may present lower quantization distortion, it will suffer more channel-induced distortion.

We first consider the study of VQ for noisy channels presented in [5]. Here, the source to be encoded is assumed to be a real-valued, discrete-time, continuous-amplitude, stationary process $\{X_n, n = 0, 1, \ldots\}$ with zero mean and variance σ_X^2. The source is encoded using

Table 7.2 Scalar quantizer performance results (SNR in dB) for a zero-mean unit-variance Gaussian-distributed source.

Bit/sample	Design	$p_e = 0.005$	$p_e = 0.01$	$p_e = 0.05$	$p_e = 0.1$
1	DNCNBC	4.25	4.10	3.15	2.27
1	DNCFBC	4.25	4.11	3.15	2.27
1	LMNBC	4.25	4.10	3.09	2.09
1	LMFBC	4.25	4.10	3.09	2.09
1	OPTA	5.75	5.54	4.31	3.22
2	DNCNBC	8.50	7.88	5.20	3.63
2	DNCFBC	8.52	7.88	5.20	3.63
2	LMNBC	8.47	7.77	4.41	2.22
2	LMFBC	8.52	7.85	4.58	2.41
2	OPTA	11.49	11.07	8.59	6.40
3	DNCNBC	11.88	10.49	6.47	4.67
3	DNCFBC	12.04	10.50	6.47	4.67
3	LMNBC	11.61	9.85	4.28	1.56
3	LMFBC	11.99	10.36	4.93	2.18
3	OPTA	17.24	16.60	12.89	9.59
4	DNCNBC	14.15	12.30	7.81	5.60
4	DNCFBC	14.14	12.30	7.81	5.60
4	LMNBC	12.92	10.33	3.79	0.93
4	LMFBC	13.84	11.36	4.84	1.89
4	OPTA	23.00	22.14	17.19	12.79

a k-dimensional vector quantizer. Recall from Chapter 2 that a k-dimensional VQ takes a block of k samples from a source, interprets it as a vector $\mathbf{x} = [x_1, x_2, \dots, x_k]$, and maps it into a code-vector $\mathbf{y} = q(\mathbf{x})$. This code-vector represents a volume in the k-dimensional space that includes the vector $\mathbf{x}$. The design of the VQ involves finding a partition of the k-dimensional space into a finite number of volumes. Because the number of volumes is finite, the set of all code-vectors has a finite number M of elements. This set is called the codebook $\mathfrak{C} = \{\mathbf{c}_1, \mathbf{c}_2, \dots, \mathbf{c}_M\}$. While the output of the VQ is a $\log_2(M)$-bit word, the code-vectors are also the reconstruction values used at the receiver. Consequently, a vector code will have a rate

$$R = \frac{\log_2(M)}{k} \text{ (bits/sample)},$$

and will be completely described by specifying the codebook $\mathfrak{C} = \{\mathbf{c}_1, \mathbf{c}_2, \dots, \mathbf{c}_M\}$ and the set of limits for each of the volumes partitioning the k-dimensional space,

Table 7.3 Scalar quantizer performance results (SNR in dB) for a zero-mean unit-variance Laplacian-distributed source.

Bit/sample	Design	$p_e = 0.005$	$p_e = 0.01$	$p_e = 0.05$	$p_e = 0.1$
1	DNCNBC	2.93	2.84	2.25	1.67
1	DNCFBC	2.93	2.84	2.25	1.67
1	LMNBC	2.92	2.84	2.22	1.55
1	LMFBC	2.92	2.84	2.22	1.55
1	OPTA	6.42	6.21	4.89	3.72
2	DNCNBC	6.81	6.26	4.05	2.80
2	DNCFBC	6.94	6.44	4.05	2.80
2	LMNBC	6.75	6.08	2.84	0.73
2	LMFBC	6.93	6.40	3.56	1.52
2	OPTA	12.13	11.70	9.23	7.05
3	DNCNBC	9.94	8.68	5.21	3.62
3	DNCFBC	10.49	9.17	5.18	3.62
3	LMNBC	9.30	7.44	1.76	−0.93
3	LMFBC	10.36	8.87	3.61	0.82
3	OPTA	17.87	17.23	13.52	10.23
4	DNCNBC	12.17	10.49	6.32	4.45
4	DNCFBC	12.76	11.03	6.82	4.79
4	LMNBC	9.78	7.11	0.52	−2.29
4	LMFBC	12.13	9.69	3.13	0.07
4	OPTA	23.65	22.77	17.82	13.42

$P = \{A_1, A_2, \ldots, A_M\}$, where the mapping $q(\cdot)$, a source sample vector, $\mathbf{x}$, a code-vector $\mathbf{c}_i$, and the corresponding partitioning volume are related through the expression:

$$q(\mathbf{x}) = \mathbf{c}_i, \text{ if } \mathbf{x} \in A_i, \quad i = 1, 2, \ldots, M.$$

Since a code-vector $\mathbf{c}_i$ is the reconstruction value for all vectors spanning the volume A_i, the average quantizer distortion per sample is

$$D_S(q) = \frac{1}{k} \sum_{i=1}^{M} \int_{A_i} f_{\mathbf{X}}(\mathbf{X}) d(\mathbf{x}, \mathbf{c}_i) d\mathbf{x}, \tag{7.22}$$

where $f_{\mathbf{X}}(\mathbf{X})$ is the joint pdf for the k random variables $x_1, x_2, \ldots, x_k$.

As mentioned earlier, and as is usually the case in digital communications, while the output of the mapping $q(\cdot)$ is a code-vector, the value transmitted is a word that identifies the code-vector. In other words, before transmission, the code-vectors are mapped into binary codewords with $\log_2 M = kR$ bits. These binary codewords take values $0, 1, \ldots, M - 1$ and are denoted $b(\mathbf{c}_i)$, with $i = 1, 2, \ldots, M$. In this case, $b(\mathbf{c}_i)$ represents the codeword and $b(\cdot)$ is the mapping relating code-vectors with codewords. The specific mapping $b(\cdot)$ is open

Table 7.4 Scalar quantizer performance results (SNR in dB) for a zero-mean unit-variance uniformly distributed source.

Bit/sample	Design	$p_e = 0.005$	$p_e = 0.01$	$p_e = 0.05$	$p_e = 0.1$
1	DNCNBC	5.77	5.53	4.06	2.84
1	DNCFBC	5.77	5.53	4.06	2.84
1	LMNBC	5.77	5.53	3.98	2.60
1	LMFBC	5.77	5.53	3.98	2.60
1	OPTA	6.44	6.29	4.91	3.60
2	DNCNBC	10.93	10.06	6.51	4.50
2	DNCFBC	10.92	10.06	6.51	4.50
2	LMNBC	10.90	10.04	6.02	3.59
2	LMFBC	10.71	9.69	5.47	3.03
2	OPTA	12.68	12.14	9.59	7.21
3	DNCNBC	14.67	13.00	8.12	5.87
3	DNCFBC	14.68	13.00	8.12	5.87
3	LMNBC	14.52	12.60	6.13	3.88
3	LMFBC	13.98	11.93	5.94	3.15
3	OPTA	18.60	17.97	14.05	10.69
4	DNCNBC	17.01	14.87	9.74	7.03
4	DNCFBC	17.02	14.86	9.74	7.03
4	LMNBC	16.23	13.60	6.92	3.95
4	LMFBC	15.42	12.72	6.07	3.18
4	OPTA	24.43	23.57	18.55	13.95

to several possible combinations and subject to a design choice. When transmitted, the binary codewords are sent through a channel that has as possible input and output any of the codewords $b(\mathbf{c}_i)$ and that will be assumed memoryless. The effects of channel errors will be modeled through the probabilities $P\left(b(\mathbf{c}_j)|b(\mathbf{c}_i)\right), j, i = 1, 2, \ldots, M$, that represent the probability that a codeword $b(\mathbf{c}_j)$ is received when the codeword $b(\mathbf{c}_i)$ was transmitted. For example, for a BSC with crossover probability p_e, the probabilities $P\left(b(\mathbf{c}_j)|b(\mathbf{c}_i)\right)$ follow an expression similar to Eq. (7.11) considering now that the binary codewords have kR bits:

$$P\left(b(\mathbf{c}_j)|b(\mathbf{c}_i)\right) = p_e^{d_H(b(\mathbf{c}_j),b(\mathbf{c}_i))}(1 - p_e)^{(kR - d_H(b(\mathbf{c}_j),b(\mathbf{c}_i)))}.\tag{7.23}$$

Then, for a given mapping $b(\cdot)$, the average distortion per source sample due to channel communication errors is

$$D_C(b) = \frac{1}{k}\sum_{i=1}^{M}\sum_{j=1}^{M}d(\mathbf{c}_i, \mathbf{c}_j)P\left(b(\mathbf{c}_j), b(\mathbf{c}_i)\right)$$

$$= \frac{1}{k}\sum_{i=1}^{M}\sum_{j=1}^{M}d(\mathbf{c}_i, \mathbf{c}_j)P\left(b(\mathbf{c}_j)|b(\mathbf{c}_i)\right)P(\mathbf{c}_i),\tag{7.24}$$

where $P\left(b(\mathbf{c}_j), b(\mathbf{c}_i)\right)$ is the joint probability distribution for codewords $b(\mathbf{c}_j)$ and $b(\mathbf{c}_i)$, and $P(\mathbf{c}_i) = P(b(\mathbf{c}_i))$ is the a priori probability for codeword $b(\mathbf{c}_i)$ (the equality here is because the mapping $b(\cdot)$ is one-to-one).

Considering the end-to-end distortion, and assuming that the mappings $q(\cdot)$ and $b(\cdot)$ have already been designed, the overall distortion from both the VQ and the channel errors is

$$D(q, b) = \frac{1}{k}\sum_{i=1}^{M}\sum_{j=1}^{M}\left(\int_{A_i} f_{\mathbf{X}}(\mathbf{X})d(\mathbf{x}, \mathbf{c}_i)d\mathbf{x}\right)P\left(b(\mathbf{c}_j)|b(\mathbf{c}_i)\right). \tag{7.25}$$

If the distortion measure being used is the squared error, i.e. $d(\mathbf{x}, \mathbf{y}) = \|\mathbf{x} - \mathbf{y}\|^2$, and each code-vector $\mathbf{c}_i$ is chosen equal to the centroid of its associated volume A_i, denoted as $\hat{\mathbf{c}}_i$, it is possible to show that the average end-to-end distortion per source sample can be decomposed into two terms:

$$D(q, b) = \frac{1}{k}\sum_{i=1}^{M}\int_{A_i} f_{\mathbf{X}}(\mathbf{X})\,\|\mathbf{x} - \hat{\mathbf{c}}_i\|^2 d\mathbf{x} + \frac{1}{k}\sum_{i=1}^{M}\sum_{j=1}^{M}\|\hat{\mathbf{c}}_i - \mathbf{c}_j\|^2 P\left(b(\mathbf{c}_j)|b(\mathbf{c}_i)\right)P(\mathbf{c}_i),$$

where the first term is the distortion introduced during source encoding (this can be seen by comparing the term with (7.22)) and the second term is the distortion introduced due to channel errors (this can be seen by comparing the term with (7.24)). Therefore, subject to the aforementioned conditions, the end-to-end distortion can be written as $D(q, b) = D_S(q) + D_C(b)$. This decomposition justifies a design approach that, given a quantization mapping $q(\cdot)$, focuses on minimizing the channel distortion by finding an appropriate index assignment function $b(\cdot)$.

A number of algorithms have been proposed to find the best index assignment function. In [6], the approach is to consider in an ordered way the effects on distortion of the different bit permutations in the indexes. Specifically, the development of the algorithm starts by making explicit in the average channel-induced distortion per source sample (7.24), the effects of exactly m bit errors:

$$D_C(b) = \frac{1}{k}\sum_{i=1}^{M}\sum_{m=1}^{kR}\sum_{j\in C_{i,m}}P\left(b(\mathbf{c}_j)|b(\mathbf{c}_i)\right)P(\mathbf{c}_i)d(\mathbf{c}_i, \mathbf{c}_j), \tag{7.26}$$

where $C_{i,m}$ is the set of codewords with index assignment that differ in m bits from the assignment for codeword $\mathbf{c}_i$. From this expression, it is possible to isolate the contribution to the average channel-induced distortion from a given codeword $\mathbf{c}_i$: $C_{\mathbf{c}_i}(b)$,

$$C_{\mathbf{c}_i}(b) = \sum_{m=1}^{kR}\sum_{j\in C_{i,m}}P\left(b(\mathbf{c}_j)|b(\mathbf{c}_i)\right)P(\mathbf{c}_i)d(\mathbf{c}_i, \mathbf{c}_j). \tag{7.27}$$

Consequently, the design problem becomes that of finding a permutation function $f(\cdot)$, that permutes the code-vectors in the codebook, so that the average channel-induced distortion per source sample is minimized:

$$\min_{f} D_C(b) = \frac{1}{k}\min_{f}\sum_{i=1}^{M}C_{\mathbf{c}_i}(b).$$

Algorithm 7.2 Optimal code-vector permutation function search algorithm

---

1: Choose an initial codebook.
2: Create a list with the code-vectors sorted by decreasing average channel-induced distortion from a given codeword $\mathbf{c}_i$, $C_{\mathbf{c}_i}(b)$.
3: Set index $i = -1$.
4: Increment counter $i = i + 1$.
5: **for** $j = i$, to M **do**
6:     Let G be the maximum reduction in average channel-induced distortion per source sample resulting after switching code-vectors $\mathbf{c}_i$ and $\mathbf{c}_j$ (with the indexes from the list) in the codebook.
7:     **if** $G > 0$ **then**
8:         switch $\mathbf{c}_i$ and $\mathbf{c}_j$ in the codebook.
9:         GOTO Step 2.
10:     **else**
11:         **if** $i \neq M$ **then**
12:             GOTO Step 4.
13:         **else**
14:             STOP
15:         **end if**
16:     **end if**
17: **end for**

The reason for searching for a permutation function that permutes the code-vectors in the codebook is that this operation is equivalent to a reassignment of indices between two codewords. Algorithm 7.2 is the one presented in [6] to find the permutation.

The idea behind Algorithm 7.2 is to search for the best switch between two code-vectors. The best switch is the one that reduces the average channel-induced distortion per source sample the most. The algorithm stops when it is not possible to find any switch resulting in a reduction of distortion for any codeword. The search is conducted one code-vector at a time, following an order of codewords from the one with largest associated average channel-induced distortion from a given codeword to the one with smallest such average distortion.

Algorithm 7.2 results in a function $b(\cdot)$ that yields a local minimum average channel-induced distortion per source sample. To address the issue of a locally, rather than globally, optimal solution, an algorithm based on simulated annealing is discussed in [5]. Simulated annealing belongs to a family of optimization algorithms that find solutions to NP-complete combinatorial optimization problems, such as the one at hand, by introducing randomization into the search. These algorithms allow, at times, temporary consideration of solutions that perform contrary to the optimization goal. This type of algorithm has been considered in many disciplines. For example, the same approach is followed in some learning algorithms where there are two phases. A randomized solution phase, where a solution contrary to the optimization goal may be temporarily accepted, is called an *exploration phase*, and a phase following standard optimization rules (orderly searching for an optimal solution) is called an *exploitation phase*.

Algorithm 7.3 shows the application of simulated annealing to find the near-optimal index assignment [5]. The algorithm starts by introducing an *effective temperature*, which represents the likelihood of taking a step against the optimization goal. A step against the optimization goal is the choice of a mapping $b(\cdot)$ that results in a larger average channel-induced distortion per source sample than the one attained with the previous choice of mapping $b(\cdot)$. When the effective temperature is high, the probability of taking a step against the optimization goal is also high. This probability decreases with temperature following an appropriately chosen function ($\exp(-\Delta D_C/T)$ for Algorithm 7.3). Consequently, in the initial iterations of the algorithm, it will be quite likely to take steps against the optimization goal. As the algorithm progresses, the effective temperature is gradually reduced, making it less and less likely to take steps against the optimization goal. The reason for allowing steps against the optimization goal is to allow the algorithm to "backtrack" from a sequence of optimization decisions that eventually would lead to a local optimum and explore other sequences of optimization solutions. The flexibility of the algorithm to explore different optimization sequences is high at the beginning and very low toward the end of the iterations. The effective temperature is also used to control when the algorithm stops. The conditions for stopping are that either a stable solution has been reached (i.e. the solution does not change for a number of iterations) or the effective temperature has decreased beyond a threshold T_f set to a small value. Algorithm 7.3 does not specify how the effective temperature is gradually reduced. There are different possible choices, related to different trade-offs between convergence speed and backtracking flexibility. It has been shown that if the effective temperature is decreased sufficiently slowly, the simulated annealing algorithm will converge to a global optimum. A function to reduce the effective temperature that satisfies the conditions for global optimum convergence is $T_k = c/\log(k+1)$, where c is a constant and k is the counter of the number of times the effective temperature has been reduced. Nevertheless, this slow reduction of temperature results in an excessively long convergence time. A typical choice for the function is $T_k = \alpha T_{k-1}$, where $0 < \alpha < 1$.

Recall that the earlier assumptions that the distortion measure is the squared error and that each code-vector is chosen equal to the centroid of its associated volume allow for the average end-to-end distortion to be decomposed into two terms: one due to the VQ distortion and a second term related to distortion introduced by channel errors. In turn, this decomposition allows for a design approach that separately designs the VQ and the code-vector index assignment function. So far, we have discussed the design of the second of these components. The design of the VQ can follow a channel-optimized philosophy as discussed in Section 7.1 for scalar quantizers. Continuing this line of thought, the index assignment function design just discussed can be considered as the design of a zero-redundancy channel encoder, as highlighted in the title of [6]. As for the quantizer design, not surprisingly, the design is the same as the one discussed in Section 7.1, extended to the number of dimensions in the VQ. Figure 7.4 shows the work flow for all the components of the VQ design for noisy channels. In the design, the codebook $\mathfrak{C} = \{c_1, c_2, \ldots, c_M\}$ and the par-

---

Algorithm 7.3 Use of simulated annealing to find the near-optimal code-vector index assignment

---

1: Define an index assignment mapping $\tilde{b} = \{b(\mathbf{c}_1), b(\mathbf{c}_2), \ldots, b(\mathbf{c}_M)\}$.

2: Define an *effective temperature*, T, and set it to a large initial value $T = T_0$.

3: Choose an initial mapping $\tilde{b} = \tilde{b}_0$.

4: **while** $T \geq T_f$ OR current solution for $\tilde{b}$ is not stable **do**

5:     Choose a new candidate solution $\tilde{b}'$ by randomly switching two components of the current solution $\tilde{b}$.

6:     Compute the change in average channel-induced distortion per source sample
$\Delta D_C = D_C(\tilde{b}') - D_C(\tilde{b})$.

7:     **if** $\Delta D_C < 0$ **then**

8:         Replace $\tilde{b}$ by $\tilde{b}'$.

9:     **else**

10:         Replace $\tilde{b}$ by $\tilde{b}'$ with probability $\exp(-\Delta D_C/T)$.

11:     **end if**

12:     **if** in Step 4 the number of successive cases with $\Delta D_C > 0$ or the number of successive cases with $\Delta D_C < 0$ exceeds a preset limit **then**

13:         Lower effective temperature T.

14:     **end if**

15: **end while**

16: Solution is the index assignment function $\tilde{b}$.

---

titions $\mathcal{P} = \{\mathcal{A}_1, \mathcal{A}_2, \ldots, \mathcal{A}_M\}$ are calculated by direct extension of (7.6) and (7.13) to the multiple dimension case:

$$\mathbf{c}_j = \frac{\sum_{i=1}^{M} P\left(b(\mathbf{c}_j)|b(\mathbf{c}_i)\right) \int_{\mathcal{A}_i} \mathbf{x}\, f_{\mathbf{X}}(\mathbf{X}) d\mathbf{x}}{\sum_{i=1}^{M} P\left(b(\mathbf{c}_j)|b(\mathbf{c}_i)\right) \int_{\mathcal{A}_i} f_{\mathbf{X}}(\mathbf{X}) d\mathbf{x}}, \quad j = 1, 2, \ldots, M, \tag{7.28}$$

$$\mathcal{A}_i = \left\{ \mathbf{x} : \sum_{j=1}^{M} P\left(b(\mathbf{c}_j)|b(\mathbf{c}_i)\right) \|\mathbf{x} - \mathbf{c}_j\|^2 \leq \sum_{j=1}^{M} P\left(b(\mathbf{c}_j)|b(\mathbf{c}_k)\right) \|\mathbf{x} - \mathbf{c}_j\|^2, \forall k \right\},$$
$$i = 1, 2, \ldots, M. \tag{7.29}$$

Next, we assume that p_e takes very small values, so that $p_e \ll 1/\log_2(M)$, or equivalently $p_e \log_2(M) \ll 1$. For example, if the codebook has $M = 256$ code-vectors, then we are assuming that $p_e \ll 0.125$, a reasonable assumption for most practical wireless channels. With this assumption for p_e, the probabilities $P\left(b(\mathbf{c}_j)|b(\mathbf{c}_i)\right)$ can be approximated as:

$$P\left(b(\mathbf{c}_j)|b(\mathbf{c}_i)\right) = p_e^{d_H(b(\mathbf{c}_j), b(\mathbf{c}_i))} (1 - p_e)^{(kR - d_H(b(\mathbf{c}_j), b(\mathbf{c}_i)))}$$
$$\approx p_e^{d_H(b(\mathbf{c}_j), b(\mathbf{c}_i))}$$
$$= p_e,$$

when $d_H(b(\mathbf{c}_j), b(\mathbf{c}_i)) = 1$,

$$P\left(b(\mathbf{c}_j)|b(\mathbf{c}_i)\right) \approx 0,$$

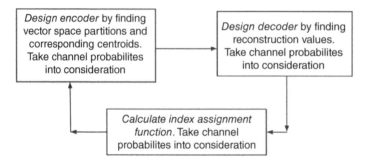

Figure 7.4 Work flow for all the components of the vector quantizer design for noisy channels.

when $d_H(b(\mathbf{c}_j), b(\mathbf{c}_i)) > 1$, and

$$P\left(b(\mathbf{c}_j)|b(\mathbf{c}_i)\right) = (1 - p_e)^{kR}$$
$$\approx 1 - kRp_e,$$

when $b(\mathbf{c}_j) = b(\mathbf{c}_i)$, this is $d_H(b(\mathbf{c}_j), b(\mathbf{c}_i)) = 0$. This approximation was used in [5] to slightly simplify the measurement of performance by expressing the probabilities as a linear function of p_e. From (7.26),

$$D_C(b) = \frac{1}{k}\sum_{i=1}^{M}\sum_{j\in C_{i,m}} p_e P(\mathbf{c}_i)d(\mathbf{c}_i, \mathbf{c}_j),$$

which allows us to write the average channel distortion normalized to the channel BER:

$$\frac{D_C(b)}{p_e} = \frac{1}{k}\sum_{i=1}^{M}\sum_{j\in C_{i,1}} P(\mathbf{c}_i)d(\mathbf{c}_i, \mathbf{c}_j). \tag{7.30}$$

This equality will hold as a reasonably good approximation as long as p_e remains small. Under this condition, the average channel distortion normalized to the channel BER does not depend on p_e (the BER), and so it is convenient as a single benchmark number to evaluate the performance of a design algorithm as a function of other design variables. This approach was used to evaluate the simulated annealing algorithm (Algorithm 7.3). Table 7.5 shows performance results obtained from minimizing the right side in (7.30) using Algorithm 7.3. The results show the average MSE distortion normalized to the channel BER, as long as the BER is kept small. The results are shown for a first-order autoregressive Gaussian source X_n (n being the discrete-time index) that behaves according to

$$X_n = \rho X_{n-1} + W_n, \quad n = 1, 2, \ldots,$$

where W_n is the noise sequence, modeled as a discrete-time, zero-mean Gaussian random process with unit variance, and ρ controls the correlation between two successive samples. Table 7.5 shows results for a memoryless Gaussian source ($\rho = 0$) and for strong correlation between successive samples ($\rho = 0.9$). While the VQ rate is fixed at $R = 1$ bit/sample, Table 7.5 shows results for four different VQ dimensions: $k = 2$, 4, 6, and 8. The quantizer itself was designed using the Linde–Buzo–Gray (LBG) algorithm (see Algorithm 2.2 in Chapter 2). For comparison, the table also shows in the last row the expected performance

Table 7.5 The mean squared channel distortion normalized to the BER for the simulated annealing algorithm and its comparison with the average performance for all possible codewords assignments, representative of a "random" codeword assignment.

	$\rho = 0.0$				$\rho = 0.9$			
	$k = 2$	$k = 4$	$k = 6$	$k = 8$	$k = 2$	$k = 4$	$k = 6$	$k = 8$
Run #1	2.56	3.56	4.94	6.73	4.35	5.42	5.86	6.00
Run #2	2.56	3.56	4.81	6.70	4.35	5.34	5.48	5.97
Run #3	2.57	3.54	5.07	6.68	4.35	5.40	5.44	5.82
Run #4	2.56	3.57	4.92	6.63	4.35	5.42	5.45	5.88
Run #5	2.56	3.52	4.73	6.72	4.35	5.40	5.51	6.20
Random	3.39	5.68	8.40	11.41	5.29	10.60	15.06	20.15

from a random codeword assignment, estimated as the average performance for all $M!$ possible codeword assignments.

The results in Table 7.5 show little difference between runs, a sign that the simulated annealing algorithm converges in all cases to a near-optimal solution. The different performances between the solutions calculated in different runs become slightly more obvious as k increases, indicating that the design is more likely to converge to various local optimum solutions as VQ dimensions increase. The results in Table 7.5 also clearly show the performance advantage of simulated annealing over randomly choosing a codeword assignment function. This performance improvement increases with VQ dimension.

The analysis so far has assumed a BSC, but the lessons learned can be applied to other setups. Consider the system studied in [7] and shown in Figure 7.5, in which a discrete-time source is transmitted through a Rayleigh fading additive white Gaussian noise (AWGN) channel using a channel-optimized vector quantization (COVQ), a binary phase shift keying (BPSK) modulator (other modulation schemes can be used) and a symbol-by-symbol maximum a posteriori (MAP) hard-decision demodulator. The system design involves finding the partitions, the codebook, and the index assignment for the COVQ and, for the MAP decoder, the mapping from a received real-valued block of symbols at the channel output, into blocks of binary words. Since the input to the modulator and the output from the MAP decoder are blocks of binary words, the tandem subsystem composed of the modulator, channel and MAP decoder can be seen as a discrete memoryless channel. If the distribution of indexes at the input of the modulator follows a uniform distribution, the equivalent discrete memoryless channel is a BSC. With this, it is possible to apply the design techniques seen in this chapter in the following iterative procedure:

- Design the VQ optimized to the equivalent BSC given by the modulator-channel-MAP demodulator tandem, e.g. using (7.28) and (7.29). The equivalent BSC crossover probability can be obtained from simulations of the modulator-channel-MAP demodulator tandem system. During the first iteration, when the MAP decoder is not yet defined, the modulator-channel-demodulator tandem system is initialized with a maximum likelihood or a simple matched filter demodulator.
- The MAP decoder is designed given the COVQ.

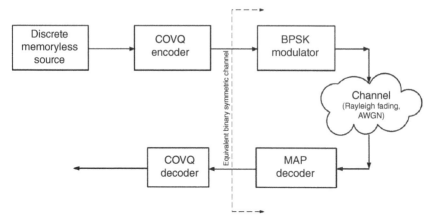

Figure 7.5 Communication system with channel-optimized vector quantizer and MAP decoder.

The techniques explained in this section can also be directly extended to channel models other than the binary symmetric one, specially when the channel remains modeled with binary inputs and outputs. In these cases, what is required is to revise the expression for $P\left(b(\mathbf{c}_j)|b(\mathbf{c}_i)\right)$ in (7.23). For example, the case for binary Markov channels was studied in [8]. The binary Markov channel is a discrete channel subject to additive noise with memory. For binary input U_t at discrete time t, and binary noise sample Z_t, the output V_t is given by $V_t = U_t \oplus Z_t$, $t = 1, 2, 3, \ldots$, where $\oplus$ refers to modulo-2 addition, that is, an error will be introduced to U_t whenever $Z_t = 1$. The channel is called a binary Markov channel because the noise samples Z_t have the Markov property $P\left(Z_t = \epsilon_t | Z_1 = \epsilon_1, Z_2 = \epsilon_2, \ldots, Z_{t-1} = \epsilon_{t-1}\right) = P\left(Z_t = \epsilon_t | Z_{t-M} = \epsilon_{t-M}, Z_{t-M-1} = \epsilon_{t-M-1}, \ldots, Z_{t-1} = \epsilon_{t-1}\right)$, for $t \geq M+1$ and Markov process order M. Furthermore, it was assumed that the process Z_t is generated by a finite-memory contagion urn model studied in [9]. The study in [9] provides an expression for $P\left(b(\mathbf{c}_j)|b(\mathbf{c}_i)\right)$ (Eq. (9)) that allows for the design of the COVQ through the conditions (7.28) and (7.29).

Before ending this section, it is worth mentioning the work in [10], where with the goal of obtaining a more compact expression for the average end-to-end distortion, it was assumed that the assignment of indexes to code-vectors was done at random with a uniform distribution. With this approach, (7.25) with squared error, can be expressed as:

$$D = \left(1 - \frac{M\bar{P}_e}{M-1}\right) D_S + \frac{M\bar{P}_e}{M-1}\sigma_X^2 + \frac{M\bar{P}_e}{M-1}S_Q, \tag{7.31}$$

where

$$\bar{P}_e = \frac{1}{M}\sum_{\substack{j=1 \\ i \neq j}}^{M}\sum_{i=1}^{M}P\left(b(\mathbf{c}_j)|b(\mathbf{c}_i)\right),$$

is the average of symbol channel error probability, D_S is the source quantization distortion given in (7.22), σ_X^2 is the source variance, and

$$S_Q = \frac{1}{kM}\sum_{j=1}^{M}|\mathbf{c}_j|^2,$$

was presented in [10] as a structural factor called the *scatter factor of the vector quantizer*. The design of the COVQ can now be formulated as finding the codebook $\mathfrak{C} = \{\mathbf{c}_1, \mathbf{c}_2, \dots, \mathbf{c}_M\}$ and partitions $\mathcal{P} = \{\mathcal{A}_1, \mathcal{A}_2, \dots, \mathcal{A}_M\}$ that minimize the distortion in (7.31). This results in the following two design conditions akin to the more general expressions (7.23) and (7.31):

$$\mathbf{c}_j = \frac{\displaystyle\int_{\mathcal{A}_i} \mathbf{x} f_{\mathbf{X}}(\mathbf{X}) d\mathbf{x}}{\displaystyle\int_{\mathcal{A}_i} f_{\mathbf{X}}(\mathbf{X}) d\mathbf{x} + \frac{\bar{P}_e}{M - 1 - M\bar{P}_e}}, \quad j = 1, \dots, M,$$

$$\mathcal{A}_i = \left\{ \mathbf{x} : \|\mathbf{x} - \mathbf{c}_i\|^2 \leq \|\mathbf{x} - \mathbf{c}_j\|^2, j \neq i \right\}, \quad i = 1, 2, \dots, M.$$

7.4 Channel Mismatch Considerations

After inspecting the design techniques discussed so far, it is clear that they depend on knowing the probabilities of data errors in the channel. But what would happen if the channel changes over time and the actual statistics of the channel during operation are different from those used during design? This section will discuss the answer to this question and, in general, how the techniques discussed so far are reshaped to consider the difference between the channel statistics at design time and during actual operation. Because they are a more general case, the discussion will focus on VQ. Also, the discussion will focus on BSCs that introduce errors with a probability equal to the channel BER (assuming ergodic conditions).

The problem of channel mismatch arises from the fact that the end-to-end performance of a COVQ depends on two channel BERs: an *actual channel BER* (p_a), which is the one experienced during the operation of the system, and a *design BER* (p_d), which is the one assumed at design time. While the design BER p_d is a deterministic value chosen at design time, the actual BER p_a is a random variable or process (if the channel changes over time). This will make the end-to-end distortion also a random variable or process.

The design of COVQs in the presence of channel mismatch was studied in detail in [11]. This work includes the results in Figure 7.6 showing the quality of the 512×512 grayscale image Lena processed through a COVQ. The table reports quality measured through the peak signal-to-noise ratio (PSNR), as defined in (1.61).

The results in Figure 7.6 are good examples of the performance trends arising from channel mismatch. The most negative effect from channel mismatch is observed when the actual channel introduces errors at a higher rate than the design assumption. In contrast, the performance of the COVQ is not severely reduced, and even in some cases improves slightly, when the design BER is larger than the one for the actual channel. (Of course, the performance would be even better if the COVQ had been designed with this more favorable channel BER in mind.) A good example for this behavior is the curve corresponding to actual channel BER $p_a = 0.03$. When the actual channel introduces very few errors, the performance of the COVQ is also degraded when the design BER is larger than that of the actual channel. Nevertheless, the quality degradation in this case is not as negative as when $p_d < p_a$ because it occurs when the overall quality is very good. This is exemplified by the

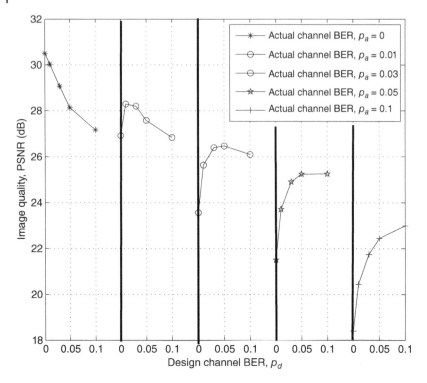

Figure 7.6 Quality (measured as PSNR in dB) for the Lena image coded with a COVQ at 0.5 bits per pixel and dimension 4×4.

curves with actual channel BER $p_a = 0$ and $p_a = 0.01$. These considerations suggest a choice of the design BER. When there is no information about the actual channel statistics other than the range of possible BERs, the best choice for design BER is likely to be the largest possible actual BER (the *worst case*) that can be expected from the channel.

Using as design BER the largest possible actual BER that can be expected from the channel is a choice arising from lack of knowledge of the channel statistics. Yet, if the largest possible actual BER is such that $p_a \ll 1/\log_2(M)$, or equivalently $p_a \log_2(M) \ll 1$, it is possible to make a more educated choice of the design BER. As discussed in Section 7.3, if $p_a \ll 1/\log_2(M)$ it is possible to approximate

$$P\left(b(\mathbf{c}_j)|b(\mathbf{c}_i)\right) \approx \begin{cases} p_a, & \text{when } d_H(b(\mathbf{c}_j), b(\mathbf{c}_i)) = 1 \\ 0, & \text{when } d_H(b(\mathbf{c}_j), b(\mathbf{c}_i)) > 1, \\ 1 - kRp_a, & \text{when } d_H(b(\mathbf{c}_j), b(\mathbf{c}_i)) = 0. \end{cases}$$

This approximation can now be used in (7.25) to obtain a simplified expression for the end-to-end distortion $D(p_a, p_d)$ (now depending on p_d and p_a):

$$D(p_a, p_d) = \frac{1 - kRp_a}{k} \sum_{i=1}^{M} \int_{\mathcal{A}_i} f_{\mathbf{X}}(\mathbf{X}) \parallel \mathbf{x} - \mathbf{c}_i \parallel^2 d\mathbf{x} + \frac{p_a}{k} \sum_{i=1}^{M} \sum_{j \in C_{i,1}} \int_{\mathcal{A}_i} f_{\mathbf{X}}(\mathbf{X}) \parallel \mathbf{x} - \mathbf{c}_j \parallel^2 d\mathbf{x}$$

where $C_{i,1}$ is, as in the previous section, the set of codewords with index assignments that differ in 1 bit from the assignment for codeword $\mathbf{c}_i$. Rearranging the terms in this expression to emphasize that it is an affine function of p_a yields

$$D(p_a, p_d) = p_a \sum_{i=1}^{M} \left(\frac{1}{k} \sum_{j \in C_{i,1}} \int_{A_i} f_\mathbf{X}(\mathbf{X}) \parallel \mathbf{x} - \mathbf{c}_j \parallel^2 d\mathbf{x} - R \int_{A_i} f_\mathbf{X}(\mathbf{X}) \parallel \mathbf{x} - \mathbf{c}_i \parallel^2 d\mathbf{x} \right)$$
$$+ \sum_{i=1}^{M} \int_{A_i} f_\mathbf{X}(\mathbf{X}) \parallel \mathbf{x} - \mathbf{c}_i \parallel^2 d\mathbf{x}.$$

This last equation expresses $D(p_a, p_d)$ as an affine function of p_a and leads to $\bar{D}(p_d) = E[D(p_a, p_d)] = D(\bar{p}_a, p_d)$, where $\bar{p}_a = E[p_a]$. Consequently, $D(\bar{p}_a, p_d) \geq D(\bar{p}_a, \bar{p}_a)$ because distortion is minimized when the design channel BER is matched to the actual channel BER. This result implies that the design choice should be $p_d = \bar{p}_a$. This choice for p_d follows from the assumption of a small p_a, which leads to expressing the average end-to-end distortion as an affine function of p_a.

When the pdf $f_b(\cdot)$ of the channel BER is known, it is possible to adjust these design methods to account for the randomness in the actual channel BER. The design goal becomes to minimize the average end-to-end distortion, now also averaged over the distribution for the channel BER:

$$\bar{D} = \int_B D(p_a) f_b(p_a) dp_a, \tag{7.32}$$

where $D(p_a)$ is the end-to-end average distortion given an actual channel BER p_a and B is the support of the actual channel BER. Writing $D(p_a)$ using (7.25), with squared error distortion, (7.32) becomes

$$\bar{D} = \frac{1}{k} \int_B \sum_{i=1}^{M} \sum_{j=1}^{M} P\left(b(\mathbf{c}_j)|b(\mathbf{c}_i)\right) \left(\int_{A_i} f_\mathbf{X}(\mathbf{X}) \parallel \mathbf{x} - \mathbf{c}_j \parallel^2 d\mathbf{x} \right) f_b(p_a) dp_a$$
$$= \frac{1}{k} \sum_{i=1}^{M} \sum_{j=1}^{M} \left(\int_B P\left(b(\mathbf{c}_j)|b(\mathbf{c}_i)\right) f_b(p_a) dp_a \right) \left(\int_{A_i} f_\mathbf{X}(\mathbf{X}) \parallel \mathbf{x} - \mathbf{c}_j \parallel^2 d\mathbf{x} \right)$$
$$= \frac{1}{k} \sum_{i=1}^{M} \sum_{j=1}^{M} \bar{P}(j, i) \left(\int_{A_i} f_\mathbf{X}(\mathbf{X}) \parallel \mathbf{x} - \mathbf{c}_j \parallel^2 d\mathbf{x} \right), \tag{7.33}$$

where, by also recalling (7.23),

$$\bar{P}(j, i) = \int_B P\left(b(\mathbf{c}_j)|b(\mathbf{c}_i)\right) f_b(p_a) dp_a \tag{7.34}$$
$$= \int_B p_a^{d_H(b(\mathbf{c}_j), b(\mathbf{c}_i))} (1 - p_a)^{(kR - d_H(b(\mathbf{c}_j), b(\mathbf{c}_i)))} f_b(p_a) dp_a.$$

This expression for expected end-to-end distortion is simply (7.25) taking into consideration the randomness of p_a through the expected probability of the channel altering the transmitted $b(\mathbf{c}_i)$. Then, a vector quantizer can be designed through a modified Lloyd's algorithm (Algorithm 2.1 in Chapter 2) where the codebook $\mathfrak{C} = \{\mathbf{c}_1, \mathbf{c}_2, \ldots, \mathbf{c}_M\}$ and the partitions $\mathcal{P} = \{\mathcal{A}_1, \mathcal{A}_2, \ldots, \mathcal{A}_M\}$ are calculated by adapting the corresponding equations that ignored the channel mismatch (7.28) and (7.29) as follows:

$$c_j = \frac{\sum_{i=1}^{M} \bar{P}(j,i) \int_{\mathcal{A}_i} x f_X(X) dx}{\sum_{i=1}^{M} \bar{P}(j,i) \int_{\mathcal{A}_i} f_X(X) dx}, \quad j = 1, 2, \ldots, M, \tag{7.35}$$

$$\mathcal{A}_i = \left\{ x : \sum_{j=1}^{M} \bar{P}(j,i) \, \|x - c_j\|^2 \leq \sum_{j=1}^{M} \bar{P}(j,k) \, \|x - c_j\|^2, \, \forall k \right\}, \quad i = 1, \ldots, M. \tag{7.36}$$

Many communication systems include techniques that allow the receiver to estimate the actual channel BER and to communicate this information to the transmitter. This information is useful for implementing adaptive modulation and coding techniques (also known as "link adaptation") as well as for scheduling transmissions in multiuser systems. By knowing the actual channel BER, it would be possible to avoid channel mismatch by using a VQ for which $p_d = p_a$. While exciting, this idea is not practical. As p_a will change over time and take values over a continuous support, this would require that the transmitter and receiver store an infinite number of codebooks and partition tables. A solution is to divide the range of p_a values into N_{p_a} segments, and then design N_{p_a} VQs, each according to the statistics of p_a over one segment. To simplify the scheme, let us assume that there is only one encoder and N_{p_a} decoders, i.e. the actual channel BER is only needed at the decoder side. The average end-to-end distortion now becomes

$$\bar{D} = \frac{1}{k} \sum_{k=1}^{N_{p_a}} \int_{B_k} \sum_{i=1}^{M} \sum_{j=1}^{M} P\left(b(c_j)|b(c_i)\right) \left(\int_{\mathcal{A}_i} f_X(X) \, \|x - c_{j,k}\|^2 dx \right) f_b(p_a) dp_a$$

$$= \frac{1}{k} \sum_{k=1}^{N_{p_a}} \sum_{i=1}^{M} \sum_{j=1}^{M} \bar{P}_k(j,i) \left(\int_{\mathcal{A}_i} f_X(X) \, \|x - c_{j,k}\|^2 dx \right),$$

where

$$\bar{P}_k(j,i) = \int_{B_k} P\left(b(c_j)|b(c_i)\right) f_b(p_a) dp_a,$$

and B_k defines the kth channel BER segment.

These equations are similar to (7.33) and (7.34), so we expect that the VQs will have partitions $\mathcal{P} = \{\mathcal{A}_1, \mathcal{A}_2, \ldots, \mathcal{A}_M\}$ and codebook $\mathfrak{C}_k = \{c_{1,k}, c_{2,k}, \ldots, c_{M,k}\}$ determined by:

$$c_{j,k} = \frac{\sum_{i=1}^{M} \bar{P}_k(j,i) \int_{\mathcal{A}_i} x f_X(X) dx}{\sum_{i=1}^{M} \bar{P}_k(j,i) \int_{\mathcal{A}_i} f_X(X) dx}, \quad j = 1, 2, \ldots, M, \tag{7.37}$$

$$\mathcal{A}_i = \left\{ x : \sum_{k=1}^{N_{p_a}} \sum_{j=1}^{M} \bar{P}_k(j,i) \, \|x - c_{j,k}\|^2 \leq \sum_{k=1}^{N_{p_a}} \sum_{j=1}^{M} \bar{P}(j,l) \, \|x - c_{j,k}\|^2, \forall l \right\},$$

$$i = 1, \ldots, M, \tag{7.38}$$

Table 7.6 PSNR results for 512×512 grayscale Lenna image coded with CPVQs under channel mismatch

Rate (bpp)	Dimension	COVQ	$p_d = \bar{p}_a$	BER pdf	DA $K = 2$	CA
2	2×2	24.29	26.33	27.00	27.27	26.05
1	4×2	23.96	25.80	26.25	26.46	25.43
0.5	4×4	23.54	25.14	25.59	25.84	24.99

with $k = 1, 2, \ldots, N_{p_a}$. As before, the encoder and N_{p_a} decoders can be designed using a modified Lloyd's algorithm.

The scheme just described has been called *decoder-adaptive quantization* [11]. In [12] and [13], *codebook-adaptive quantization* was presented that used a single encoder (frequently designed as the encoder for a noiseless channel) and a decoder that uses an estimate of p_a to calculate a new codebook using (7.28) and (7.23). The complexity associated with computing (7.28) can be reduced by prestoring in the decoder the M results $\int_{A_i} \mathbf{x} f_{\mathbf{X}}(\mathbf{X}) d\mathbf{x}, i = 1, 2, \ldots, M$. The advantage of this scheme is that it does not require knowledge of channel statistics. However, as shall be seen next, the use of an encoder designed for a noiseless channel negatively affects the performance.

Table 7.6 shows results from [11] comparing the different schemes discussed in this section. The results show the quality of a 512×512 grayscale Lenna image. In the table, COVQ refers to a CPVQ designed using the simulated annealing method discussed in the previous section, "$p_d = \bar{p}_a$" refers to the approach that uses the average actual channel BER for the design, "BER pdf" refers to the approach that uses the pdf of the channel BER (Eqs. (7.35) and (7.36)), DA is the decoder-adaptive scheme and CA refers to the codebook-adaptive scheme. The randomness of the channel is introduced through a log-normal fading with average carrier SNR of 10 dB (which could model fading due to shadowing).

7.5 Structured Vector Quantizers

A practical issue when using VQs is codebook design complexity. When designing a VQ using the LBG algorithm (see Algorithm 2.2 in Chapter 2), the complexity grows exponentially with kR, where k is the vector dimension and R is the code rate. Approaches to reduce VQ design complexity are frequently based on introducing some structure into the codebook to reduce the dimensionality of the codebook design. Tree-structured vector quantization (TSVQ), multistage VQ (MSVQ), product VQ, and lattice VQ are some of the proposed VQ designs following this approach.

TSVQs and multistage VQs have similarities. In both, complexity is reduced by breaking down the code design into stages, each associated with a VQ code of smaller dimensionality than the full-rate code (presumably designed using an algorithm such as the LBG). Specifically, consider that a real-valued, discrete-time, continuous-amplitude source, characterized by the stationary process $\{X_n, n = 0, 1, ...\}$ with zero mean and variance σ_X^2, is encoded with k-dimensional VQ. The operation of each stage of the TSVQ and MSVQ remains the same as with any VQ, by taking a block of k input samples and mapping them into a code-vector. Each code-vector corresponds to one of the partitions of the k-dimensional space. We keep the earlier assumption that a standard VQ partitions the k-dimensional space into M volumes. In TSVQ and MSVQ, while the complete encoding operation still partitions the space into M volumes, the first stage of encoding is based on a partition of the k-dimensional space into $M_1 < M$ volumes, corresponding to an encoding rate $R_1 = \frac{1}{k}\log_2(M_1) < \frac{1}{k}\log_2(M) = R$. Therefore, the output from the first stage is a code-vector that only approximates the code-vector that would be output by the full-rate encoder. For the first stage, denoting the encoding mapping from the input vector into a codeword as $b_1(\mathbf{x}) = i - 1$, with the mapping result $i \in \{1, 2, ..., M_1\}$, and the decoding mapping as $g_1(i) = \mathbf{c}_{1,i}$, with $\mathbf{c}_{1,i}$ being in the reproduction codebook $\mathfrak{C}_1 = \{\mathbf{c}_{1,1}, \mathbf{c}_{1,2}, ..., \mathbf{c}_{1,M_1}\}$, the encoding–decoding VQ operation can be written as:

$$\mathbf{c}_{1,i} = q_1(\mathbf{x}) = g_1(b_1(\mathbf{x})).$$

The second and possibly succeeding stages refine the quantization performed in previous stages, gradually approaching encoding at full rate. With an MSVQ, a second VQ stage operates on the first-stage quantization error $\mathbf{e}_1 = \mathbf{x} - q_1(\mathbf{x})$, with encoding–decoding operation:

$$\mathbf{c}_{2,i} = q_2(\mathbf{e}_1) = g_2(b_2(\mathbf{e}_1)).$$

Note that implied in this expression is that the encoding mapping $b_2(\cdot)$ is a function of $\mathbf{e}_1$. For TSVQ, the second stage quantizes $\mathbf{e}_1 + q_1(\mathbf{x})$, which ideally matches $\mathbf{x}$, with encoding–decoding operation:

$$\mathbf{c}_{2,i} = q_2(\mathbf{e}_1 + q_1(\mathbf{x})) = g_2(b_2(\mathbf{e}_1 + q_1(\mathbf{x}))),$$

and the encoding mapping $b_2(\cdot)$ is a function of $\mathbf{e}_1$ and $b_1(\cdot)$. To simplify the study of MSVQ and TSVQ, it is possible to translate the definition for the encoding–decoding functions into a common framework by viewing the second stage for both encoders as operating on the first-stage error $\mathbf{e}_1$.

Generalizing, any stage s of the total S stages in a TSVQ or MSVQ quantizer can be characterized by the encoding mapping $b_s(\cdot)$, with partition set $P_s = \{\mathcal{A}_{s,1}, \mathcal{A}_{s,2}, ..., \mathcal{A}_{s,M_s}\}$, the decoding mapping $g_s(\cdot)$, the codebook $\mathfrak{C}_s = \{\mathbf{c}_{s,1}, \mathbf{c}_{s,2}, ..., \mathbf{c}_{s,M_s}\}$, and the encoder–decoder quantizing function q_s. The rate for stage s is $R_s = \frac{1}{k}\log_2(M_s)$. At the output of the overall TSVQ or MSVQ encoders, the codeword to be transmitted is formed by concatenating the codewords obtained from each stage encoder. The $\log_2(M_1)$ bits from the first stage occupy the most significant bit positions, the $\log_2(M_2)$ bits from the second stage occupy the next most significant positions, and so on until the $\log_2(M_S)$ bits from the last stage occupy the least significant positions. Mathematically, this structure for the overall encoder mapping function can be written as:

$$b(\mathbf{x}) = \sum_{s=1}^{S} \Upsilon_s b_s(\mathbf{e}_{s-1}),$$

where $\Upsilon_s = \prod_{i=s+1}^{S} M_i$, for $s = 1, 2, \ldots, S - 1$, $\Upsilon_S = 1$, $\mathbf{e}_0 = \mathbf{x}$, and from now on it is assumed that all values M_s are chosen so that $\log_2(M_s)$ is an integer. The overall decoder output is calculated by adding the output from each stage decoder:

$$\mathbf{c}_i = \sum_{s=1}^{S} g_s \left(\left\lfloor \frac{i}{\Upsilon_s} \right\rfloor \bmod M_s \right),$$

and the overall encoder–decoder operation is

$$q(\mathbf{x}) = \sum_{s=1}^{S} g_s(b_s(\mathbf{e}_{s-1})).$$

Consequently, the overall rate R, combining the operation of all stages, is $R = \sum_{s=1}^{S} R_s$.

As the overall transmitted codeword is formed by concatenating the encoder output from all the stages, when the channel is memoryless, the output from each encoding stage will be affected by the channel independently from the others. Nevertheless, the effects of the channel on overall distortion will be different depending on which stage an affected codeword belongs to. Each stage incrementally improves the coding resolution from the previous stages. This means that an error in the first-stage codeword will affect the overall distortion in a similar way as an error in the codeword of a standard VQ encoding at low rate. An error in the second-stage codeword will affect the quality improvement provided by this stage but will not affect the basic quality provided by the first stage. Consequently, to consider channel effects in the design of TSVQ and MSVQ, we will assume that the encoder and decoder for the first stage are already given and we concentrate on second-stage design.

We start by assuming a BSC and introducing notation to characterize channel effects. The encoder output is a binary word in the range from 0 to $M_1 - 1$ for the first stage and from 0 to $M_2 - 1$ for the second stage. Let the codeword from the first stage be $b_1(\mathbf{x}) = i - 1$, $i = 1, 2, \ldots, M_1$, and the one from the second stage be $b_2(\mathbf{x}) = u - 1$, $u = 1, 2, \ldots, M_2$. The effects of the channel on the transmission of i can be modeled with a transition probability matrix with elements $P_1(j|i)$, where $j = 1, 2, \ldots, M_1$ is the first-stage codeword at the output of the channel. Similarly, the effects of the channel on the second-stage codeword can be modeled with a transition probability matrix with elements $P_2(v|u)$, where $v = 1, 2, \ldots, M_2$ is the second-stage codeword at the output of the channel.

Consider first the design of the channel-matched TSVQ given in [14]. The difference between the TSVQ and the MSVQ is that the second stage operates with M_1 different partition sets and associated codebooks. The partition set used by the encoder depends on the output of the first stage. This means, for example, that if the codeword at the output of the first-stage encoder is i, the partition set used in the second stage is $\mathcal{P}_2^{(i)} = \{A_{2,1}^{(i)}, A_{2,2}^{(i)}, \ldots, A_{2,M_2}^{(i)}\}$, and the codebook is $\mathfrak{C}_2^{(i)} = \{\mathbf{c}_{2,1}^{(i)}, \mathbf{c}_{2,2}^{(i)}, \ldots, \mathbf{c}_{2,M_2}^{(i)}\}$. The presence of channel errors that make the transmitted codeword i be received as j will lead to the decoder using the partition set $\mathcal{P}_2^{(j)} = \{A_{2,1}^{(j)}, A_{2,2}^{(j)}, \ldots, A_{2,M_2}^{(j)}\}$, and the codebook $\mathfrak{C}_2^{(j)} = \{\mathbf{c}_{2,1}^{(j)}, \mathbf{c}_{2,2}^{(j)}, \ldots, \mathbf{c}_{2,M_2}^{(j)}\}$. Then, the end-to-end distortion can be calculated as:

$$D = \frac{1}{k} \sum_{i,u} \sum_{j,v} \left\{ P_1(j|i) P_2(v|u) \int_{A_i^u} f_{\mathbf{X}}(\mathbf{X}) d\left(\mathbf{x}, \mathbf{c}_{1,j} + \mathbf{c}_{2,v}^{(j)}\right) d\mathbf{x} \right\},$$

where $f_{\mathbf{X}}(\mathbf{X})$ is the joint pdf for the k source samples random variables $x_1, x_2, \dots, x_k$, and $\mathcal{A}_i^u$ is the partition created by the two encoding stages: $\mathcal{A}_i^u = \mathcal{A}_{1,i} \cap \mathcal{A}_{2,u}^{(i)}$. Note how in calculating the distortion, the reconstruction of the source vector is done by adding the output from the two decoding stages $\mathbf{c}_{1j} + \mathbf{c}_{2,v}^{(j)}$. Working on the expression for the end-to-end distortion, the integral and summation on the codeword indexes at the channel output can be interchanged:

$$D = \frac{1}{k} \sum_{i,u} \int_{\mathcal{A}_i^u} \left\{ \sum_{j,v} P_1(j|i) P_2(v|u) d\left(\mathbf{x}, \mathbf{c}_{1j} + \mathbf{c}_{2,v}^{(j)}\right) \right\} f_{\mathbf{X}}(\mathbf{X}) d\mathbf{x}. \tag{7.39}$$

For the expression within the braces, the values for i and u are fixed (as they are given by the first sum), which also implies that $\mathcal{A}_i^u$ is given. Then, $\sum_{j,v} P_1(j|i) P_2(v|u) d\left(\mathbf{x}, \mathbf{c}_{1j} + \mathbf{c}_{2,v}^{(j)}\right)$ can be interpreted as a distortion for which the source encoding result is given. We denote it as $d_{i,u}(\mathbf{x})$.

Minimizing the end-to-end distortion is accomplished by minimizing the distortion $d_{i,u}(\mathbf{x})$. Thus, the design for the second coding stage can be formalized as seeking to minimize $d_{i,u}(\mathbf{x})$. The design of each coding stage follows the same approach as seen earlier, where the optimum partition is found under the assumption of a fixed codebook, and the codebook is calculated assuming a fixed partition. For the second-stage design, calling the output from the first stage $b_1(\mathbf{x}) = i - 1$, the optimum partition $\mathcal{P}_2^{(j)*} = \{\mathcal{A}_{2,1}^{(i)*}, \mathcal{A}_{2,2}^{(i)*}, \dots, \mathcal{A}_{2,M_2}^{(i)*}\}$ must follow the condition:

$$\mathcal{A}_{2,u}^{(i)*} = \{\mathbf{x} : d_{i,u}(\mathbf{x}) \le d_{i,u'}(\mathbf{x}) \ \forall u' = 1, 2, \dots, M_2, \quad u' \ne u\}.$$

To find the optimum codebook design, $\mathfrak{C}_2^{(j)*} = \{\mathbf{c}_{2,1}^{(j)*}, \mathbf{c}_{2,2}^{(j)*}, \dots, \mathbf{c}_{2,M_2}^{(j)*}\}$, the partition is fixed and we use the squared error distortion measure: $d(\mathbf{x}, \mathbf{y}) = \|\mathbf{x} - \mathbf{y}\|^2$. Recalling that in a vector space we can write $\|\mathbf{x} - \mathbf{y}\|^2 = \|\mathbf{x}\|^2 + \|\mathbf{y}\|^2 - 2 < \mathbf{x}, \mathbf{y} >$, where $< \mathbf{x}, \mathbf{y} >$ is the inner product between the two vectors, the distortion $d_{i,u}(\mathbf{x})$ can be expanded as:

$$d_{i,u}(\mathbf{x}) = \sum_{j,v} \|\mathbf{x} - (\mathbf{c}_{1j} + \mathbf{c}_{2,v}^{(j)})\|^2 P_1(j|i) P_2(v|u)$$

$$= \sum_{j,v} \left(\|\mathbf{x}\|^2 + \|\mathbf{c}_{1j} + \mathbf{c}_{2,v}^{(j)}\|^2 - 2 < \mathbf{x}, \mathbf{c}_{1j} + \mathbf{c}_{2,v}^{(j)} > \right) P_1(j|i) P_2(v|u)$$

$$= \|\mathbf{x}\|^2 + \sum_{j,v} \|\mathbf{c}_{1j} + \mathbf{c}_{2,v}^{(j)}\|^2 P_1(j|i) P_2(v|u) - 2 \sum_{j,v} < \mathbf{x}, \mathbf{c}_{1j} + \mathbf{c}_{2,v}^{(j)} > P_1(j|i) P_2(v|u),$$

where the last equality follows from the property that $\sum_{j,v} P_1(j|i) P_2(v|u) = 1$. Next, the term with $\|\mathbf{c}_{1j} + \mathbf{c}_{2,v}^{(j)}\|^2$ can be further expanded in the same way as:

$$\sum_{j,v} \|\mathbf{c}_{1j} + \mathbf{c}_{2,v}^{(j)}\|^2 P_1(j|i) P_2(v|u) = \sum_{j,v} \left(\|\mathbf{c}_{1j}\|^2 + \|\mathbf{c}_{2,v}^{(j)}\|^2 + 2 < \mathbf{c}_{1j}, \mathbf{c}_{2,v}^{(j)} > \right) P_1(j|i) P_2(v|u),$$

and

$$\sum_{j,v} \|\mathbf{c}_{1j}\|^2 P_1(j|i) P_2(v|u) = \sum_j \|\mathbf{c}_{1j}\|^2 P_1(j|i) \sum_v P_2(v|u)$$

$$= \sum_j \|\mathbf{c}_{1j}\|^2 P_1(j|i),$$

because $\sum_v P_2(v|u) = 1$. To ease the presentation, we introduce the notation:

$$\bar{\mathbf{c}}_{1,j} = \sum_{j,v} \mathbf{c}_{1,j} P_1(j|i) P_2(v|u)$$

$$= \sum_j \mathbf{c}_{1,j} P_1(j|i),$$

(again using the fact that $\mathbf{c}_{1,j}$ does not depend on v and that $\sum_v P_2(v|u) = 1$) and

$$\bar{\mathbf{c}}_{2,v}^{(j)} = \sum_{j,v} \mathbf{c}_{2,v}^{(j)} P_1(j|i) P_2(v|u).$$

Furthermore, using the linearity property for inner products, we can write

$$\sum_{j,v} <\mathbf{c}_{1,j} + \mathbf{c}_{2,v}^{(j)}> P_1(j|i) P_2(v|u) = <\bar{\mathbf{c}}_{1,j}, \bar{\mathbf{c}}_{2,v}^{(j)}>,$$

and

$$\sum_{j,v} <\mathbf{x}, \mathbf{c}_{1,j} + \mathbf{c}_{2,v}^{(j)}> P_1(j|i) P_2(v|u) = <\mathbf{x}, \bar{\mathbf{c}}_{1,j} + \bar{\mathbf{c}}_{2,v}^{(j)}>.$$

Combining all parts for the expansion of $d_{i,u}(\mathbf{x})$ yields

$$d_{i,u}(\mathbf{x}) = \|\mathbf{x}\|^2 - 2 <\mathbf{x}, \bar{\mathbf{c}}_{1,j} + \bar{\mathbf{c}}_{2,v}^{(j)}> + \sum_j \|\mathbf{c}_{1,j}\|^2 P_1(j|i)$$

$$+ 2 <\bar{\mathbf{c}}_{1,j}, \bar{\mathbf{c}}_{2,v}^{(j)}> + \sum_{j,v} \|\mathbf{c}_{2,v}^{(j)}\|^2 P_1(j|i) P_2(v|u).$$

Using again the linearity property for inner products results in

$$d_{i,u}(\mathbf{x}) = \underbrace{\|\mathbf{x}\|^2 - 2 <\mathbf{x}, \bar{\mathbf{c}}_{1,j}> + \sum_j P_1(j|i) \|\mathbf{c}_{1,j}\|^2}_{\text{First-stage distortion}}$$

$$\underbrace{- 2 <\mathbf{x} - \bar{\mathbf{c}}_{1,j}, \bar{\mathbf{c}}_{2,v}^{(j)}> + \sum_{j,v} \|\mathbf{c}_{2,v}^{(j)}\|^2 P_1(j|i) P_2(v|u).}_{\text{Second-stage distortion}} \qquad (7.40)$$

The first three terms in (7.40) are the average distortion from the first encoding stage for fixed i and u and partition $\mathcal{A}_i^u$. This is because, by following the previous steps in reverse order, they can be combined as $\sum_j P_1(j|i)(\|\mathbf{x}\|^2 - 2 <\mathbf{x}, \mathbf{c}_{1,j}> + \|\mathbf{c}_{1,j}\|^2) = \sum_j \| \mathbf{x} - \mathbf{c}_{1,j}\|^2 P_1(j|i)$. Also note in (7.40) that the last two terms are the distortion associated with the second stage (also for fixed i and u and partition $\mathcal{A}_i^u$) because they are the difference between the overall distortion and that from the first stage. When combined with the other elements used to compute distortion present in (7.39), the distortion associated with the second stage can be seen as that resulting from a regular VQ when the source is $\mathbf{x} - \bar{\mathbf{c}}_{1,j}$. Consequently, by straightforward extension of (7.6) to the vector case, the optimum codebook is

$$\mathbf{c}_{2,v}^{(j)*} = \frac{\displaystyle\sum_{i,u} P_1(j|i) P_2(v|u) \int_{\mathcal{A}_i^u} (\mathbf{x} - \mathbf{c}_{1,j}) f_{\mathbf{X}}(\mathbf{X}) d\mathbf{x}}{\displaystyle\sum_{i,u} P_1(j|i) P_2(v|u) \int_{\mathcal{A}_i^u} f_{\mathbf{X}}(\mathbf{X}) d\mathbf{x}}.$$

With this result, it is possible to set an iterative algorithm that computes the optimum codebook assuming a fixed partition and then uses this result to calculate the optimum partition while the codebook is kept fixed. This design procedure really focuses on the second stage, with the design of the first stage assumed to be given, because the analysis showed that there is no difference between the first stage for the TSVQ and a standard VQ. Details of the design can be found in [14].

The design of the MSVQ is a simplified case of the TSVQ, as the second coding stage uses a single design irrespective of the encoding result from the first stage. Thus, the second-stage decoding depends only on the output v from the second channel. As mentioned earlier, some variables related to the second stage of quantization depended on the output of the first-stage channel, j, but now this dependence is gone. This is, the end-to-end distortion can still be written as:

$$D = \frac{1}{k} \sum_{i,u} \int_{A_i^u} d_{i,u}(\mathbf{x}) f_{\mathbf{X}}(\mathbf{X}) d\mathbf{x},$$

but $d_{i,u}(\mathbf{x})$ is now $d_{i,u}(\mathbf{x}) = \sum_{j,v} P_1(j|i) P_2(v|u) d\left(\mathbf{x}, \mathbf{c}_{1,j} + \mathbf{c}_{2,v}\right)$ and can be written as:

$$d_{i,u}(\mathbf{x}) = \|\mathbf{x}\|^2 - 2 < \mathbf{x}, \bar{\mathbf{c}}_{1,j} > + \sum_j P_1(j|i) \|\mathbf{c}_{1,j}\|^2$$

$$-2 < \mathbf{x} - \bar{\mathbf{c}}_{1,j}, \bar{\mathbf{c}}_{2,v}^{(j)} > + \sum_v \|\mathbf{c}_{2,v})\|^2 P_2(v|u), \tag{7.41}$$

with

$$\bar{\mathbf{c}}_{2,v} = \sum_v \mathbf{c}_{2,v} P_2(v|u).$$

These differences in the MSVQ formulation make the design process simpler than for TSVQ. Still, the design procedure for both is similar: The first-stage design is done separately following the procedures discussed earlier, and is assumed as given, with an output $b_1(\mathbf{x}) = i - 1$, for the second-stage design, which is in turn designed by finding the optimum partition under the assumption of a fixed codebook and calculating the codebook assuming a fixed partition. As such, the optimum partition $\mathcal{P}_2^* = \{A_{2,1}^*, A_{2,2}^*, \ldots, A_{2,M_2}^*\}$ is given by the condition:

$$A_{2,u}^* = \{\mathbf{x} : d_{i,u}(\mathbf{x}) \le d_{i,u'}(\mathbf{x}) \quad \forall u' = 1, 2, \ldots, M_2, \quad u' \ne u\},$$

with $d_{i,u}(\mathbf{x})$ given by (7.41). Moreover, following the aforementioned differences between the MSVQ and TSVQ, the optimum codebook design for MSVQ follows the same procedure as for the TSVQ case, but now it is only necessary to find one optimum codebook $\mathfrak{C}_2^* = \{\mathbf{c}_{2,1}^*, \mathbf{c}_{2,2}^*, \ldots, \mathbf{c}_{2,M_2}^*\}$ that is applicable to all values of j. Consequently, the optimum codebook is given by the equation:

$$\mathbf{c}_{2,v}^* = \frac{\sum_{i,u} P_2(v|u) \int_{A_i^u} (\mathbf{x} - \mathbf{c}_1) f_{\mathbf{X}}(\mathbf{X}) d\mathbf{x}}{\sum_{i,u} P_2(v|u) \int_{A_i^u} f_{\mathbf{X}}(\mathbf{X}) d\mathbf{x}}.$$

The results presented in [14] for the channel-optimized MSVQ and TSVQ show improvements in SNR of up to 4.9 dB when the channel is very noisy. Here, the SNR is essentially a measure of the average end-to-end squared error distortion, D, relative to the source variance, σ^2, SNR= $10\log_{10}\sigma^2/D$ (when measuring using dB). As the channel quality improves, the advantage of a channel-matched design is progressively reduced relative to a design not matched to the channel. Eventually, when the channels have high SNR, the non-channel-matched designs show a negligible performance advantage compared to the channel-matched designs. Also, the results in [14] show that the channel-matched MSVQ is more robust to a noisy channel environment than TSVQ. This is contrary to the error-free case, where TSVQ outperforms MSVQ. Both results are easily justified as the TSVQ utilizes a different codebook in the second stage for each first-stage output. This improves the coding efficiency of TSVQ compared with MSVQ but when considering transmission over a noisy channel, with MSVQ an error in the code-vector at a given coding stage will only affect the decoding of that stage, but with TSVQ it will affect that stage and all successive stages. In other words, TSVQ is more efficient in performing source coding than MSVQ, but is also more sensitive to channel errors, even when resorting to channel-matched designs. Also, Phamdo et al. [14] note a difference in performance behavior for the VQs depending on the source characteristics. For a memoryless Gaussian source, after fixing the overall coding rate and the number of coding bits used in the first stage, the performance of the channel-matched VQs does not increases monotonically with the dimension of the vector space and even decreases in several cases. This is not the case when the source is a Gauss–Markov one, for which performance always improved as the vector space dimensions increased.

Finally, a fairly robust performance was observed in the presence of a mismatch between the design and the actual channel BER for both MSVQ and TSVQ. These results are similar to those previously observed for standard VQs. Of course, the issues associated with channel mismatch can be addressed using the same techniques discussed in the previous section.

References

1 Kurtenbach, A. and Wintz, P. (1969). Quantizing for noisy channels. *IEEE Transactions on Communication Technology* 17 (2): 291–302.

2 Totty, R. and Clark, G. Jr. (1967). Reconstruction error in waveform transmission (corresp.). *IEEE Transactions on Information Theory* 13 (2): 336–338.

3 Farvardin, N. and Vaishampayan, V. (1987). Optimal quantizer design for noisy channels: an approach to combined source-channel coding. *IEEE Journal on Selected Areas in Communications* 33 (6): 827–838.

4 Fine, T. (1964). Properties of an optimum digital system and applications. *IEEE Transactions on Information Theory* 10 (4): 287–296.

5 Farvardin, N. (1990). A study of vector quantization for noisy channels. *IEEE Transactions on Information Theory* 36 (4): 799–809.

6 Zeger, K.A. and Gersho, A. (1987). Zero redundancy channel coding in vector quantization. *IEEE Electrononics Letters* 23 (12): 654–655.

7 Saffar, H.E., Alajaji, F., and Linder, T. (2009). COVQ for MAP hard-decision demodulated channels. *IEEE Transactions Communications Letters* 13 (1): 28–30.

8 Phamdo, N., Alajaji, F., and Farvardin, N. (1997). Quantization of memoryless and Gauss–Markov sources over binary Markov channels. *IEEE Transactions on Communications* 45 (6): 668–675.

9 Alajaji, F. and Fuja, T. (1994). A communication channel modeled on contagion. *IEEE Transactions on Information Theory* 40 (6): 2035–2041.

10 Yu, X., Wang, H., and Yang, E.-H. (2010). Design and analysis of optimal noisy channel quantization with random index assignment. *IEEE Transactions on Information Theory* 56 (11): 5796–5804.

11 Jafarkhani, H. and Farvardin, N. (2000). Design of channel-optimized vector quantizers in the presence of channel mismatch. *IEEE Transactions on Communications* 48 (1): 118–124.

12 Hung, A.C. and Meng, T.H.-Y. (1993). Adaptive channel optimization of vector quantized data. *Proceedings of the Data Compression Conference (DCC)*, pp. 282–291, Snowbird, Utah.

13 Skoglund, M., Hagen, P., and Hedelin, P. (1995). Fixed and adaptive decoding in robust LPC quantization. *IEEE Speech Coding Workshop*, pp. 71–72.

14 Phamdo, N., Farvardin, N., and Moriya, T. (1993). A unified approach to tree-structured and multistage vector quantization for noisy channels. *IEEE Transactions on Information Theory* 39 (3): 835–850.

8

Error-Resilient Source Coding

As discussed in Chapter 5, different parts of a source-encoded stream have different sensitivity to channel errors and so require different levels of error protection. In some parts, an error during transmission would lead to high distortion, and such important parts would require more error protection. In other parts, an error would have little effect on end-to-end distortion, and so less error protection is needed. Chapter 5 discussed unequal error protection (UEP) techniques, where forward error correction (FEC) redundancy is allocated to match the different effects of channel errors. This chapter will discuss another approach to handle unequal bit stream sensitivity, called *error-resilient source coding*. Here, the source encoder is designed so that its output presents an inherent resiliency against channel errors. For example, since predictive coding can propagate channel errors (see Chapter 5), an error-resilient source coder might limit the use of predictive coding. This idea is used in video coding, since it is possible to change the intra-coding rate based on channel error rate. As the channel error rate increases, the use of predictive coding is reduced by increasing the frequency of I-frames. In the extreme, it is possible to encode the video sequence using only I-frames, avoiding all temporal prediction coding. This solution is called motion-JPEG when JPEG image compression is used to encode the I-frames. In this chapter, we will cover different techniques to increase the resiliency of the source-encoded stream to channel errors.

Lack of resiliency in a coded bit stream can sometimes be addressed with a mechanism to retransmit missing information, an approach studied in Chapter 6. Retransmissions require a feedback channel, which may not always be available. Even when a feedback channel is available, its use may not be advisable. For example, for multicast transmission, feedback channels present a problem when some receivers signal successful reception and others indicate failure. Also for real-time transmission, the delay and possible unreliability of the feedback channel may preclude its use. The techniques discussed next, instead of relying on retransmission, directly improve the resiliency of a coded bit stream.

8.1 Multiple-Description Coding

In multiple-description source coding (MDSC), the bit stream at the output of the source encoder contains multiple representations. Figure 8.1 illustrates a dual-description

Joint Source-Channel Coding, First Edition. Andres Kwasinski and Vinay Chande.
© 2023 John Wiley & Sons Ltd. Published 2023 by John Wiley & Sons Ltd.

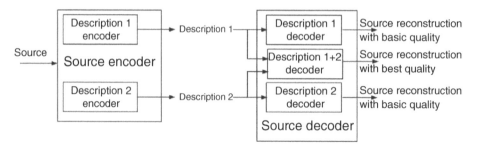

Figure 8.1 A dual-description encoder and decoder.

encoder and decoder, with two encoded descriptions of the source. As depicted, the two descriptions are generated by separate encoders. This is not always the case, but it simplifies the presentation. Each description can be decoded independently of the others and results in a source reconstruction that has a *basic quality*. The reconstruction quality of each description may be the same as or different from the quality of other descriptions. When receiving multiple descriptions, the decoder can combine them to obtain a reconstruction with better quality than the basic one obtained from individual descriptions. The decoder which combines descriptions is known as the combined or central decoder, and the decoders above and below it which handle only a single description are known as individual or side decoders. Descriptions that are combined need to have been received error-free.

Historically, MDSC studies originated with scenarios where multiple communication channels are available to be used in parallel, and it was useful if the source encoder output could be split and transmitted through the available channels. MDSC provided a solution where each description can be communicated through a different independent link, affected by independent channel impairments. Application examples are transmission over packet-switched networks and fading multipath channels which frequently resort to diversity techniques. With MDSC, to obtain at least a basic quality at the decoder, at least one coded description must be correctly received. As illustrated in Figure 8.1, if more than one description are received error-free, they can be combined at the decoder to reconstruct the source with better quality. This possibility of improving the reconstruction by combining correctly received descriptions is why MDSC is often used to leverage transmit diversity schemes, where transmit paths are independently affected by channel impairments.

The dual-description codec of Figure 8.1 has been well studied for the memoryless Gaussian source. This is the only case for which the multiple-description rate-distortion region is completely known. The set of achievable mean squared distortions for a Gaussian source with zero mean and variance σ^2 was given in [1] by the union of all triplets d_0 (distortion at the central decoder), d_1 (distortion at one side decoder), and d_2 (distortion at the other side decoder), such that

$$d_1 \geq \sigma^2 2^{-2x_1}, \tag{8.1a}$$

$$d_2 \geq \sigma^2 2^{-2x_2}, \tag{8.1b}$$

$$d_0 \geq \frac{\sigma^2 2^{-2(x_1+x_2)}}{1 - \left(\left|\sqrt{(1-\frac{d_1}{\sigma^2})(1-\frac{d_2}{\sigma^2})} - \sqrt{\frac{d_1 d_2}{\sigma^4} - 2^{-2(x_1+x_2)}}\right|^+\right)^2},$$ (8.1c)

where x_1 and x_2 are the source encoding rates associated with each of the side descriptions and the function $|u|^+$ stands for

$$|u|^+ = \begin{cases} u, & \text{if } u > 0 \\ 0 & \text{else.} \end{cases}$$

The function $|u|^+$ in (8.1c) only becomes equal to zero when $d_1 + d_2 > \sigma^2(1 + 2^{-2(x_1+x_2)})$, which corresponds to high-side distortion conditions. This condition is not active in the distortion-rate region boundary given by the inequalities (8.1).

The distortion-rate region given by (8.1) can be studied in a simple case that provides insight into MDSC operation. Assume that the source has unit variance, that the two side encoders are symmetric (perform with the same rate-distortion curve), and that they encode at the same individual source coding rate $x_1 = x_2 = x$ with distortion $d_1 = d_2 = 2^{-2x} = d$ given by the distortion-rate curve. For this setup, the side distortion-rate curve is shown in Figure 8.2 labeled as *Individual Decoder*, and the combined distortion-rate performance given by (8.1c) becomes

$$d_0 \geq \frac{d^2}{1 - \left(\left|(1-d) - \sqrt{d^2 - d^2}\right|^+\right)^2} = \frac{d^2}{1-(1-d)^2} = \frac{d^2}{2d - d^2} = \frac{d}{2-d}.$$ (8.2)

This result implies that the achievable combined distortion is no better than half the achievable distortion on the individual channels. Figure 8.2 shows the combined distortion

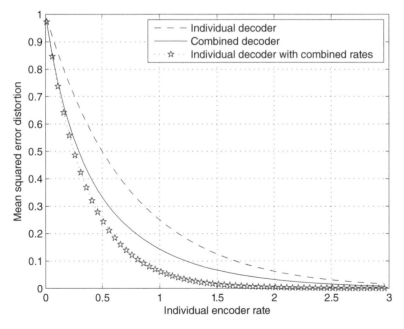

Figure 8.2 The distortion associated with a symmetric dual-description codec.

bound (8.2) with equality using the label *Combined Decoder*. Comparing the *individual decoder* and *combined decoder* curves verifies that the ratio d_0/d_1 is never less than one half; it goes from a value of 1 when the individual encoder rate is zero to asymptotically approaching 1/2 at large individual encoder rates (for example, the ratio is 0.66 at rate 0.5, 0.57 at rate 1, and 0.53 at rate 1.5).

Figure 8.2 also shows a curve labeled *individual decoder with combined rates*. This curve corresponds to a codec that is designed to provide the best possible distortion when using the combined encoding rate $2x$ (this is, the distortion is 2^{-4x}). Comparing this curve with the *Combined Decoder* curve shows that MDSC introduces some performance loss relative to single-description encoding, because of the introduction of redundancy across the multiple descriptions. However, this redundancy provides the error resiliency characteristic of MDSC.

Continuing with the example of a unit-variance source with symmetric side decoders each operating at rate x, (8.1) can be rewritten as:

$$d_0 = \begin{cases} \dfrac{2^{-4x}}{1-\left(1-d_1-\sqrt{d_1^2-2^{-4x}}\right)^2}, & \text{if } 2^{-2x} \le d_1 < \frac{1}{2}(1+2^{-4x}) \\ 2^{-4x}, & \text{if } d_1 \ge \frac{1}{2}(1+2^{-4x}). \end{cases} \tag{8.3}$$

This expression shows that when the central distortion is minimum (equal to 2^{-4x}), the side decoders operate with their maximum distortion level. Also the central distortion is maximum (equal to $1/(2^{2x+1}-1)$), when the side decoders achieve the minimum distortion bound. Because of this, a multiple-description codec could be designed to optimize the performance of the central decoder, resulting in worse side decoding performance, or it could be designed to optimize the performance of the side decoders, resulting in a worse central decoder performance. The overall performance of a dual-description codec can be calculated from the average end-to-end distortion considering the central and side distortions as well as the distortion associated with zero correct descriptions:

$$D = d_0 p_2 + d_1 p_1 + p_0, \tag{8.4}$$

where p_2 is the probability of receiving both descriptions, p_1 is the probability of receiving one of the two descriptions, and p_0 is the probability of not receiving any useful description. In this equation, we do not need to multiply p_0 with the distortion associated with zero correct descriptions, because the assumption of a unit-variance source means the distortion associated with zero correct descriptions is equal to 1.

From examining the average end-to-end distortion for the dual-description codec, and as discussed in [2], if the transmitter knew that the channel presents a high probability of receiving both descriptions (p_2 close to 1), it should optimize the multiple-description codec for best central distortion. When the channel is known to be degraded to the extent that it is highly unlikely that both descriptions will be received, it would be better to configure the dual-description codec to achieve the best-side distortion. In between the best and worst channel conditions, the configuration should balance optimization for the central and side decoding performance. With this idea, [2] studied how to find the best operating point for the dual-description codec with a bit rate constrained to $2x$ and when the probabilities p_2,

p_1, and p_0 are known at the transmitter. This is achieved by first combining (8.3) and (8.4), leading to

$$D = \frac{p_2 2^{-4x}}{1 - \left(1 - d_1 - \sqrt{d_1^2 - 2^{-4x}}\right)^2} + d_1 p_1 + p_0, \tag{8.5}$$

for $2^{-2x} \leq d_1 < (1 + 2^{-4x})/2$. As there is a fixed total bit rate constraint, there is no point in designing for a side distortion larger than $(1 + 2^{-4x})/2$ because over this range the central distortion is constant and equal to 2^{-4x}. Over $2^{-2x} \leq d_1 < (1 + 2^{-4x})/2$, the average end-to-end distortion is a convex function. Consequently, the optimal operating point can be found by setting the derivative of D to zero, which results in the operating condition [2]:

$$\frac{\partial d_0}{\partial d_1} = \frac{-p_1}{p_2}. \tag{8.6}$$

The interpretation of this condition is that, for a given coding rate budget x for each side encoding and given the channel condition through p_1 and p_2, the central and side multiple-description codecs should have the operating point such that in the range $2^{-2x} \leq d_1 < (1 + 2^{-4x})/2$, the slope of the curve $2^{-4x}/(1 - (1 - d_1 - \sqrt{d_1^2 - 2^{-4x}})^2)$ is $-p_1/p_2$.

An example of this solution was presented in [2] for a two-state Gilbert–Elliot channel which models successful reception of a transmitted packet as one state (good) and failed reception as another state (bad). The probability of a transition from the good state to the bad state is p, and the probability of a transition from the bad state to the good state is q. It can be shown that, in the steady state, the probability that the channel is in the good state is $\Pi_0 = q/(p + q)$, while the probability of being in the bad state is $\Pi_1 = p/(p + q)$. With this model, when the descriptions are transmitted over independent Gilbert–Elliot channels with the same statistics, the probability of successfully receiving the two descriptions equals the probability that both channels are in the good state:

$$p_2 = \Pi_0^2 = \frac{q^2}{(p + q)^2}.$$

The probability of receiving successfully only one of the two descriptions equals the probability that one channel is in the good state and the other is in the bad state:

$$p_1 = \Pi_0 \Pi_1 + \Pi_1 \Pi_0 = \frac{2pq}{(p + q)^2}.$$

Applying the expressions for p_2 and p_1 in (8.6), the condition to choose the best operating mode for the dual-description codec becomes

$$\frac{\partial d_0}{\partial d_1} = \frac{-2p}{q}.$$

In the Gilbert–Elliot channel, as the probabilities p and $1 - q$ become larger, the channel is more degraded, so the slope of the curve d_0 as a function of d_1 increases or, equivalently, the multiple-description codec becomes more optimized for better side decoding distortion.

The operating point choice just described depends on the transmitter having information of the channel statistics. This may not be the case in many application scenarios. When the transmitter does not know the statistics of the channels, the question then becomes how to define an optimal multiple-description codec. The problem of optimal design of a multiple-description scalar quantizer was studied in [3], where optimality was taken to mean minimizing the central average distortion subject to a constraint on the maximum side average distortions. For a dual-description quantizer, this means minimizing $E[d_0]$ subject to $E[d_1] < D_1$ and $E[d_2] < D_2$.

There exist several approaches to MDSC design. One approach is to design a multiple-description scalar quantizer. Single-description scalar quantizers were reviewed in Section 2.3.1. In multiple-description scalar quantization, the encoder first performs a standard scalar quantization operation, as shown in Figure 8.3a, where a range of real numbers from $-A$ to A is divided into quantization intervals of equal length. Each interval is assigned a distinct index number i. The encoding operation maps the input sample value x (where $-A \leq x \leq A$) into the index i corresponding to the quantization interval which includes x. This operation is denoted $i = Q(x)$. The operation is turned into a multiple

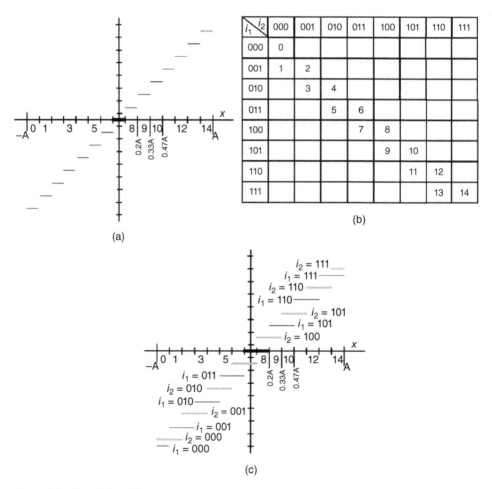

Figure 8.3 The design of a multiple-description scalar quantizer: (a) a four-bit standard scalar quantizer, (b) mapping from the standard scalar quantization index i into indices i_1 and i_2 for two descriptions, and (c) relation between the two description indices.

description encoding by taking the index i and mapping it into a pair of indices (i_1, i_2), which are transmitted as different descriptions. This mapping should be such that when both descriptions are successfully received (indices i_1 and i_2 received with no errors), it is possible to uniquely determine the original scalar quantizer index i. At the same time, when receiving only one of the two indices, it should be possible to estimate x with reasonable precision. Note that the standard scalar quantization operation shown in Figure 8.3a is particular in that it has fifteen quantization intervals. A scalar quantizer as reviewed in Section 2.3.1 would have sixteen intervals, as these would result in the finest partition of the range $-A \le x \le A$ that can be encoded into four bits, thus resulting in the lowest distortion. The scalar quantizer in Figure 8.3a is restricted to 15 intervals due to the need to be able to define pair of indexes i_1 and i_2 that together can uniquely determine an index i from the scalar quantizer, while they each individually can index a set of intervals spanning the range $-A \le x \le A$ using a smaller number of bits than the scalar quantizer. This qualitative observation previews the fact that the central decoder will exhibit some performance loss compared to a scalar quantizer designed to operate as an individual codec. This observation will be numerically confirmed soon.

Figure 8.3b from [4] shows one possible mapping from the standard scalar quantization index i into indices for two descriptions. In the figure, the possible values that i can take are shown as base-10 numbers inside the array. The corresponding values for index i_1 are the row numbers of the array, shown as binary numbers; the binary column numbers are the index i_2. Figure 8.3c shows the correspondence between indices and quantization intervals for both the standard scalar quantizer and both descriptions. For example, if the value of the input sample is $x = 0.22A$, the standard scalar quantizer index would be $i = 9$, and the multiple-description indices would be $i_1 = 101$ and $i_2 = 100$, as can be seen from the corresponding quantization levels. If both indices $i_1 = 101$ and $i_2 = 100$ are successfully received, the decoder can trace the cell in the array in Figure 8.3b where column i_2 and row i_1 intersect and find that the standard scalar quantization index was $i = 9$ and the value of the input sample is estimated to be $\hat{x} = 0.27A$. This result arises from using the combined multiple-description decoder. If only the index $i_1 = 101$ is received error-free, the standard scalar quantization index could have been $i = 9$ or $i = 10$ (see Figure 8.3b, c). In this case, the estimated input sample is the centroid of the region corresponding to $i_1 = 101$, which is $\hat{x} = 0.33A$.

There is increased expected distortion when using only one individual description as compared to the performance of combining the two descriptions into a single estimate of the input sample. As can be seen in Figure 8.3b,c, the combination of the two descriptions results in an uniform scalar quantizer with fifteen levels and quantization step size $d = 2A/15$. The quantization error signal $e(x) = x - q_i$, where q_i is the reconstruction value corresponding to quantization index (and quantization interval) i, will be in the range $-d/2 \le e(x) \le d/2$. This error can be thought of as a quantization noise that additively affects the signal being quantized. We assume that the input signal samples x follow a random process with zero mean and variance σ_x^2. The value A determining the range of input signal that can be handled by the quantizer without being clipped is assumed to be chosen as $A = 4\sigma_x$. The rationale for this choice is that the probability of clipping the input signal should be small. If the number of quantization levels is large with respect to the value of A, in the sense that the quantization step size d is relatively small, it can be assumed that the quantization noise signal $e(x)$ is approximately uniformly distributed within the range $-d/2 \leq e(x) \leq d/2$ (the assumption of large number of quantization levels may not be the case rigorously in the present example, but it allows for a simple approximate quantitative analysis). The quality of the quantization process can be measured by calculating the signal-to-quantization noise ratio:

$$SNR = \frac{\sigma_x^2}{\sigma_e^2}. \tag{8.7}$$

Following the assumption that $e(x)$ is uniformly distributed in $-d/2 \leq e(x) \leq d/2$, the variance σ_e^2 is as follows:

$$\sigma_e^2 = \int_{-d/2}^{d/2} x^2 \frac{1}{d} dx$$
$$= \frac{d^2}{12}.$$

When using the central decoder, the combination of the two descriptions results in a uniform scalar quantizer with fifteen levels, and thus the quantization step size is $d = 2A/15$. Therefore, for the central decoder, the signal-to-quantization noise ratio is as follows:

$$SNR_0 = \frac{\sigma_x^2}{\sigma_e^2} = \frac{A^2/4^2}{4A^2/(15^2 12)} = 42.2, \tag{8.8}$$

or $SNR_0 \approx 16.3$ dB. As indicated earlier, if the standard uniform scalar quantizer where to have been used as an individual codec, it would have had sixteen quantization intervals with step size $d = 2A/16$. Redoing the calculations for this case yields an signal-to-noise ratio (SNR) of 16.8 dB, illustrating the loss of 0.5 dB associated with the multiple-description design for the central decoder in this example.

When only one description is received, it can be seen in Figure 8.3b and c that the side decoder behaves as a uniform quantizer with eight quantizer levels (although this is an approximation for the intervals with index $i_1 = 000$ and $i_2 = 111$), and thus, $d = 2A/8$ and $\sigma_e^2 = 4A^2/(8^2 12)$. Consequently, the signal-to-quantization noise ratio for the side decoder in this example is as follows:

$$SNR_1 = \frac{\sigma_x^2}{\sigma_e^2} = 12, \tag{8.9}$$

or $SNR_1 \approx 10.8$ dB. The central decoder outperforms the side decoder by 5.5 dB in signal-to-quantization noise ratio.

There are multiple options on how to choose the mapping from the standard scalar quantizer into the indices for the multiple descriptions. The choice shown in the example in Figure 8.3b, c is just one possibility. Figure 8.4 shows another choice which is more complicated, but the principles of operation are the same. Compared with Figure 8.3b, c, the most evident change is that the central decoder in Figure 8.4 has twenty-two quantization levels instead of fifteen. This would improve the SNR by a little more than 3 dB. At the same time, as seen in Figure 8.4, there is more ambiguity in the index of the standard scalar quantizer when only one description is successfully received. This larger ambiguity translates into higher-side distortion, so there is a trade-off between central and side distortion.

i_1\\i_2	000	001	010	011	100	101	110	111
000	0	1						
001	2	3	5					
010		4	6	7				
011			8	9	11			
100				10	12	13		
101					14	15	17	
110						16	18	19
111							20	21

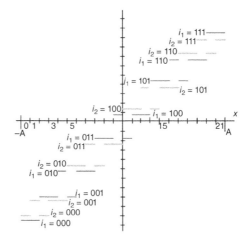

Figure 8.4 An alternative multiple-description scalar quantization design with lower central decoder distortion but higher-side decoder distortion compared to the design in Figure 8.3.

Because some of the most important applications for video communication are multicast transmission and video over the Internet, MDSC was adopted in relatively early video coding standards, such as H.263+, [5]. Under the name *video redundancy coding* (VRC), MDSC was included as a technique that implements multiple-coded descriptions through the use of more than one reference frame and more than one prediction thread. It creates a minimum of two coded bit streams. As exemplified in Figure 8.5, each bit stream carries a thread of inter-predicted P-frames that depend on earlier frames belonging to the same thread and bit stream, but not on any information contained in the other coded bit streams. The original video frames are assigned to the threads in an interleaved fashion, and they all start and end with common reference frames. These common starting and ending frames, which usually are I-frames, do not follow the precise definition for MDSC but are nonetheless used as synchronization frames for all threads. With two threads, irreparable

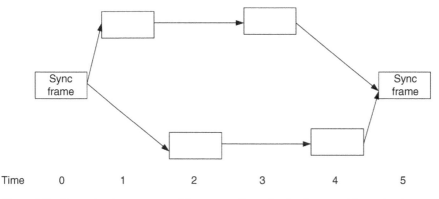

Figure 8.5 Example of a sequence of frames configured to implement video redundancy coding.

damage to one of the coded bit streams results in a decoded video sequence that is played back using only the frames encoded in the thread received correctly. Because frames are assigned to threads in an interleaved fashion, damage to one thread results in a video playback at half the frame rate. This playback at half the frame rate starts at the time associated with the damaged frame and lasts until the next correctly received synchronization frame.

With the introduction of 3D or stereoscopic video, MDSC was extended in the new dimension of extra video views [6]. The first scheme in [6], called *multistate stereo-MDC*, extends video redundancy to two views. As shown in Figure 8.6, the scheme generates two descriptions. One description encodes the odd-numbered frames from both views. In this description, the first frame from the left view is intra-coded and subsequent odd-numbered frames in the left view are predicted from the previous odd-numbered frame of the left view. The right-view frames in that description are predicted using as reference the preceding right-view odd-numbered frames (motion compensation) or the left-view frames at the corresponding time (disparity compensation). The second description in this scheme is constructed as a complement of the first, using only the even-numbered frames, and with an anchor intra-frame in the right view, rather than the left, as shown in Figure 8.6 This MDSC scheme maintains the same functionality as VRC, extended to stereoscopic video. In the event of successful reception of only one of the two descriptions, the stereoscopic video is reproduced at half the frame rate and with approximately the same quality for both views.

In the second scheme presented in [6], called *spatial scaling stereo-MDC*, loss of a description does not cause a reduction in frame rate but rather results in lower quality for one view. Assuming two views, in spatial scaling stereo-MDC (Figure 8.7), one description is formed from the left view encoded as a regular single-view video sequence (that is, encoding a group of frames starting with an I-frame, followed by a sequence of temporally inter-predicted frames), and the right view spatially downsampled and then predictively encoded using either motion compensation or disparity compensation (that is, using either

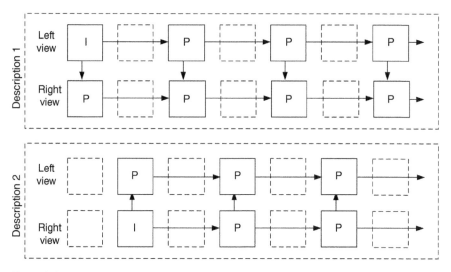

Figure 8.6 Multistate stereo-multiple-description coding.

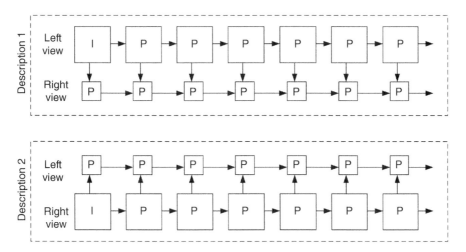

Figure 8.7 Spatial scaling stereo-multiple-description coding.

the previous frame in the same view or the frame from the other view at the same time). The second description reverses the roles of the left and right views. If only one description is successfully received after transmission, one of the two views is reproduced with a reduced spatial resolution because of the downsampling. This reduced quality is to some extent masked because of *binocular suppression*, an effect of the human visual system in which blurring effects in one stereo view are often imperceptible if the other stereo view presents high quality. Of course, when receiving the two descriptions, the video can be recovered with the least possible distortion (there still remains the typical quantization distortion) by using both full spatial resolution views.

Another technique to create multiple descriptions is to apply a linear transform to sets of source symbols that have already been quantized. Frequently. the transform is applied to pairs of quantized source symbols, resulting in a dual-description codec. The goal for applying the transform is to introduce some redundancy, or correlation, between the two descriptions that can be used to reduce the side distortion by better estimating a missing description from a correctly received one. Because of the common use of two input variables that are transformed to introduce correlation between them, this technique has received the name of *multiple-description coding from correlated pairwise transform* [7].

Figure 8.8 shows the basic block diagram for this technique. The input source symbols A and B are quantized and the resulting variables A_q and B_q are linearly transformed:

$$\begin{bmatrix} C_q \\ D_q \end{bmatrix} = \mathbf{T} \begin{bmatrix} A_q \\ B_q \end{bmatrix},$$

$$(8.10)$$

where C_q and D_q are the transformed variables and $\mathbf{T}$ is a 2×2 matrix:

$$\mathbf{T} = \begin{bmatrix} a & b \\ c & d \end{bmatrix}.$$

The design of the matrix $\mathbf{T}$ controls the correlation between C_q and D_q.

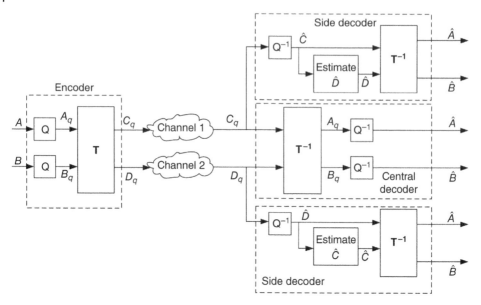

Figure 8.8 Block diagram to implement multiple-description coding from correlated pairwise transform.

We use $\hat{A}$ and $\hat{B}$ to denote the reconstructed values for A and B. After transmission of C_q and D_q over separate independent channels, if both C_q and D_q are correctly received, $\hat{A}$ and $\hat{B}$ can be computed by applying the inverse transform:

$$\begin{bmatrix} A_q \\ B_q \end{bmatrix} = \mathbf{T}^{-1} \begin{bmatrix} C_q \\ D_q \end{bmatrix},$$ (8.11)

to obtain A_q and B_q, followed by inverse quantization of A_q and B_q to obtain $\hat{A}$ and $\hat{B}$.

When only one of the two transmissions is received successfully, for example C_q, the originally encoded symbols A and B can be reconstructed (with a higher distortion than that obtained when both transmissions are received successfully) by first performing inverse quantization on C_q to obtain $\hat{C}$, and then using a linear estimator to estimate an inverse quantized version $\hat{D}$ of the lost D_q:

$$\hat{D} = a_C \hat{C}.$$ (8.12)

This is followed by application of the inverse transform:

$$\begin{bmatrix} \hat{A} \\ \hat{B} \end{bmatrix} = \mathbf{T}^{-1} \begin{bmatrix} \hat{C} \\ \hat{D} \end{bmatrix}.$$ (8.13)

Suppose now that the two inputs A and B are independent Gaussian random variables with zero mean and variances σ_A^2 and σ_B^2, respectively. In a simple approach, the linear estimator used in (8.12) can be designed to minimize the mean squared error between D and $\hat{D}$:

$$E[(D - \hat{D})^2] = E[(D - a_C \hat{C})^2]$$

$$= \sigma_D^2 - 2a_C \, \text{Cov}(\hat{C}, D) + a_C^2 \sigma_{\hat{C}}^2.$$

Taking the derivative of this error with respect to a_C and equating it to zero results in the linear estimator coefficient:

$$a_C = \frac{\text{Cov}[\hat{C}, D]}{\sigma_{\hat{C}}^2}$$

$$= \frac{\text{E}[C, D]}{\sigma_C^2 + \sigma_{q_C}^2},$$

where $\text{E}[C, D]$ is the correlation between C and D and $\sigma_{q_C}^2$ is the variance of the quantization error associated with $\hat{C}$. In the aforementioned expression, to calculate the covariance $\text{Cov}[\hat{C}, D]$ between $\hat{C}$ and D, and the variance $\sigma_{\hat{C}}^2$, the inverse quantized version of C_q is expressed as $\hat{C} = C + q_C$, where q_C is the quantization error. Assuming high-rate quantization, C and the quantization noise q_C become uncorrelated and can be treated as being independent. They are also zero mean, leading to $\sigma_{\hat{C}}^2 = \sigma_C^2 + \sigma_{q_C}^2$. The high-rate quantization assumption also leads to

$$\text{Cov}[\hat{C}, D] = \text{E}[\hat{C}, D] - \text{E}[\hat{C}]\text{E}[D]$$

$$= \text{E}[C + q_C, D]$$

$$= \text{E}[C, D] + \text{E}[q_C, D]$$

$$= \text{E}[C, D],$$

where these equalities use the facts that D is zero mean, that $\hat{C} = C + q_C$, and that q_C and D are independent. The correlation $\text{E}[C, D]$ is controlled by the transform $\mathbf{T}$ and can be expressed as $\text{E}[C, D] = \sigma_C \sigma_D \cos \phi$ by introducing a correlation angle ϕ between C and D. Then the coefficient of the linear estimator in (8.12) that minimizes the mean squared error is as follows:

$$a_C = \frac{\sigma_C \sigma_D \cos \phi}{\sigma_C^2 + \sigma_{q_C}^2}.$$

The correlation angle between C and D can be related to the parameters of the transform $\mathbf{T}$ by considering that $C = aA + bB$, $D = cA + dB$ and that any cross-correlation $\text{E}[A, B]$ is equal to zero (because it is assumed that A and B are independent and zero mean). Then the covariance matrix for the random variables C and D can be calculated to be

$$\mathbf{R}_{CD} = \begin{bmatrix} a^2\sigma_A^2 + b^2\sigma_B^2 & ac\sigma_A^2 + bd\sigma_B^2 \\ ac\sigma_A^2 + bd\sigma_B^2 & c^2\sigma_A^2 + d^2\sigma_B^2 \end{bmatrix}.$$

From here

$$\cos \phi = \frac{ac\sigma_A^2 + bd\sigma_B^2}{\sqrt{a^2\sigma_A^2 + b^2\sigma_B^2}\sqrt{c^2\sigma_A^2 + d^2\sigma_B^2}}.$$

While useful for design purposes, these expressions lend little insight into analytical results. It is more practical to write the covariance matrix for the random variables C and D in the simple form:

$$\mathbf{R}_{CD} = \begin{bmatrix} \sigma_C^2 & \sigma_C \sigma_D \cos\phi \\ \sigma_C \sigma_D \cos \phi & \sigma_D^2 \end{bmatrix},$$

and the covariance matrix for the random variables A and B as:

$$\mathbf{R}_{AB} = \begin{bmatrix} \sigma_A^2 & 0 \\ 0 & \sigma_B^2 \end{bmatrix}.$$

Since $\mathbf{R}_{CD} = \mathbf{T}^T \mathbf{R}_{AB} \mathbf{T}$, taking the determinants on both sides yields

$$\sigma_C^2 \sigma_D^2 \sin^2 \phi = \sigma_A^2 \sigma_B^2, \tag{8.14}$$

which allows calculation of the redundancy introduced by the transformation. This redundancy is the extra coding bit rate needed to achieve a given value of distortion (denoted as D_0 below). Let R^* be the minimum encoding rate for the source needed to achieve D_0, which is given by the rate-distortion relation:

$$R^* = \frac{1}{2} \log_2 \frac{\sigma_A \sigma_B}{D_0} + K,$$

where K is a constant. After applying the transform $\mathbf{T}$, the rate needed to achieve a distortion D_0 is as follows:

$$R = \frac{1}{2} \log_2 \frac{\sigma_C \sigma_D}{D_0} + K.$$

Therefore, the redundancy introduced by the transform is as follows:

$$\rho = R - R^* = \frac{1}{2} \log_2 \frac{\sigma_C \sigma_D}{D_0} - \frac{1}{2} \log_2 \frac{\sigma_A \sigma_B}{D_0} = \frac{1}{2} \log_2 \frac{\sigma_C \sigma_D}{\sigma_A \sigma_B}.$$

Using (8.14), the redundancy can be written as:

$$\rho = -\frac{1}{2} \log_2 \sin \phi.$$

Figure 8.9 shows the relation between the redundancy and the correlation angle. A consequence of this result is that it is completely equivalent to talk about redundancy or correlation angle. Since $E[C, D] = \sigma_C \sigma_D \cos \phi$, there is a proportional relation between the correlation between C and D and $\cos \phi$. Figure 8.9 illustrates this relation showing how the redundancy changes with $\cos \phi$ (or equivalently, the correlation between C and D). As the correlation increases, so does the coding redundancy.

A common way to write transforms such as the ones considered in this technique is with a scaling-rotation form:

$$\mathbf{T} = \begin{bmatrix} r_2 \cos \theta_2 & -r_2 \sin \theta_2 \\ -r_1 \cos \theta_1 & r_1 \sin \theta_1 \end{bmatrix}.$$

It is also common practice to seek a transform that has determinant equal to 1 because it allows for lossless implementation of the transform and its inverse with fixed point arithmetic. Applying this condition results in $\sin \Delta_\theta = 1/(r_1 r_2)$, where $\Delta_\theta = \theta_1 - \theta_2$. Using this relation, it is possible to derive for this form of transform matrix a trigonometric relation between the correlation angle ϕ and the angle Δ_θ given the angle θ_1 as a fixed parameter, [7],

$$\cot \Delta_\theta = \frac{2r \cot \phi - (r^2 - 1) \sin(2\theta_1)}{2r^2 \cos^2 \theta_1 + 2 \sin^2 \theta_1},$$

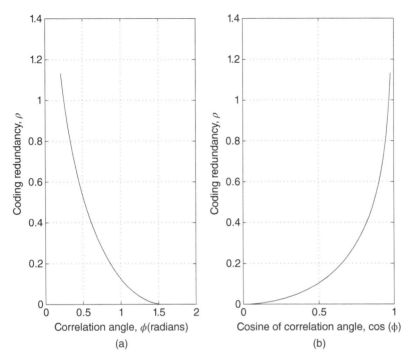

Figure 8.9 (a) Relation between coding redundancy ρ and correlation angle ϕ. (b) Relation between ρ and $\cos \phi$.

where $r = \sigma_A/\sigma_B$. Following some algebraic operations, it is possible to derive an expression for the side distortion as a function of the transform parameters and its corresponding correlation angle (or equivalently, the coding redundancy) [7]:

$$D_1 = \frac{(\sigma_A^2 - \sigma_B^2)((\sigma_A^2\cos^2\theta_1 - \sigma_B^2\sin^2\theta_1)\sin^2\phi - \sigma_A\sigma_B \sin(2\theta_1)\sin\phi\cos\phi) + 2\sigma_A^2\sigma_B^2}{4(\sigma_A^2\cos^2\theta_1 + \sigma_B^2\sin^2\theta_1)}.$$

Figure 8.10 illustrates the behavior of this result with $\sigma_A = 1$, $\sigma_B = 0.3$, and $\theta_1 = \pi/5$. The addition of redundancy (or correlation, when considering Figure 8.9) results in a lower-side decoder distortion.

Sometimes, one joint source–channel coding (JSCC) technique inspires development of another technique, and their difference may be subtle. This happened with a multiple-description technique in [8] related to Priority Encoding Transmission discussed in Chapter 5. Called *multiple-description coding using FEC codes*, this technique generates many descriptions from a scalable (also layered or progressive) source-encoded bit stream. The multiple-description bit stream using FEC codes transforms an error-sensitive progressive bit stream into one more robust against errors, a kind of transcoding operation.

In the following, we use n to denote the number of descriptions to be generated, and $D(k)$ denotes the distortion when receiving k out of the n descriptions. The input bit stream is partitioned into n layers with n distortion values $D(k)$. The error correcting code is designed so

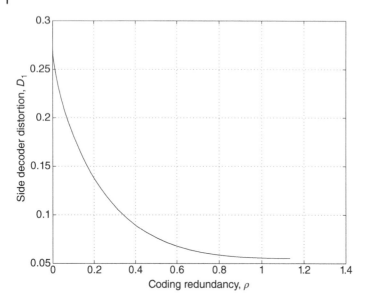

Figure 8.10 Side decoder distortion as a function of the coding redundancy ρ, with $\sigma_A = 1$, $\sigma_B = 0.3$, and $\theta_1 = \pi/5$.

that the ith description is successfully received when there are no more than $n - i$ erasures during the communication process. The ith layer can be decoded when the ith description is successfully received. This behavior is implemented with Reed–Solomon block channel codes. The ith layer is separated into i channel code symbols and is encoded using an (n, i) Reed–Solomon code. Recall that this code will be able to recover from the loss of any $n - i$ symbols. The channel coding operation from every one of the n layers will result in n symbols that are placed each in a different description. Then, each description will have one symbol from the output of channel encoding each layer. This operation is illustrated in Figure 8.11.

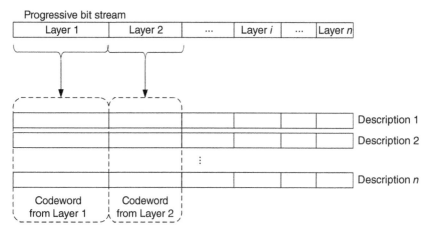

Figure 8.11 The transcoding operation in the multiple-description source coding using forward error correction codes.

To complete the scheme described earlier, we must specify how to separate the progressive input bit stream into layers. Since the partition results in n distortion values $D(k)$, finding the partition of the progressive input bit stream involves finding the rate partition of the input bit stream, or equivalently, how many bits are assigned to each layer. Since by virtue of the multiple-description transcoding scheme, the n distortion values $D(k)$ correspond to the n side distortion values for the n descriptions, the rate partition of the input stream also determines the distortion corresponding to receiving any number of descriptions. The partition needs to be calculated with the goal of minimizing the expected distortion. The problem is reminiscent of the allocation of bit rates to different source encoders subject to a total coding bit rate constraint. It can be solved in a similar way using Lagrange multipliers. An algorithm based on Lagrange multipliers was presented in [8].

8.2 Error-Resilient Coded Bit Streams

Previous chapters discussed various embedded codecs; an embedded codec allows for progressive transmission and reception of the source (i.e. successive refinement). Some embedded codecs are also attractive because they provide high performance and low complexity. Examples are the embedded zerotree wavelet (EZW) coder [9] and the set partitioning in hierarchical trees (SPIHT) coder, [10], described in Chapter 2. However, a progressively encoded bit stream is typically more sensitive to transmission noise, because an error may cause misinterpretation of later bits, leading to error propagation and a possible loss in synchronization. After the first error has affected an embedded bit stream, the progressive property is lost as the bits following the error may not improve the reconstruction quality; in fact, they might degrade it.

Earlier chapters discussed some techniques to address this error sensitivity, e.g. UEP can be used to protect the earlier part of the bit stream more than later parts. Bit streams with high sensitivity to communication errors that lead to error propagation and loss of synchronization between encoder and decoder states are not exclusive to embedded bit streams. The use of variable length coding also increases the sensitivity to transmission errors and can lead to loss of synchronization. Coded bit streams that result from predictive coding, commonly found in video or speech encoding, also suffer from similar sensitivity to communication errors. For these, some of the techniques that are applicable to embedded bit streams are also applicable, while other techniques are of specific use with predictive coding. All of these techniques present a trade-off between increasing the resilience of the source-encoded bit stream and decreasing the coding efficiency. These trade-offs need to be considered from the perspective of end-to-end distortion, in which case the reduction in distortion provided by the error resilience could outpace the increase in distortion from the reduced source encoding efficiency.

8.2.1 Robust Entropy Coding

Entropy coding usually results in source codewords of variable length. An error during transmission can cause the decoder to lose synchronization with the encoder. For example, a single-bit error in a long codeword may make its beginning portion look like a short codeword. In this case, not only will the affected codeword be wrong, but the decoder will attempt to start reading the next codeword in the wrong place. It may be impossible to decode the rest of the transmitted data block after the corrupted source codeword. One solution to increase the resiliency against transmission errors is to insert resynchronization markers at regular intervals within the transmitted bit stream. A resynchronization marker is a sequence of bits with a known pattern that allows the decoder to identify them within the received bit stream, providing an anchor point from where to resume decoding. It is common practice to include, immediately after the resynchronization marker, some metadata for the subsequent content of the bit stream. For example, after the marker, the metadata could indicate which video frame the next content refers to. The addition of resynchronization markers reduces coding efficiency, because those bits do not carry source information. More frequent resynchronization markers make the transmitted bit stream more error-resilient, so there is a trade-off between the improvement in resiliency and the loss in encoding efficiency.

Since resynchronization markers are used to contain errors within sections of the bit stream by demarcating sections of encoded data in order to use them effectively it is important that there is no dependency between the data delimited by a marker and the data demarcated by another marker. Otherwise, damage in data delimited by one resynchronization marker affects data demarcated by another marker, and the error will affect more than the bit stream section where it occurred.

Figure 8.12 illustrates another technique for robust entropy coding. Called *reversible variable length coding (RVLC)*, it is used in combination with resynchronization markers to reduce error damage in a stream of variable length codewords. An RVLC encoder generates codewords of variable length that can be read in the standard forward direction (say, from left to right, as in Figure 8.12), or in the reverse direction (from right to left). When an error is encountered in the RVLC stream, the decoder can stop forward decoding, search forward in the bit stream for a resynchronization marker, and from this point, decode in the reverse direction until an error is encountered (Figure 8.12). Reversible variable length codes are defined by having the prefix property (see Chapter 2) hold in both the forward and reverse decoding directions.

Tables 8.1 and 8.2 show examples of variable length codes modified, so they can be decoded in the reverse direction. Table 8.1 shows a Hamming weight variable length code. The RVLC version of this code, from [11], is constructed by adding a fixed length bit sequence as prefix and suffix to each codeword. The idea here is to ensure that the prefix

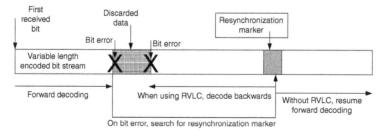

Figure 8.12 Illustration of the use of resynchronization markers and reversible variable length coding (RVLC).

Table 8.1 A Hamming weight variable length code and its equivalent reversible variable length code (RVLC).

Hamming weight variable length code	RVLC
1	111
01	1011
001	10011
0001	100011

property of the code holds in both the forward and reverse decoding directions. For Table 8.1, the prefix or suffix sequence is just the single bit "1". In this case, the decoding can be done in either direction by counting the number of zeros until three ones have been encountered, which would signal the end of a codeword.

Table 8.2 exemplifies another approach, from [12], that can be used to construct an RVLC. The variable length Golomb–Rice code in Table 8.2 has suffixes of constant length (one bit), while the prefixes have variable length. The RVLC is constructed by replacing the prefixes with others that have the same lengths but are palindromes. A palindrome is a word or sentence that reads the same way in both the forward and reverse directions, e.g. "radar," "level," and "never odd or even." In this case, this is done by starting and ending the prefix with a "1" and adding in-between as many bits "0" as needed. To decode in the reverse direction, the decoder would record the first bit as the suffix and then would complete reading the codeword by using the prefix property in the prefix.

While in general an RVLC has lower coding efficiency than a comparable variable length coding (VLC), there exist practical implementations where the efficiency loss is minimal. For example, for video coding, training sequences can be used so that the reversible

Table 8.2 Golomb–Rice code and its equivalent reversible variable length code (RVLC).

Source symbol	Non-reversible Golomb–Rice code			Reversible Golomb–Rice code		
	Prefix	Suffix	Codeword	Prefix	Suffix	Codeword
0	0	0	00	0	0	00
1	0	1	01	0	1	01
2	10	0	100	11	0	110
3	10	1	101	11	1	111
4	110	0	1100	101	0	1010
5	110	1	1101	101	1	1011
6	1110	0	11100	1001	0	10010
7	1110	1	11101	1001	1	10011

variable length codewords are designed to closely match the statistics of the discrete cosine transform (DCT) coefficients in the video texture representation [11]. Because of this, RVLC has been included in video coding standards such as Moving Picture Experts Group (MPEG)-4.

Another idea to increase the resiliency of variable length coding is reorganizing the codewords into data blocks with a size known in advance. The key property of the reorganization is that each data block contains the beginning of only one variable length codeword at the start of the data block itself. The goal of this operation is to avoid the use of resynchronization markers. This idea was presented in [13] under the name of *error-resilient entropy code* (EREC). The variable length codewords are organized into a frame with size L bits. The frame is divided into N time slots, each with length l_i bits. The receiver must know in advance the values of L, N and l_i for $i = 1, 2, \ldots, N$. To reduce the overhead, N can be fixed in advance and the l_i can be determined as a function of L, in which case only the value of L needs to be exchanged. When possible, it is advantageous to operate with all frame slots having the same length $l_i = L/N$, for all i. Figure 8.13a shows an example of this case. The EREC frame is depicted as a 2D structure; each row is one frame slot with length $l_i = L/N$ and $N = 8$.

Given this configuration, EREC can be used to transmit N codewords per frame. The ith variable length codeword has b_i bits. The bit reorganization starts by placing one variable length codeword at the start of each frame slot (Figure 8.13b,c). When doing this, some codewords will fit perfectly into the frame slot (that is, $b_i = l_i$) as with the second and sixth frame slot in Figure 8.13c. Usually, most codewords are either too long ($b_i > l_i$), as with the third, fourth, and last slots in Figure 8.13c, or too short ($b_i < l_i$), as with the first, fifth, and seventh slots. The EREC algorithm iterations systematically reorganize the bits from the long codewords into the available space in slots with short codewords. At iteration n, the bits from a codeword in slot i that exceed l_i are allocated to the slot $i + \phi_n$ mod N, where ϕ_n is an offset value that can be made to change at each iteration. The sequence is predetermined and known by the encoder and decoder. According to [13], a pseudorandom sequence ϕ_n provides better error-resilient properties.

Figure 8.13 shows an example with four iterations of the algorithm. In the first iteration, shown in Figure 8.13d, the offset value is $\phi_1 = 4$. The bits of the codeword in the third slot that cannot fit are moved to the seventh slot, where some of them fit but some do not. In the second iteration (Figure 8.13e), with $\phi_2 = 1$, the bits that did not fit in the seventh slot are moved to the eighth slot, which has no available space. So the assignment of these bits is not accomplished yet. The remaining bits from the codeword originally assigned to the third slot are finally allocated in the third iteration with $\phi_3 = 5$ (Figure 8.13f), finding available space in the fifth time slot. Note that this slot contains bits from three different codewords by the end of the third iteration. It contains bits from four different codewords after the fourth iteration shown in Figure 8.13g. Having slots that contain bits from more than one codeword is in fact the goal of the EREC algorithm. This is not a problem for decoding as the sequence ϕ_n is known at the receiver and also the end of each codeword can be determined as part of the standard decoding process (assuming that there is no corruption due to communication errors). The final result of the bit reorganization is shown in Figure 8.13g. Some space in the EREC frame (in the fifth slot) is not used. This unused space (bit positions) reduces the coding efficiency of the original variable length code.

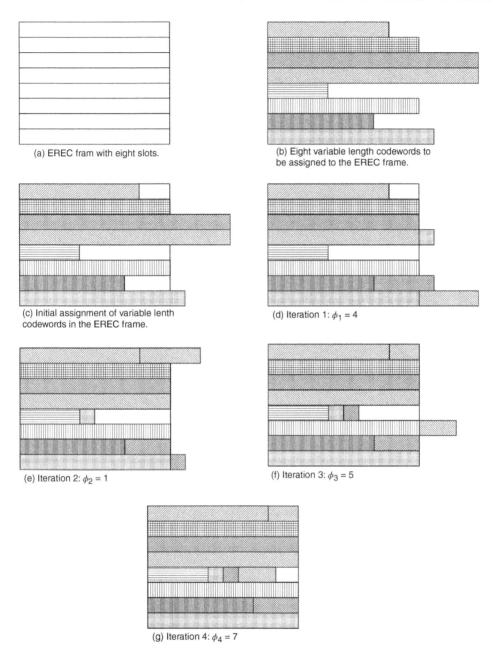

(a) EREC fram with eight slots.

(b) Eight variable length codewords to be assigned to the EREC frame.

(c) Initial assignment of variable lenth codewords in the EREC frame.

(d) Iteration 1: $\phi_1 = 4$

(e) Iteration 2: $\phi_2 = 1$

(f) Iteration 3: $\phi_3 = 5$

(g) Iteration 4: $\phi_4 = 7$

Figure 8.13 Example of the operation of the EREC bit allocation algorithm.

The goal of EREC is to reorganize the bits from a variable length code into a fixed known structure that implicitly determines the start of the codewords. In this way, the reorganization addresses the reason for the lack of resiliency in variable length coding: errors in the bit stream make the decoder lose synchronization, and all data after the first error must be discarded. A similar problem occurs when transmitting progressive bit streams. In this case, the decoder depends on earlier parts of the bit stream to decode later parts. Once an error occurs, the rest of the information in the bit stream cannot be used. The SPIHT and EZW coders have high coding efficiency, and for cases where progressivity is not important but other properties are, [14] presented a simple scheme to improve error resiliency for the SPIHT- and EZW-coded streams. The scheme addressed the central reason why the SPIHT- and EZW-coded streams show sensitivity to channel errors, namely its data organization. The scheme in [14] increased the robustness to errors by reorganizing the information in the encoded stream at the cost of sacrificing its progressivity.

SPIHT decomposes an image into wavelet coefficients across spatial frequencies. The encoding operation efficiently represents the tree structures that are created between wavelet coefficients, starting each tree from one of the coefficients in the low–low spatial frequency band and growing toward the three children coefficients in each of the next directional bands. Then, from each of the three children coefficients, the tree grows by spawning four branches at each subband decomposition. For example, in a 512×512 pixel image, with four levels of subband decompositions, there are 1024 "root" coefficients in the low–low spatial frequency band and each will have $3(1 + 4 + 16 + 64) = 255$ descendants in its tree. Once the tree structure has been created, the structure is encoded progressively by transmitting the representation bits in order of significance. It is this transmission in order of significant bits that enables the embedded coding property and also what makes the transmission more susceptible to channel errors.

The scheme in [14] reorganizes the structure into sub-streams, where each sub-stream corresponds to one tree rooted in a wavelet coefficient in the low–low band. Using again the case of a 512×512 pixel image, with four levels of subband decompositions, there are 1024 sub-streams. Transmission is implemented using fixed-size packets. Each packet will carry a certain number of sub-streams. The bits corresponding to one sub-stream are transmitted in only one packet, but packets can carry more than one sub-stream. The constraint in the assignment of sub-streams to packets follows the recurring idea in many error resilience schemes: the decoding of information in one packet is independent of the information contained in any other packet. In this way, when a packet is lost, the information of the sub-streams within is lost, but all the sub-streams in the successfully received packets can be recovered without seeing any effect from the lost packets. Moreover, the wavelet coefficient trees that are successfully recovered can be used to conceal the information lost in the coefficient trees that were not received.

To assign sub-streams to fixed-size packets, the sub-streams are placed in a predetermined order known by both transmitter and receiver. Following this order, sub-streams are assigned to a packet by placing as many sub-streams as possible in one packet such that they fit completely within. The packet will need to carry some extra overhead bits to identify the first sub-stream in the packet. Unused space left in the packet can be fully utilized by extending the sub-streams with more levels of significant bits. To avoid extra overhead to communicate when each sub-stream ends, the sub-streams can be interleaved within the packet. The interleaving still requires overhead to indicate the number of sub-streams carried in the packet, but this overhead is small and has a very small impact on performance.

8.2.2 Predictive Coding Mode Selection

As mentioned in Chapter 2, predictive coding is a popular technique widely used for coding multimedia sources. But predictive coding can cause error propagation and distort the reconstructed source for as long as the erroneous sample is used in the prediction of later source samples. Another class of error-resilient coding techniques aims at limiting these negative effects. One can limit the extent of error propagation by regularly disabling the use of predicted coding and, instead, communicating a coded representation of the source that does not use prediction. Because this technique has been very popular in video coding – it has been incorporated into video coding standards for decades – it is frequently named using video coding terminology as "Intra/Inter mode selection," "adaptive Intra refresh" or "Intra mode refresh." These differ in details but share the same core approach. For example, the work in [15] discussed the problem of choosing when to use intra- or inter-prediction when encoding video for transmission over a network subject to packet dropping.

Joint source-channel coding solutions often occur when considering a source coding design problem from an expanded end-to-end perspective that includes the complete communication processing chain, from input source samples to the reconstruction at the receiver. In this context, consider the traditional problem of choosing the temporal prediction mode during video source encoding:

$$\min_{M_i^n} D_q(M_i^n) \qquad \text{subject to } R(M_i^n) \le R_b, \tag{8.15}$$

where M_i^n is the temporal prediction mode used during source encoding the video macroblock at position i in frame n, D_q is the quantization distortion, $R(M_i^n)$ is the number of bits used for quantization of the macroblock, and R_b is the bit budget available for this quantization. For simplicity, we consider only two modes here: no temporal prediction and prediction based on the preceding video frame. In [15], this problem was studied for a packet erasure channel (e.g. Internet) with distortion calculated end to end. Now, the problem (8.15) is written as:

$$\min_{M_i^n} D(M_i^n) \qquad \text{subject to } R(M_i^n) \le R_b, \tag{8.16}$$

where $D(M_i^n)$ is the end-to-end distortion for the video macroblock at position i in frame n when using temporal prediction mode M_i^n. The value of this distortion depends on source encoding characteristics, the quality of the communication channel, and the error concealment procedure applied at the receiver.

The effects of the source encoder on the distortion $D(M_i^n)$ go beyond the quantization operation. The distortion also depends on how the quantized data is organized within the bit stream and into packets for transmission. The case studied in [15] is video transmission over the Internet where an IP default packet size (576 bytes) far exceeds the maximum size of a macroblock (90 bytes). A video bit stream is usually organized into data sets larger than a macroblock, such as a slice or a group of blocks (GOBs). Slices and GOBs are similar concepts, both being a collection of macroblocks, but they differ in that a GOB has a fixed number of macroblocks from a fixed row-by-column configuration in the video frame, while the slice can contain a variable number of consecutive macroblocks from the video frame. To limit the effect of a packet loss to only one GOB or slice, the encoded bit stream should be divided for transmission in such a way that a GOB or slice is not separated into more than one packet. Given the typical sizes for a GOB or slice, each packet will contain no more than one GOB or slice.

The quality of the communication channel of course affects the end-to-end distortion. For Internet communication, the Gilbert–Elliot channel model is a handy tool, modeling the channel as a Markov chain with two states: packet loss and packet successfully delivered [15].

The end-to-end distortion also depends on the operations performed at the receiver to conceal the errors created by a lost packet. The subject of error concealment is especially rich for video communication. In [15], the error concealment scheme is a simple common one: when a macroblock is lost, its motion vector is estimated to equal that of one of the neighboring macroblocks. If these motion vectors are not available, the estimated motion vector is taken to equal zero. The lost macroblock is then replaced by the colocated macroblock from the previous video frame.

The problem (8.16) can now be restated to consider the source encoding characteristics, the quality of the communication channel and the error concealment procedure applied at the receiver. To consider the source encoding packetization characteristics, it is necessary to solve problem (8.16) across a complete GOB/slice. Let a GOB or slice be a collection of macroblocks using the notation $B_g^n = \{B_g^n, B_{g+1}^n, \ldots, B_{g+N_G-1}^n\}$, where B_g^n is the macroblock from frame n with index g and B_g^n is the GOB or slice containing the N_G macroblocks from B_g^n through $B_{g+N_G-1}^n$. Under these settings, problem (8.16) becomes that of jointly finding the set of N_G predictive coding modes $\mathcal{M}_g^n = \{M_g^n, M_{g+1}^n, \ldots, M_{g+N_G-1}^n\}$ for all the macroblocks in the GOB or slice that minimizes the end-to-end distortion subject to a bit budget $R(B_g^n, \mathcal{M}_g^n)$, for the GOB or slice:

$$\min_{\mathcal{M}_g^n} D(B_g^n, \mathcal{M}_g^n) \quad \text{subject to } R(B_g^n, \mathcal{M}_g^n) \leq R_B. \tag{8.17}$$

When the distortion measure is additive, this problem can be simplified by using Lagrange multipliers to convert it to an unconstrained minimization problem:

$$\min_{\mathcal{M}_g^n} J(B_g^n, \mathcal{M}_g^n) = \min_{\mathcal{M}_g^n} \sum_{i=g}^{g+N_G-1} D(B_i^n, \mathcal{M}_g^n) + \lambda R(B_i^n, \mathcal{M}_g^n),$$

where λ is the Lagrange multiplier. This problem can be further simplified by making the assumption that the mode chosen for macroblock B_i^n is the only one that affects its distortion and rate (this is not the case in general but simplifies the problem by assuming negligible effects between macroblocks). With this assumption, the mode selection problem can be written as:

$$\sum_{i=g}^{g+N_G-1} \min_{M_i^n} J(B_i^n, M_i^n) = \sum_{i=g}^{g+N_G-1} \min_{M_i^n} D(B_i^n, M_i^n) + \lambda R(B_i^n, M_i^n). \tag{8.18}$$

At this point, the mode selection problem for the GOB/slice has become one that is common in rate-distortion optimization. The solution includes finding the value for the Lagrange multiplier λ. One approach is to regard this as rate-distortion optimization for a set of heterogeneous source encoders that are all subject to a common bit rate budget. The solution to this problem was presented in [16]. In Chapter 11, we will see that this has applications in the solution of other JSCC problems. Nonetheless, the approach taken in [15] to compute λ is a simpler one, based on considering the state of the buffer that holds the video frames during encoding. At the end of frame n, λ_n is recomputed for frame $n + 1$ as:

$$\lambda_{n+1} = \frac{2\Gamma_n + (L_\Gamma - \Gamma_n)}{\Gamma_n + 2(L_\Gamma - \Gamma_n)}\lambda_n,$$

where Γ_n is the buffer occupancy state at the end of frame n and L_Γ is the buffer size. In this formulation, as the buffer gets full ($\Gamma_n \to L_\Gamma$) λ_{n+1} gets updated to be close to $2\lambda_n$, thereby significantly increasing the penalty on rate in the Lagrangian and dictating lower encoding rates. Conversely, as the buffer gets empty ($\Gamma_n \to 0$) λ_{n+1} gets updated to be close to $0.5\lambda_n$, thereby significantly decreasing the penalty on rate and allowing for higher encoding rates.

The computation of distortion $D(B_i^n, M_i^n)$ in the aforementioned equations still remains. It depends on the characteristics of the macroblock and on whether predictive coding is used. The computation can be split into two calculations, one when no prediction is used and the other when prediction is used. In the present setup, the formulas that model distortion also need to consider the use of error concealment. The calculation of macroblock distortion for both prediction modes starts by modeling the distortion as the average distortion between the pixels comprising the macroblock and their corresponding expected reconstructed values. It is in these expected reconstructed values where the considerations of source encoding characteristics, the quality of the communication channel, and the error concealment procedure come into play.

For a successfully received intra-coded macroblock, the expected reconstructed distortion equals the quantization distortion. When the data for the macroblock is received with errors, the distortion will become, in general, the average difference between the pixels comprising the macroblock and the expected reconstructed value for the macroblock in the previous video frame that is pointed to by the estimated motion vector (for example, the motion vector of the macroblock directly above the lost macroblock). If this motion vector has not been successfully received, its value is assumed to equal zero. The motion vector is also set to zero when the lost macroblock is at the top of the frame and does not have another macroblock above. In calculating the expected values for the distortion, it is necessary to include the probabilities for each event, which follow from the probabilities that a packet is received or lost, determined by the state probabilities in the Gilbert–Elliot model (appropriate to model Internet communication in simple terms, as mentioned earlier). The computation of distortion for an inter-coded macroblock is similar to the intra-coded case, with the difference that the expected reconstructed value for an inter-coded macroblock equals the expected reconstructed value of the macroblock in the previous video frame plus the expected reconstructed prediction error.

References

1 Ozarow, L. (1980). On a source-coding problem with two channels and three receivers. *The Bell System Technical Journal* 59 (10): 1909–1921.

2 Kim, M.Y. and Kleijn, W.B. (2006). Comparative rate-distortion performance of multiple description coding for real-time audiovisual communication over the internet. *IEEE Transactions on Communications* 54 (4): 625–636.

3 Vaishampayan, V.A. (1993). Design of multiple description scalar quantizers. *IEEE Transactions on Information Theory* 39 (3): 821–834.

4 Goyal, V. (2001). Multiple description coding: compression meets the network. *IEEE Signal Processing Magazine* 18 (5): 74–93.

5 Wenger, S. (1997). Video redundancy coding in H.263+. *Proceedings of AVSPN*, Volume 97.

6 Norkin, A., Aksay, A., Bilen, C. et al. (2006). Schemes for multiple description coding of stereoscopic video. In: *Multimedia Content Representation, Classification and Security*, (eds. Bilge Gunsel, Anil K. Jain, A. Murat Tekalp, Bülent Sankur) 730–737. Springer.

7 Wang, Y., Orchard, M.T., Vaishampayan, V., and Reibman, A.R. (2001). Multiple description coding using pairwise correlating transforms. *IEEE Transactions on Image Processing* 10 (3): 351–366.

8 Puri, R. and Ramchandran, K. (1999). Multiple description source coding using forward error correction codes. *Conference Record of the 33rd Asilomar Conference on Signals, Systems, and Computers, 1999*, Volume 1, pp. 342–346. IEEE.

9 Shapiro, J.M. (1993). Embedded image coding using zerotrees of wavelet coefficients. *IEEE Transactions on Signal Processing* 41 (12): 3445–3462.

10 Said, A. and Pearlman, W.A. (1996). A new, fast, and efficient image codec based on set partitioning in hierarchical trees. *IEEE Transactions on Circuits and Systems for Video Technology* 6 (3): 243–250.

11 Talluri, R. (1998). Error-resilient video coding in the ISO MPEG-4 standard. *IEEE Communications Magazine* 36 (6): 112–119.

12 Wen, J. and Villasenor, J.D. (1998). Reversible variable length codes for efficient and robust image and video coding. *Data Compression Conference, 1998. DCC'98. Proceedings*, pp. 471–480. IEEE.

13 Redmill, D.W. and Kingsbury, N.G. (1996). The EREC: an error-resilient technique for coding variable-length blocks of data. *IEEE Transactions on Image Processing* 5 (4): 565–574.

14 Rogers, J.K. and Cosman, P.C. (1998). Robust wavelet zerotree image compression with fixed-length packetization. *Data Compression Conference, 1998. DCC '98. Proceedings*, pp. 418–427.

15 Wu, D., Hou, Y.T., Li, B. et al. (2000). An end-to-end approach for optimal mode selection in internet video communication: theory and application. *IEEE Journal on Selected Areas in Communications* 18 (6): 977–995. and

16 Shoham, Y. and Gersho, A. (1988). Efficient bit allocation for an arbitrary set of quantizers. *IEEE Transactions on Acoustics, Speech, and Signal Processing* 36 (9): 1445–1453.

9

Analog and Hybrid Digital–Analog JSCC Techniques

A communication process eventually leads to having an analog signal (a signal that is continuous in time and amplitude) that can propagate through a physical medium. Since the joint source–channel coding (JSCC) problem is at its essence one of mapping a message into this analog signal sent through the physical channel, it makes sense to consider JSCC schemes where the coding operation is done all at once, mapping the source output directly onto an analog signal. In this chapter, we study this type of JSCC, which entails mapping either analog or digital signals from the source directly into an analog signal that can propagate through the physical channel. The study will take us back to the very beginning of JSCC to learn about schemes first proposed by Claude Shannon in his seminal information theory papers. The chapter will also discuss recent advances in artificial neural networks to implement both analog and hybrid digital–analog JSCC.

9.1 Analog Joint Source–Channel Coding Techniques

9.1.1 Analog Joint Source–Channel Coding in Vector Spaces

Claude Shannon's seminal work on information theory set the mathematical foundation for the formal study of the communication of information [1], paving the way for the development of digital communications. The remarkable growth in digital communications that followed Shannon's work is reflected in many of the preceding chapters by considering JSCC techniques that operate on digital signals. But communication over a physical channel is done by propagating analog waves. Also, many sources, particularly those related with human senses (e.g. sound), originate as analog sources and are digitized for communication. So we consider communication systems where analog sources are transmitted over the channel as analog signals with no separation between source and channel coding, an approach named *Analog Source–Channel Coding*. The transmitter conditions the analog signal generated at the source to a signal matching the channel characteristics and requirements.

The first important ideas related to analog JSCC techniques can be traced back to Claude Shannon [2]; the presentation seamlessly links the sampling theorem with the communication capacity of a channel. The central concept employed by Shannon was to represent signals, messages, the distortion introduced by the channel, the transmitter

Joint Source-Channel Coding, First Edition. Andres Kwasinski and Vinay Chande.

and the receiver, and their interrelation – in essence, the components of a communication process – as entities in vector spaces. We begin by explaining Shannon's ideas from [2].

Let us assume that the communication of the source is implemented through the transmission of signals (the signals that are determined from the modulation operation) with a duration T. During transmission, these signals are limited to the channel bandwidth W. The sampling theorem tells us that ideally each of the transmitted signals can be represented without loss of any information by a sequence of samples that are $1/(2W)$ seconds apart. Since the duration of each signal is constrained to T seconds, each of the transmitted signals can be represented without loss of information with a set of $2WT$ samples. We can regard the $2WT$ samples that represent a signal as the coordinates of a vector in a space of dimension $2WT$. The signal being represented by the $2WT$ samples is a point in a space, that we call the *signal space*, of dimension $2WT$.

Shannon considered a geometrical representation of signals because it becomes possible to abstract key concepts in signal and communications theory as geometric entities. A case in point are the distances in the signal space. Assuming that the $2WT$ signal coordinates are indicated with respect to axes that are orthogonal to each other, the distance from the origin to a point with coordinates $(x_1, x_2, \ldots, x_{2WT})$ (or equivalently, the length of the vector associated with a signal with samples $\{x_n\}_1^{2WT}$) is as follows:

$$d = \sqrt{\sum_{n=1}^{2WT} x_n^2}.$$

The sampling theorem tells us that a signal $f(t)$ can be reconstructed from its $2WT$ samples through the addition of time-shifted sinc functions:

$$f(t) = \sum_{n=1}^{2WT} x_n \frac{\sin(2W\pi t - n\pi)}{\pi(2Wt - n)}.$$

Using the fact (shown using the Fourier transform) that

$$\int_{-\infty}^{\infty} \frac{\sin(2W\pi t - n\pi)}{\pi(2Wt - n)} \frac{\sin(2W\pi t - m\pi)}{\pi(2Wt - m)} dt = \begin{cases} \frac{1}{2W}, & m = n, \\ 0, & m \neq n, \end{cases}$$

we have that the signal energy E is as follows:

$$E = \int_{-\infty}^{\infty} f(t)^2 dt = \frac{1}{2W} \sum_{n=1}^{2WT} x_n^2 = \frac{d^2}{2W}.$$

Therefore, the square of the distance from the center of coordinates to a point is $2W$ times the energy E of the signal represented at the corresponding point. That is, $d^2 = 2WE = 2WTP$, where P is the average power of the signal. This observation can be extended to other geometric setups in the signal space. For example, the distance between two points in the signal space is equal to $\sqrt{2WT}$ times the root mean square (RMS) difference between the signals corresponding to the points. Also, all the signals whose average power is less than P will correspond to points inside a sphere of radius $r = \sqrt{2WTP}$.

The effect of additive noise experienced by a signal during communication can also be thought of in geometric terms. If a signal is affected by additive noise, the point representing the signal is moved by a distance that is proportional to the RMS value of the noise.

Following the same thinking, the messages generated by a source can be thought of as points in a *message space*. Consider a speech source where the messages are sounds limited to a duration of T_m seconds and a maximum frequency W_m. Similar to the signal space, the message space will have $2W_m T_m$ dimensions. There is a key conceptual difference between the message and signal spaces, as for the message space, multiple points may represent the same message in what concerns the receiver. For example, sounds with different phases may be perceived as the same by humans. Since all points that convey the same message can be treated as one, the effective dimensionality of the message space may be less than what comes from duration and maximum frequency. As will be discussed more later on, this characteristic of the message space is exploited by many source coding techniques, but dimensionality reduction does not happen in all applications and is not the general case.

Given a signal space and message space, it becomes natural to interpret the operation of the transmitter and the receiver from a geometric perspective. Since the input to the transmitter is a message and the output is a signal, the transmitter performs a mapping from the message space into the signal space. Conversely, the receiver maps the signal space into the message space. As a summary of the discussion so far, the correspondences between communication systems components and geometric entities are listed in Table 9.1 (from [2]).

The mapping operation at the transmitter can be formalized mathematically in the case of a discrete-time system, by assuming that the messages from the source are samples and that the transmitter operates on a set of M such source samples. Continuing with Shannon's vector space correspondences, this input can be modeled as a vector $\vec{x}$ with real-valued components $x_1, x_2, \ldots, x_M$, thus, representing the input to the transmitter as an M-dimensional vector. Similarly, the signals at the transmitter output, which can be assumed to be a set of K channel symbols, can be modeled as a vector $\vec{y}$ in a K-dimensional space with real-valued

Table 9.1 Correspondences between communication systems components and geometric entities.

Communication system	Geometric entity
A particular signal	A point in a vector space called *signal space*
The set of possible signals	A vector space of $2TW$ dimensions
The average power of the signal	The square of the distance from the origin to the point (signal), divided by $2TW$
The set of signals of power P	The set of points on a sphere of radius $\sqrt{2TWP}$
Additive noise in the channel	Displacement of the point representing a signal
A message	A point in a vector space called *message space*
The set of possible messages	A vector space of $2T_m W_m$ dimensions
The set of actual messages distinguishable by the destination	A space of D dimensions obtained by removing from the message space those messages that the source cannot produce and regarding all equivalent messages as one point in the message space
The transmitter	A mapping from the message space into the signal space
The receiver	A mapping from the signal space into the message space

components $y_1, y_2, \ldots, y_K$. Since the sizes of the source sample and channel symbol sets need not be the same, the mapping operation that is implemented in the transmitter may be between spaces with the same dimensions or may increase or decrease the dimension from the message space to the signal space. If it is assumed that the duration of signals and messages is the same ($T = T_m$, and $T = T_m = 1$ for a discrete-time system), the change of space dimensions corresponds to a bandwidth expansion (with $2W_m T_m < 2WT$ or $K > M$) or compression (with $2W_m T_m > 2WT$ or $K < M$) during the operation of the transmitter.

Depending on the mapping function, the system will be able to accommodate more users on a physical channel, through a process of bandwidth compression, or will be able to increase the level of protection against noise in the channel through a process of bandwidth expansion. Overall, the mapping function establishes a clear pathway for designing a joint source–channel codec through designing a function that in one operation maps the M-dimensional source sample space into the K-dimensional channel symbol space.

When Shannon suggested mapping from the source space into the channel space in [2], he presented a mapping that expands the dimension ($K > M$). Figure 9.1 illustrates this mapping for the case when the signal space has a single dimension and the channel space had two dimensions. The mapping is implemented by considering the source sample amplitude to be equal to the length along the curve from some reference point. If this reference point is at the curve's midpoint, the length along the curve is taken in one direction (to the right in the figure) for positive source samples and is taken in the opposite direction (to the left in the figure) for negative source samples. Once the point on the curve that corresponds to the source sample has been determined, the coordinates of this point

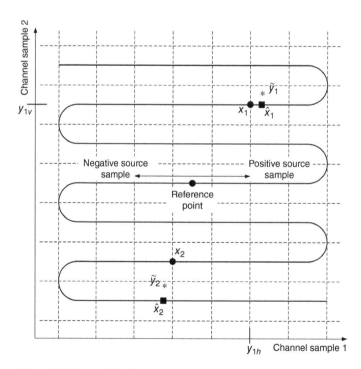

Figure 9.1 A mapping from a single-dimensional source space to a two-dimensional channel space.

become the two channel samples that are transmitted. Figure 9.1 shows two source sample examples, located on the curve based on their sign and magnitude. In the figure, source sample x_1 has a positive value and sample x_2 has negative value. At the receiver, the source sample is recovered by performing the reverse operations: the coordinates represented by the channel samples are mapped on the plane, and the closest point on the curve is taken to calculate the sign and amplitude of the reconstructed source sample.

Channel noise, which we denote as the random variable N, may affect the value of the channel samples during transmission. Moderate values of channel noise will result in a small displacement of the original point on the curve and, consequently, a small distortion of the transmitted source sample. Figure 9.1 exemplifies this case with the transmission of sample x_1, shown in the figure as a point on the wandering line where the value of x_1 equals the distance from the reference point measured along the line. In the transmitter, point x_1 is mapped to a channel sample (signal) y_1, represented as a vector with coordinates (y_{1h}, y_{1v}) in the horizontal and vertical Cartesian coordinate system shown in Figure 9.1. After adding channel noise to the transmitted signal, the received signal is represented by the Cartesian coordinates corresponding to the point $\tilde{y}_1$. In the geometric representation of Figure 9.1, the noise is represented by the vector from the coordinate (y_{1h}, y_{1v}) to the point $\tilde{y}_1$. The reconstructed value for x_1 is the value $\hat{x}_1$, measured as the distance along the wandering line from the reference point to the point $\hat{x}_1$ on the curve closest to $\tilde{y}_1$. In this case, the difference between x_1 and $\hat{x}_1$ is small. A more significant distortion of the reconstructed source sample occurs when the noise is large enough to cause the reconstructed point to be off of the original segment of the mapping curve, and on a neighboring parallel segment. This case is exemplified in Figure 9.1 with sample x_2. At the receiver, the noise results in a displacement of the point corresponding to x_2 into the received value $\tilde{y}_2$. The reconstructed value for x_2 is the point $\hat{x}_2$ that lies on a segment of the curve that is closest to $\tilde{y}_2$ but is not the same one as where x_2 lies.

The Shannon mapping illustrated in Figure 9.1 can also be used in a reverse form to compress the dimensions of the spaces $(K < M)$. In an example consistent with Figure 9.1, two source samples indicate the coordinates for a point on a plane or, equivalently, a source vector. The point on the plane can be mapped to the closest point on the mapping curve. Finally, the channel symbol is calculated as the distance along the curve from a reference point which could be, for example, the initial point of the curve on the upper left corner.

To see the relation between a mapping from the space of source samples to the space of channel signals and the bandwidth compression or expansion, consider first the matching of source information rate to channel capacity. In the simplest case where the channel is discrete-time memoryless and affected by additive white Gaussian noise (AWGN) with signal-to-noise ratio (SNR) γ_c, the channel capacity, measured in bits per channel use (or equivalently bits per input symbol into the channel), is given by Shannon's formula:

$$C = \frac{1}{2}\log_2(1 + \gamma_c). \tag{9.1}$$

Assuming sampling at the Nyquist rate, if the channel has a bandwidth W, it would be possible to transmit $2W$ symbols per second, making the channel capacity measured in bits per second be

$$C = 2W\frac{1}{2}\log_2(1 + \gamma_c) = W\log_2(1 + \gamma_c). \tag{9.2}$$

At the same time, the channel is used as the medium to communicate a source that is assumed to be discrete-time, generating independent and identically distributed (i.i.d.) samples following a zero-mean Gaussian distribution with variance σ_X^2. For a mean squared error distortion D, the source coding rate, measured in bits per sample, is as follows (see (1.30)):

$$
R = \begin{cases} \frac{1}{2}\log_2\left(\frac{\sigma_X^2}{D}\right), & \text{for } 0 \leq D \leq \sigma_X^2 \\ 0, & \text{for } D \geq \sigma_X^2. \end{cases} \tag{9.3}
$$

If the source signal occupies a bandwidth B, following the Nyquist–Shannon sampling theorem, the number of transmitted samples per second is $2B$. Consequently, for a mean squared error distortion D, the source coding rate, now measured in bits per second, is as follows:

$$
R = \begin{cases} 2B\frac{1}{2}\log_2\left(\frac{\sigma_X^2}{D}\right), & \text{for } 0 \leq D \leq \sigma_X^2 \\ 0, & \text{for } D \geq \sigma_X^2. \end{cases} \tag{9.4}
$$

To achieve the smallest source distortion, the source coding rate needs to be picked as large as possible to be transmitted through the channel. Therefore, the relation between the coded source SNR and the channel SNR can be calculated by equating the source coding rate to the channel capacity as given by (9.2) and (9.4):

$$
B\log_2\left(\frac{\sigma_X^2}{D}\right) = W\log_2(1+\gamma_c). \tag{9.5}
$$

Solving for the coded source SNR results in what is known as the optimal performance theoretically attainable (OPTA) curve:

$$
\frac{\sigma_X^2}{D} = (1+\gamma_c)^{W/B}. \tag{9.6}
$$

This result depends on the ratio of channel to source bandwidth (W/B) but, because of the application of the Nyquist–Shannon sampling theorem to derive the relation, a change in bandwidth from the source to the channel implies a change in sampling rate. Because of this, the bandwidth ratio can be approximated as the ratio between the number of channel samples K and the number of source samples M that are combined or mapped into the K channel samples: $W/B \approx K/M$. Considering this, the OPTA relation can be expressed as:

$$
\frac{\sigma_X^2}{D} \approx (1+\gamma_c)^{K/M}, \tag{9.7}
$$

where the dependency on the effects of mapping from the space of source samples to the space of channel signals is made explicit. In addition, the relation $W/B \approx K/M$ shows how the mapping results in bandwidth compression or expansion. Figure 9.2 shows the OPTA curves from (9.7) for different values of the ratio $W/B \approx K/M$. It can be seen how choosing larger bandwidth expansion ratios results in better quality of the source signal.

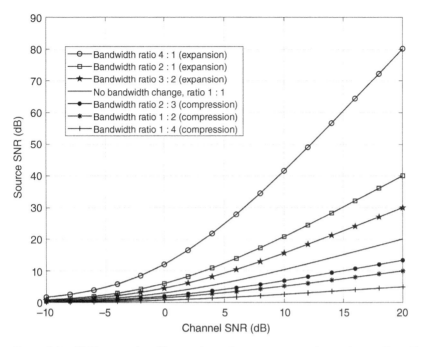

Figure 9.2 OPTA curves for different channel to source bandwidth ratios, $W/B \approx K/M$.

Also, it can be seen how bandwidth expansion implicitly increases the resiliency against channel errors, as the same source SNR can be achieved at a lower channel SNR for larger bandwidth expansion.

A design for Shannon mapping JSCC scheme with $2:1$ dimensionality compression (bandwidth ratio $W/B \approx K/M$ of $1:2$) was discussed in [3]. In this case, the input to the joint source–channel encoder consists of one sample each from two i.i.d. Gaussian-distributed sources X_1 and X_2. The two samples form the coordinates of a vector $\vec{x} = (x_1, x_2) \in \mathbb{R}^2$ or, equivalently, a point in a two-dimensional space. The dimensionality reduction is implemented by finding the point on a curve that is closest to the point $\vec{x}$. In [3], this operation is implemented by finding the point on a double Archimedes' spiral that is closest to $\vec{x}$. Given the circular symmetry of the joint probability density function for the source output values X_1 and X_2, as illustrated in Figure 9.3, it is natural to consider a coding scheme with a circularly symmetric representation, as is the case with the double Archimedes' spiral. (This intuition could be confirmed mathematically to be a close approximation of the optimal solution). The double Archimedes' spiral is shown in Figure 9.4, where the spiral arm drawn as a solid line is defined by the parametric equations:

$$x_1 = 2\Delta \frac{\theta}{2\pi} \cos(\theta),$$
$$x_2 = 2\Delta \frac{\theta}{2\pi} \sin(\theta), \tag{9.8}$$

where θ is the angle for a vector originating at the center of coordinates and ending on a point on the curve and Δ is a parameter that defines the distance between two neighboring

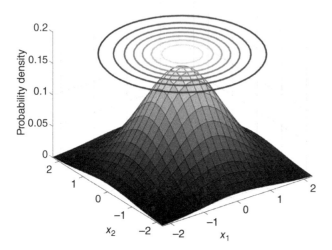

Figure 9.3 The probability density function for a two-dimensional circularly symmetric random process with zero-mean and unit-variance components and corresponding contour lines on top.

spiral arms as measured along the abscissa ($\Delta = 1$ in Figure 9.4). From (9.8), we can see that the radius or, equivalently, the distance from the center of coordinates to a point on the spiral arm drawn as a solid line is as follows:

$$r = \frac{\Delta}{\pi}\theta, \tag{9.9}$$

meaning that on the spiral the angle is proportional to the radius. In a similar fashion, we can define the spiral arm drawn as a dashed line through the parametric equations:

$$x_1 = 2\Delta\frac{\theta}{2\pi}\cos(\theta + \pi),$$
$$x_2 = 2\Delta\frac{\theta}{2\pi}\sin(\theta + \pi). \tag{9.10}$$

Clearly, the proportionality between the radius and the angle still holds for this spiral arm.

After reducing the dimensionality by finding the point on the double Archimedes' spiral that is closest to $\vec{x}$, the channel symbol z is calculated as the distance from the origin over the appropriate spiral arm (the one shown in Figure 9.4 as a solid line or the one shown as a dashed line). This operation effectively implements a $2:1$ dimensionality reduction from the two dimensions at the input of the encoder (source output) to the single dimension at the output of the encoder (channel input). The calculation of the distance from the origin to a point on the double spiral can be approximated using the operators:

$$z = T_+(\theta) = \frac{\Delta}{2\pi}\theta^2, \text{ for the spiral arm shown as a solid line,}$$

$$z = T_-(\theta) = -\frac{\Delta}{2\pi}\theta^2, \text{ for the spiral arm shown as a dashed line.} \tag{9.11}$$

As can be seen, the dimensionality reduction effectively encodes the coordinates (x_1, x_2) of the point corresponding to the source samples as the angle θ. Of course, because of the

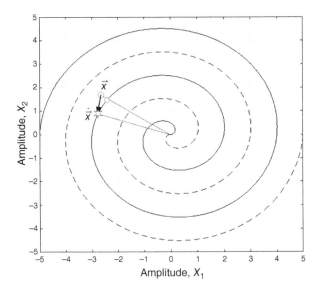

Figure 9.4 Double Archimedes' spiral used for a 2 : 1 dimensionality reduction. Source: Adapted from [3].

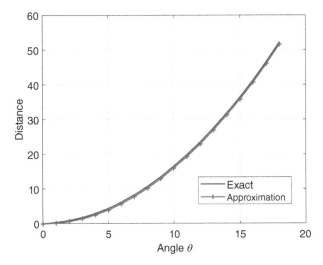

Figure 9.5 Exact distance along a branch of the double Archimedes' spiral and the approximation using the $T_+(\theta)$ operator.

proportionality between angle and radius, the same can be said of a calculation of the channel symbol based on the radius. Figure 9.5 illustrates the accuracy of the operators used to estimate the channel symbol as the length along the spiral.

During the encoding process, finding the point on the double spiral that is closest to $\vec{x}$ is an approximation operation that will introduce some error. Intuitively, this error will be smaller if the spiral arms are closer to each other and, consequently, the encoding error will depend on the design choice for the parameter Δ. Therefore, the design of the dimensionality reduction JSCC scheme entails finding the optimal value for Δ. As with other JSCC schemes, the optimal value for Δ can be found by solving an optimization problem that minimizes the end-to-end distortion per symbol, $D_{ee}(\Delta)$, subject to an average power constraint P_{avg}:

$$\Delta_{opt} = \underset{\Delta:E[Z^2]<P_{avg}}{\arg\min}\ D_{ee}(\Delta), \qquad (9.12)$$

where Z is the random variable for the transmitted channel symbol with an average power $E[Z^2]$. The end-to-end distortion per source $D_{ee}(\Delta)$ is defined as:

$$D_{ee}(\Delta) = \frac{E[\|\vec{x} - \hat{\vec{x}}\|^2]}{2}.$$

In this case, the distortion is measured as half of the mean square error, because the distortion is over two source symbols. At high quantization SNR, the spiral arms as seen in Fig. 9.4 will be more closely packed with respect to the source signal power. Under this condition, when projecting the vector of two source samples onto the spiral, the average error is approximately equal for the two source samples. This allows the assumption that the distortion can be decomposed into two components: a radial distortion, D_r, and an angular distortion, D_θ:

$$D_{ee}(\Delta) = D_r + D_\theta = \frac{E[(\|\vec{x}\| - \|\hat{\vec{x}}\|)^2]}{2} + \frac{E[\|\hat{\vec{x}}\|^2(\angle\vec{x} - \angle\hat{\vec{x}})^2]}{2}.$$

Note that the radial distortion relates to the quantization operation (because the projection onto the spiral is performed on radial directions), while the angular distortion relates to the channel noise. Resorting again to the assumption of a dense spiral relative to the source signal power, the projection onto the spiral operation can be seen as that of a standard scalar quantizer. Consequently, the quantization error can be written as $D_r = \Delta^2/12$ (see (2.5)).

Because of the average power constraint $E[Z^2] < P_{avg}$, the channel symbol needs to be multiplied by a scaling factor $1/\sqrt{\alpha}$ before being transmitted. At the receiver, the received signal is multiplied by $\sqrt{\alpha}$ to compensate for this scale factor. This operation scales the channel noise by the same factor, making its power $E[(\sqrt{\alpha}N)^2] = \alpha\sigma_N^2$. The value of α is determined by the power constraint on the transmitted symbol as:

$$E\left[\left(\frac{Z}{\sqrt{\alpha}}\right)^2\right] = \frac{1}{\alpha}E\left[Z^2\right] = \frac{\sigma_Z^2}{\alpha} = P_{avg},$$

and thus, $\alpha = \sigma_Z^2/P_{avg}$ and the angular distortion is equal to $D_\theta = \sigma_n^2\sigma_Z^2/P_{avg}$. Combining the expressions for radial and angular distortions, the end-to-end distortion can now be written as:

$$D_{ee}(\Delta) = \frac{1}{2}\left(\frac{\Delta^2}{12}\right) + \frac{1}{2}\left(\frac{\sigma_Z^2}{P_{avg}}\sigma_N^2\right). \qquad (9.13)$$

As we have indicated, the value of the channel symbol is proportional to the square of the radius. More precisely, from (9.9) and (9.11) we can write

$$z = \pm\frac{1}{2}\frac{\pi}{\Delta}r^2 = \pm\frac{1}{2}\frac{\pi}{\Delta}(x_1^2 + x_2^2). \qquad (9.14)$$

Since the two coordinates x_1 and x_2 are outcomes of the two i.i.d. Gaussian-distributed sources X_1 and X_2, after applying a transformation of random variables we can conclude that the random variable for the channel symbol Z follows a Chi-square distribution with two degrees of freedom and a variance:

$$\sigma_Z^2 = 2\left(\frac{\pi}{\Delta}\sigma_X^2\right)^2, \tag{9.15}$$

where σ_X^2 is the variance of the source. Combining this result with Eq. (9.13), we get

$$D_{ee}(\Delta) = \frac{\Delta^2}{24} + \frac{\pi^2\sigma_X^4}{\Delta^2 P_{avg}}\sigma_N^2. \tag{9.16}$$

We can now find the solution to the optimization problem (9.12) by calculating the derivative of $D_{ee}(\Delta)$ with respect to Δ and equating to zero:

$$\frac{dD_{ee}(\Delta)}{d\Delta} = \frac{\Delta}{12} - \frac{2\pi^2\sigma_X^4}{\Delta^3 P_{avg}}\sigma_N^2 = 0.$$

Solving for Δ, we get

$$\Delta_{opt} = \sigma_X \sqrt[4]{\frac{24\pi^2\sigma_N^2}{P_{avg}}}.$$

Plugging the result for Δ_{opt} into the radial and angular distortions, we see that they turn out to be equal (that is, the optimal value for Δ is the one that makes the radial and angular distortions equal):

$$D_r = D_\theta = \frac{\pi\sigma_X^2}{2}\sqrt{\frac{\sigma_N^2}{6P_{avg}}}.$$

Denoting the channel SNR P_{avg}/σ_N^2 as SNR_c, we can write the end-to-end distortion as:

$$D_{ee}(\Delta) = D_r + D_\theta = \frac{\pi\sigma_X^2}{\sqrt{6\ SNR_c}}, \tag{9.17}$$

and with the reconstructed source $SNR = \sigma_X^2/D_{ee}$ we finally obtain that, for the optimum choice of Δ, we have

$$SNR = \frac{\sqrt{6\ SNR_c}}{\pi}. \tag{9.18}$$

Considering now (9.7) for this case of a $2:1$ dimensionality reduction JSCC scheme, we see that the OPTA relation is $SNR = \sqrt{1 + SNR_c}$. Compared with the result using the double Archimedes' spiral in (9.18) for the high channel SNR regime, we can see that transmission under the OPTA conditions has a gain of approximately $\sqrt{6}/\pi$, or 1.1 dB, with respect to the code based on the double Archimedes' spiral.

9.1.2 Analog Joint Source–Channel Coding Through Artificial Neural Networks

Artificial neural networks are inspired by the human brain and are formed by the interconnection of elementary signal processing units called *artificial neurons*. This field spans decades of research and here we limit ourselves to a narrow topic (interested readers can learn more from [4] and other books dedicated to this subject). An artificial neuron is intended to behave in a way that loosely resembles the behavior of a biological neuron. An artificial neuron operates by forming a weighted linear combination of a number of inputs and applying to the result a function, called the activation function, which is usually a nonlinear operation. Mathematically, the operation of a single artificial neuron can be represented through the expression:

$$y(t) = \phi \left(b + \sum_{k=1}^{n} w_k x_k(t) \right), \tag{9.19}$$

where $x_k(t)$ are the inputs, w_k are weights, b is a bias term, $y(t)$ is the output, and $\phi(\cdot)$ is the activation function. For implementation in computing devices, the inputs and the output take, respectively, the form of a vector and a scalar of quantized values. A single artificial neuron has very limited processing capabilities, but by connecting very large numbers of single artificial neurons (usually called nodes) it becomes possible to solve complex problems. It has been shown that artificial neural networks appropriately configured with a sufficiently large number of nodes are capable of approximating any function; they are known as universal function approximators. More importantly, inspired by the human brain, artificial neural networks present a different approach to solving problems in that they are not explicitly programmed to solve the problem (e.g. by implementing an algorithm that stipulates a sequence of instructions to follow), but rather they learn to solve a problem by being presented with examples of the correct solution. Before an artificial neural network is used to solve a problem, its configuration (layer structure) is designed and it is trained to learn to solve the problem. Inputs are presented to the network at the input layer (technically not formed by artificial neurons but rather just by a node for each input). The output layer produces the network outputs, and the remaining layers between the input and output layers are called hidden layers.

Figure 9.6 illustrates an artificial neural network with two hidden layers. For a feed-forward network such as that in the figure, the outputs from the nodes in one layer form the inputs for the following layer. It is also possible to design a network where the outputs from one layer act as inputs for the same layer or previous layers, in which case the network is called a recurrent neural network (RNN). A network where a node from one layer feeds into all nodes of the following layer, as in Figure 9.6, is called a fully connected network (or layer if referring to a specific layer). While originally artificial neural networks would have one or two hidden layers (a configuration called a *shallow network*), major advances in this area are owed to configurations with many hidden layers, in what is called a *deep neural network*. The layers progressively process the input signal, gradually representing it at increasing levels of abstraction until the information from the input layer reaches a representation that allows a problem solution. For example, a neural network designed to classify different objects from an image would start at the first hidden layer by identifying very simple features (e.g. vertical, horizontal, or slanted short edges), which then becomes more complex at the following layers (e.g. corners at the second layer, squares or ellipsoids in the third layer, etc.) until the later layers (specially the output layer) are able to differentiate between different

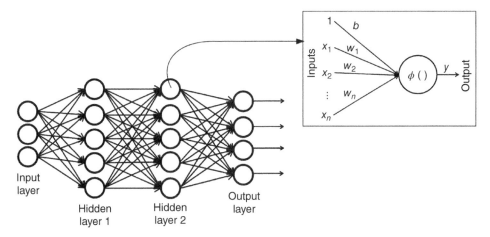

Figure 9.6 A fully connected feed-forward artificial neural network with two hidden layers. Each node in the network (with the exception of those in the input layer) represents an artificial neuron as shown in the upper right insert.

objects. For our purposes, it is important to recognize that the processing within a neural network represents information at different levels of complexity over spaces of different dimensions.

For neural networks to be able to solve a target problem, they need to first be trained so that they can learn (without explicit programming) to solve the problem. The training procedure is called *supervised learning* when it is based on presenting to the network examples of inputs and the corresponding correct outputs. In practice, the process of supervised training consists of finding the values for the nodes' weights and biases that minimize the error between the output generated by the network and the output that should be generated for the training example. During training, thousands and even millions of examples could be presented to the network.

The success of deep neural networks has resulted in the development of many different network configurations, intended for different tasks. Of interest here is a configuration called an *autoencoder*. Autoencoders are the artificial neural network architectures behind remarkable applications where a painting or a musical composition is presented as input to a trained autoencoder and the network converts the input to a painting in the style of a renaissance master or to a musical composition that sounds as if it was composed by one of the great composers of the past. The autoencoder is formed by a sequence of layers called the encoder, connected back to back with a mirror sequence of layers called the decoder. Autoencoders are trained to copy the input to the output, after the input has gone through stages of abstracting and representing the features that characterize the input signal. The encoder layers progressively extract key features of the input at increasing level of abstraction. In the neural art style transfer application, the encoder layers have learned what features characterize the style of a composer or a painter. The function of the decoder layers is to create an output that is based on, and resembles, the input in its main characteristics, but that incorporates the individualizing characteristics in the features extracted by the encoder.

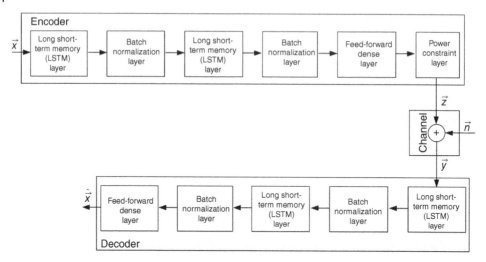

Figure 9.7 An analog joint source–channel coder implemented with an autoencoder deep neural network.

In [5], an autoencoder architecture was used to design the encoder and decoder for an analog JSCC scheme. Figure 9.7 shows the architecture. The encoder core comprises two layers of a form of RNN called a long short-term memory (LSTM) layer. The encoder also includes two batch normalization layers (which control the statistics of the hidden layer inputs to help the network learn faster), a feed-forward fully connected layer (which essentially combines the feature representations abstracted in the previous layers), and a layer to manage transmit power constraints. Compared to a traditional autoencoder, the configuration in Figure 9.7 was adapted to the JSCC purpose by opening up the traditional back-to-back connection between encoder and decoder to insert in between a layer that constrains the signal power at the encoder output and, more importantly, a layer that models the channel. Both of the inserted layers are non-trainable, meaning that they are not formed by artificial neurons but rather operate as fixed algorithms. The channel layer implements an emulation of the channel that will be present during operation (in [5], it was an AWGN channel that operates on a vector with the dimension of the encoder output). The power constraining layer converts the output vector from the last proper layer of the encoder into a zero-mean, unit-variance vector which can be scaled to the desired power (set to one in [5]). Thinking of an artificial neural network as a universal function approximator, the complete encoder can be characterized as a function $f_{w_e}(\cdot)$ that maps an input a signal $\vec{x} \in \mathbb{R}^M$ from the M-dimensional source space into a signal $\vec{z} \in \mathbb{R}^K$ from the K-dimensional channel space. The subscript emphasizes that the function $f_{w_e}(\cdot)$ is parameterized by the weights and biases of the constituent artificial neural network. The decoder side of the autoencoder architecture is a mirror image of the encoder (with respect to the LSTM layers), with layers appearing in the inverse order of the encoder with the exception that, because of its function, the feed-forward fully connected layer is moved to be the decoder's last layer (at the same time, one can consider that a fully connected layer is still present at the decoder input, since its inputs are the scaled and noisy outputs from the fully connected layer at the encoder). Similar to the encoder, the decoder can be characterized as a function $f_{w_d}(\cdot)$ that maps an input signal $\vec{y} \in \mathbb{R}^K$ from the K-dimensional channel space into the reconstructed source signal $\hat{\vec{x}} \in \mathbb{R}^M$ from the M-dimensional source space.

In supervised learning, the weights and biases of a neural network are calculated by presenting many examples to the network and adjusting the weights and biases to minimize the error between the network output and the expected output for the training example. In [5], the encoder and decoder are designed following the same principle of thinking of them as black boxes that can be trained to do joint source–channel encoding and decoding by presenting many examples of encoder inputs (an input signal) paired with the corresponding output that should be seen at the end of the decoder processing (the reconstructed signal). By presenting examples of inputs and corresponding expected outputs, the weights and biases inside the encoder and decoder black boxes are iteratively updated to minimize the error between the input signal and its reconstruction. Specifically, training in each iteration aims at minimizing the error given by the square of the Euclidean distance between the source signal and its reconstruction:

$$||\vec{x} - f_{w_d}(f_{w_e}(\vec{x}) + \vec{n})||_2^2, \tag{9.20}$$

where $\vec{n}$ is the channel noise.

The implementation of analog JSCC using an autoencoder is agnostic to whether bandwidth is expanded or compressed; either case can be designed following the same procedure as long as an appropriate set of examples is used during training. Moreover, as long as training is conducted appropriately, the resulting networks can generalize what was learned during training. This means that the autoencoder design will not be restricted to the same channel conditions seen during training and can perform effective JSCC for input signals and channel conditions that were not explicitly presented as examples during training.

Perhaps the most interesting aspect of the analog JSCC technique using an autoencoder is the ability of the artificial neural network to learn from examples solutions that rival the best known techniques obtained through analytical methods (while also adding some generalization capability). In [5], the autoencoder solution was evaluated in a 2 : 1 bandwidth compression coding configuration with an input given by two samples, one each from two i.i.d. Gaussian-distributed sources. The evaluation showed a performance (measured in terms of the source SNR vs. channel SNR) practically indistinguishable from the one obtained by the optimal design of a double Archimedes' spiral JSCC discussed earlier. Even more remarkable is that when analyzing the encoder function in the autoencoder, it can be seen that the mapping learned after training is a good approximation of the optimal double Archimedes' spiral, with the differences between the two probably being attributable to the generalization capability of the autoencoder.

9.2 Hybrid Digital–Analog JSCC Techniques

The fact that information transmission over any medium of physical existence is ultimately through an analog signal presents a natural reason to think of coding techniques in the form of mappings from the source into an analog channel symbol. However, when considering the source output (the encoder input), the decision is less natural; in our digital world, a source is often presented to the communication system as generating digital symbols. Even when the source generates analog signals (e.g. a sensor in an Internet-of-Things network), it may be a shrewd design move to digitize the signal to subsequently take advantage of efficient source compression algorithms. These cases, where an encoder directly maps from a digital source to an analog channel symbol, are known as *hybrid digital–analog* JSCC techniques.

A typical application for hybrid digital–analog JSCC is the direct mapping of a source that generates images (an array of pixels with discrete values) into a channel symbol, an analog signal from a constellation of multilevel modulation. Such a technique was presented in [6] based on a system shown in Figure 9.8. The signal first undergoes a stage of decomposition, for example using the techniques of Section 2.3.4; the purpose is to remove the correlation between the source samples, leading to a simpler mapping operation onto the channel signal space. The encoder shown reduces the source bandwidth B_s by a factor ρ to fit the channel bandwidth B_c. The codec aims to achieve bandwidth and power efficiency, as well as graceful degradation in the presence of transmission errors for the recovered source. Bandwidth efficiency can be achieved by a dimensionality reduction from the source signal space to the channel signal space. However, the encoder design needs to take into consideration that not all the samples resulting from the source decomposition have the same importance. Consequently, the least important samples may be ignored altogether (lossy coding), samples of intermediate importance may be combined and mapped to a single-channel symbol (dimensionality reduction), and the most important samples may be mapped without applying any dimensionality reduction. To achieve power efficiency, the mapping should match source symbols that are more likely (after decomposition) with channel symbols that have associated lower power. Finally, although channel errors may affect the quality of the reconstructed signal, with graceful degradation, the mapping introduces robustness against these errors by building into the design the minimization of the effects of errors.

Figure 9.9 shows the hybrid digital–analog joint source–channel encoder for images presented in [6] that develops these ideas by expanding the conceptual codec of Figure 9.8. In this encoder, an image X from the source is first decomposed through an analysis filter bank into eight-by-eight subbands $X_0, X_1, ..., X_{63}$. The output from each subband is divided into blocks of eight-by-eight pixels, and each block is subsequently classified based on its mean square value into one of five classes: Y_0 (lowest power), $Y_1, ..., Y_4$ (highest power). This classification is intended for nonuniform quantization; in the subsequent quantization stage, the blocks in class four are not quantized at all, while class zero blocks are eliminated altogether. For the other classes, the quantization stage assigns more resolution to the classes associated with larger power.

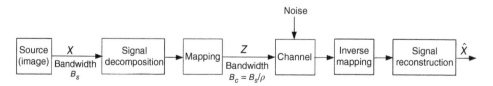

Figure 9.8 Hybrid digital–analog JSCC through signal decomposition and mapping with channel bandwidth reduction by a factor of ρ.

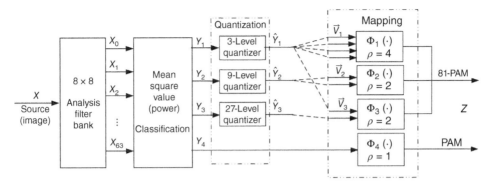

Figure 9.9 Hybrid digital–analog joint source–channel encoder for images.

The particularity in this case is that the quantizer is ternary rather than binary; the number of levels in quantizers of different resolutions is a power of 3 (see Figure 9.9). The quantized subband blocks $\hat{Y}_r$, $r = 1, 2, 3$, are organized into vector structures $\vec{V}_r$, $r = 1, 2, 3$ in such a way that each has associated components that can take 81 different combinations of values. As shown in Figure 9.9, $\vec{V}_1$ is formed by 4 input components, each with 3 quantization levels, yielding $3^4 = 81$ different combination of values, $\vec{V}_2$ is formed by 2 input components, each with 9 quantization levels, yielding $9^2 = 81$ different combinations of values, and $\vec{V}_3$ is also formed by 2 input components, but now with one component with 3 quantization levels and the other with 27 quantization levels, yielding $3 \times 27 = 81$ different combinations of values. Therefore, the definition of the three vector structures $\vec{V}_r$ defines a mapping from the quantized subband blocks into the channel space. All that remains to complete the definition of the mapping is to associate the 81 different combinations of values of each vector structure into a modulated signal. In all cases, there are 81 different combinations of values, so the association is done to an 81-PAM signal.

This mapping operation is one of the dimensionality reduction. Indeed, the mapping Φ_1 is from a four-dimensional space into a one-dimensional space ($\rho = 4$), and the mappings Φ_2 and Φ_3 are both from a two-dimensional space into a one-dimensional space ($\rho = 2$). Since the subband blocks in class four are not quantized, they are directly mapped into a pulse amplitude modulation (PAM) signal and there is no bandwidth compression or expansion in this case ($\rho = 1$).

In this case, the design of the mappings Φ_1, Φ_2, and Φ_3 involves finding the assignment from the different combinations of vector structure components at the input into 81 values at the output. Recalling the design criteria for the whole system, the mapping assignments should achieve the power efficiency and graceful degradation goals. This can be achieved by designing each mapping Φ_r based on the goal of minimizing the error:

$$\varepsilon = \sum_{k=1}^{81} P[\vec{V}_r^{(k)}] \sum_{m=1}^{81} P[\Phi_r(m)|\Phi_r(k)] d(\vec{V}_r^{(m)}, \vec{V}_r^{(k)}), \tag{9.21}$$

subject to

$$P_{avg} = \sum_{k=1}^{81} P[\vec{V}_r^{(k)}] |\Phi_r(k)|^2, \tag{9.22}$$

where $P[\vec{V}_r^{(k)}]$ is the probability of transmission for the kth value of the vector structure $\vec{V}_r$, $\Phi_r(k)$ is the mapping of the kth value of the vector structure $\vec{V}_r$ onto one of the symbols in the 81-PAM constellation, $d(\cdot, \cdot)$ is a distortion measure, and P_{avg} is a fixed average transmit power constraint.

The solution to the design problem (9.21) can be done through numerical methods. Interestingly but unsurprisingly, when there is symmetry across the dimensions of the input vector structure, the solution shows signs of association with the Shannon mappings for dimensionality reduction discussed in the previous section. Figure 9.10 shows the result from designing the mapping Φ_2. Each of the two components of the two-dimensional input that has been quantized into nine levels results in a version of the double Archimedes' spiral sampled at discrete locations. To finish the mapping, the 81 points on the new double spiral are mapped to values on the real line that correspond to 81-PAM modulation signals. The process slightly departs from that seen in the previous section in that it is not based on measuring the distance from the center along a spiral arm but, instead, on dividing the real line segment for the constellation points into 81 points with neighbors separated by the same distance. Then, starting from the center of the real line (the zero coordinate), each point along the two spiral arms, one in the positive direction and the other in the negative direction, is assigned to the sequence of points on the real line. It is not possible to visualize this phenomenon with Φ_1 (as it involves an input space of four dimensions) or Φ_3 (as there is no symmetry in the quantization levels for the components of each of the two dimensions).

For hybrid digital–analog JSCC, as with analog JSCC, there has been recent codec realizations based on deep neural networks. The work in [7] presented a deep neural network hybrid digital–analog JSCC scheme for image communication. Figure 9.11 shows the configuration for the encoder and decoder. Similar to the deep neural network analog

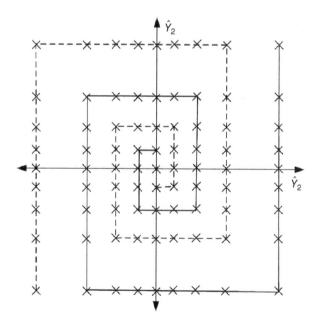

Figure 9.10 Mapping Φ_2, from two components with nine possible values each to a single-dimension value.

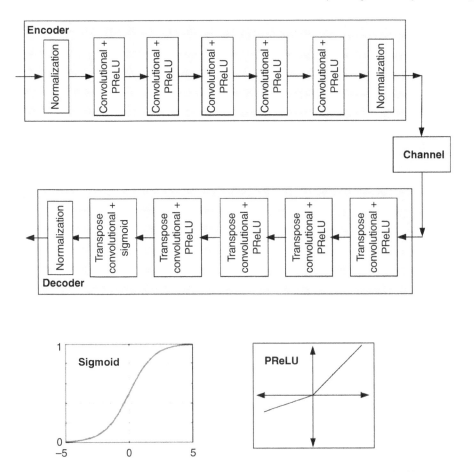

Figure 9.11 A hybrid digital–analog joint source–channel coder implemented with an autoencoder deep neural network. The sigmoid and PReLU activation functions are also shown.

JSCC discussed in the previous section, it is based on an autoencoder structure. This is a natural design choice given the direct correspondence of the encoder and decoder parts of an autoencoder with the encoder and decoder in a JSCC communication scheme. The network configuration in Figure 9.11 reveals a conceptual similarity with the encoder in Figure 9.9. The autoencoder in Figure 9.11 is configured around convolutional layers, whose operation is similar to analysis filter banks. In a traditional filter bank, the filters tend to be statically configured to analyze the input signal into frequency bands that partition the complete frequency span into bands with some regularity (e.g. octaves). However, for a convolutional layer, each of the filters that form the layer (usually from a few up to a few dozen) processes the image scanning for a specific feature. In a convolutional layer near the input of the artificial neural network, one filter may scan the image looking for a short vertical edge, another filter searches for a short horizontal edge, another filter may search for a diagonal edge, etc. Filters that are part of deeper convolutional layers search for more complex image features (e.g. a small cross, an ellipsoid, etc.).

However, the main difference between a traditional filter bank and the filters in a convolutional layer is that the coefficients of filters in a convolutional layer are not pre-calculated but instead are learned during training. A convolutional layer can be imagined as an analysis filter bank where the filters have been designed to analyze signal information in terms of specific features tailored to the task at hand. Yet, at a deep level, the concept behind the filtering in the schemes in Figures 9.9 and 9.11 is the same in analyzing the input image to do a bandwidth efficient mapping by deciding which components of the image (subbands or features) need to be represented with more fidelity and which ones are less important.

Compared to traditional filter banks, another important difference in convolutional layers is that they introduce a nonlinear processing element through the activation function in the constituent artificial neurons. The autoencoder in Figure 9.11 implements a parametric rectified linear unit (PReLU) activation function in all convolutional layers except the last one of the decoder, which uses a sigmoid activation function. The PReLU and sigmoid activation functions are shown at the bottom of Figure 9.11. The collective function of the nonlinear activation functions is to implement a classification of the features into different classes that are separated through complex nonlinear surfaces in high-dimensional spaces. This classification over multiple layers with increasing levels of abstraction eventually leads to a mapping from a signal in the source space into its representation in the channel space. Lastly, we note that the autoencoder in Figure 9.11 also has two normalization layers in the encoder and one in the decoder. Their purpose is to scale the signal, so it remains within preset power limits (and for the normalization layer at the encoder output to also meet a transmit power constraint).

In conclusion, the application of advances in artificial neural networks presents a tantalizing approach to JSCC where the traditional elements of joint codecs remain at a conceptual level but are implemented with a new paradigm of realizations that are not explicitly programmed but, instead, learned design parameters through training. This learning/training-based approach leads to signal processing structures tailored to the problem to be solved. Because the learning/training-based approach has no explicit programming based on specific models, when the training is performed appropriately the resulting neural network has the capability to generalize what it has learned to provide an appropriate response to an input and situations (e.g. channel conditions) that may not have been presented during training. It is still the task of the designer to pick a network configuration that is efficient in extracting and classifying features, which would lead to more efficient performance.

References

1 Shannon, C.E. (1948). A mathematical theory of communication. *The Bell System Technical Journal* 27: 379–423.
2 Shannon, C.E. (1949). Communication in the presence of noise. *Proceedings of the IRE* 37 (1): 10–21.
3 Hekland, F., Oien, G.E., and Ramstad, T.A. (2005). Using 2:1 Shannon mapping for joint source-channel coding. *Proceedings of the Data Compression Conference (DCC)*, pp. 223–232, Snowbird, Utah, March 2005.

4 Goodfellow, I., Bengio, Y., and Courville, A. (2016). *Deep Learning*. MIT Press. http://www.deeplearningbook.org.

5 Xuan, Z. and Narayanan, K. (2020). Analog joint source-channel coding for Gaussian sources over AWGN channels with deep learning. *2020 International Conference on Signal Processing and Communications (SPCOM)*, pp. 1–5.

6 Lervik, J.M. and Ramstad, T.A. (1996). Robust image communication using subband coding and multilevel modulation. *Visual Communications and Image Processing'96*, Volume 2727, pp. 524–535. International Society for Optics and Photonics.

7 Bourtsoulatze, E., Kurka, D.B., and Gündüz, D. (2019). Deep joint source-channel coding for wireless image transmission. *IEEE Transactions on Cognitive Communications and Networking* 5 (3): 567–579.

10

Joint Source–Channel Decoding

As discussed at different points in this book, the complexities associated with the channel encoding and source encoding operations, and the non-stationarity and delay-constrained characteristics of the source, make practical source and channel coding operations behave non-ideally. As a result, source and channel codecs do not behave as completely independent entities and can benefit from information exchange. This is one of the guiding principles in the design of joint source–channel coding (JSCC) schemes. Figure 10.1 is a modified version of Figure 5.9, as shown in Chapter 5, which was used to introduce one technique associated with this exchange of information. In Chapter 5, the exchange of source significance information from the source encoder to the channel encoder was used to design unequal error protection schemes.

Figure 10.1 differs from Figure 5.9 in that now the figure highlights the information used at the decoding side of the communication chain. This chapter will focus on JSCC techniques that use at the decoders different forms of side information, other than the source information being communicated. Source a priori information describes statistical information of the source that can improve the error correction capability of the channel decoder. This information could also be used to improve error concealment techniques at the source decoder (not depicted in the figure). The source decoder could also improve error concealment by using decoder reliability information, a set of measurements that the channel decoder could provide to indicate its level of confidence in each channel decoded bit. Also, the source decoder could provide to the channel decoder a posteriori source information, which can also improve channel decoding performance. This chapter starts by studying the use of the a priori and a posteriori source information to improve channel decoding performance.

10.1 Source-Controlled Channel Decoding

One of the first works to study a priori and a posteriori source information to improve channel decoding was [1], where the term *source-controlled channel decoding* was introduced. In this and related techniques, the information fed into one of the decoding units is a measure of the likelihood, level of confidence, or reliability for each symbol input

Joint Source-Channel Coding, First Edition. Andres Kwasinski and Vinay Chande.
© 2023 John Wiley & Sons Ltd. Published 2023 by John Wiley & Sons Ltd.

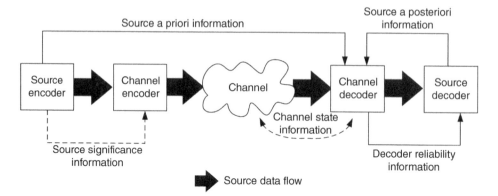

Figure 10.1 Main processing blocks in a typical digital communication system, showing the flow of source data (indicated with the bold arrows) and possible exchange of information between source and channel codecs that allow for JSCC schemes.

to the decoder. First, we consider the case of a priori source information. The intuitive idea is that, for a source generating bits with equiprobable values 0 and 1, the a priori information value should equal 0, indicating no higher likelihood of either bit value. As the probability of an output bit equating 1 increases, the a priori information should be a positive number with a larger magnitude. Conversely, the a priori information should be a negative number with increasing magnitude as the probability of the output bit equating 0 increases. A useful concept introduced in [2] that can represent this intuition for a priori information is the log-likelihood $L(x)$ of a binary random variable x defined as:

$$L(x) = \ln \frac{P(x = +1)}{P(x = -1)}, \tag{10.1}$$

where $P(x = +1)$ denotes the probability that x is equal to 1. Figure 10.2 illustrates the relation between $P(x = +1)$ and the log-likelihood $L(x)$.

The log-likelihood can also be defined for a continuous random variable that is conditioned on a binary outcome:

$$L(x) = \ln \frac{p(y|x = +1)}{p(y|x = -1)}, \tag{10.2}$$

where $p(y|x = +1)$ denotes the probability density function (PDF) of the random variable y conditioned on $x = 1$. The Viterbi algorithm, reviewed in Chapter 3, is particularly well suited to be used in combination with log-likelihood values, so the source-controlled channel decoding technique uses convolutional channel codes with Viterbi decoding. The good integration between the Viterbi algorithm and log-likelihood functions (involving expressing the algorithm's metrics in terms of log-likelihood functions) has resulted in their use in applications beyond source-controlled channel decoding.

Consider a system setup as in Figure 10.1 and similar to the one used in Section 3.2.3. The output from the source encoder is a sequence of bits $\vec{u}$. The discrete time in the model corresponds to the duration of a source-encoded bit. At time k, the input to the convolutional channel encoder is a bit u_k and the corresponding output is the vector $\vec{x}_k = \{x_{k,1}, x_{k,2}, \ldots, x_{k,N}\}$ of the N channel-encoded bits. The sequence of all the bits that results from channel encoding the sequence $\vec{u}$ is the sequence $\vec{x}$. The channel-encoded bits

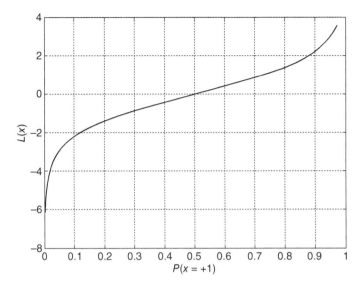

Figure 10.2 The log-likelihood of a binary random variable as a function of the probability of one of the random variable outcomes.

are transmitted through a channel and processed at the receiver's demodulator, resulting in a sequence $\vec{y}$, where the element corresponding to time k is represented as the vector $\vec{y}_k = \{y_{k,1}, y_{k,2}, \ldots, y_{k,N}\}$. For convenience, all bits are encoded to take values $+1$ or -1. After decoding, the original sequence of bits $\vec{u}$ is estimated as the sequence $\hat{\vec{u}}$. The function of all decoders is to use the sequence $\vec{y}$ to calculate $\hat{\vec{u}}$ as the estimate for $\vec{u}$.

The Viterbi algorithm decodes by finding among all the possible sequences $\hat{\vec{u}}$ the one that yields the maximum probability distribution $p(\vec{y}|\hat{\vec{u}})$. The Viterbi algorithm achieves the seemingly daunting task of computing all the possible probability distributions $p(\vec{y}|\hat{\vec{u}})$ and then finding the maximum, by mapping the sequence of estimated bits $\hat{\vec{u}}$ into a sequence of estimated channel code states $\hat{\vec{S}}_k = (s_0, s_1, \ldots, s_k)$ that are subsequently visualized as a path in the code's trellis diagram. Since at any given time k there exist multiple possible paths through the trellis diagram, we will use an index i to denote an specific trellis path with sequence of states $\vec{S}_k^{(i)}$. With the translation of possible bit sequences $\hat{\vec{u}}$ into trellis paths, the Viterbi algorithm decodes by finding the sequence of code states (trellis path) that has the maximum probability given the sequence $\vec{y}$. In other words, the Viterbi algorithm performs decoding by maximizing the a posteriori probability $P(\vec{S}_k^{(i)}|\vec{y})$:

$$\hat{\vec{S}}_k = \arg\max_i P(\vec{S}_k^{(i)}|\vec{y}), \tag{10.3}$$

Using Bayes' rule, the maximization problem can be rewritten as:

$$\hat{\vec{S}}_k = \arg\max_i P(\vec{S}_k^{(i)}|\vec{y})$$

$$= \arg\max_i \frac{P(\vec{S}_k^{(i)}, \vec{y}_k)}{p(\vec{y})}$$

$$\overset{+}{=} \arg\max_i P(\vec{S}_k^{(i)}, \vec{y})$$

$$= \arg\max_i P(\vec{S}_k^{(i)}, \vec{y})$$

$$= \arg\max_i p(\vec{y}|\vec{S}_k^{(i)})P(\vec{S}_k^{(i)}),$$

where step ✚ follows from the observation that $p(\vec{y})$ does not affect the maximization because it is common for all state sequences. Next, after denoting as $u_k^{(i)}$ the bit that is input at time k which results in the final state of the sequence $\vec{S}_k^{(i)}$, the probability $P(\vec{S}_k^{(i)})$ can be written as:

$$P(\vec{S}_k^{(i)}) = P(\vec{S}_{k-1}^{(i)})P(u_k^{(i)}), \tag{10.4}$$

and the maximization problem can be expressed as:

$$\hat{\vec{S}}_k = \arg\max_i p(\vec{y}|\vec{S}_k^{(i)})P(\vec{S}_{k-1}^{(i)})P(u_k^{(i)}). \tag{10.5}$$

Considering now the PDF $p(\vec{y}|\vec{S}_k^{(i)})$, we see that it can be expanded into an expression using distributions conditioned on transitions from one state to the immediately following one, that is, single-stage transitions in the code trellis:

$$p(\vec{y}|\vec{S}_k^{(i)}) = p(\vec{y}_k|\vec{S}_k^{(i)}, \vec{y}_{0,k-1})p(\vec{y}_{k-1}|\vec{S}_k^{(i)}, \vec{y}_{0,k-2}) \dots p(\vec{y}_1|\vec{S}_k^{(i)}, \vec{y}_0)p(\vec{y}_0|\vec{S}_k^{(i)})$$
$$= p(\vec{y}_k|s_k^{(i)}, s_{k-1}^{(i)})p(\vec{y}_{k-1}|s_{k-1}^{(i)}, s_{k-2}^{(i)}) \dots p(\vec{y}_0|s_0^{(i)}), \tag{10.6}$$

where a notation $\vec{y}_{0,k-2}$ means the sequence of received bits $\vec{y}_0, \vec{y}_1, \dots, \vec{y}_{k-3}, \vec{y}_{k-2}$. The result in (10.6) follows from the observation that the received bits $\vec{y}_j$ do not depend on any other received bits, given the previous and current states $s_{j-1}^{(i)}$ and $s_j^{(i)}$. Using (10.6) in (10.5) yields

$$\hat{\vec{S}}_k = \arg\max_i P(\vec{S}_{k-1}^{(i)})P(u_k^{(i)})p(\vec{y}_k|s_k^{(i)}, s_{k-1}^{(i)})p(\vec{y}_{k-1}|s_{k-1}^{(i)}, s_{k-2}^{(i)}) \dots p(\vec{y}_0|s_0^{(i)}).$$

This expression emphasizes the nature of the problem where the goal is to find a path through the trellis, which accounts for all the transitions, that results in the largest a posteriori probability. Recall that the Viterbi algorithm calculates the path by accumulating a metric for all the possible paths through the trellis, and then choosing as solution the path with the largest such metric. Thus, the maximization problem can be rewritten as:

$$\hat{\vec{S}}_k = \arg\max_i M_k^{(i)},$$

where $M_k^{(i)}$, the metric for path i accumulated throughout k stages in the trellis, is equal to

$$M_k^{(i)} = P(\vec{S}_{k-1}^{(i)})P(u_k^{(i)})p(\vec{y}_k|s_k^{(i)}, s_{k-1}^{(i)})p(\vec{y}_{k-1}|s_{k-1}^{(i)}, s_{k-2}^{(i)}) \dots p(\vec{y}_0|s_0^{(i)}).$$

For computational purposes, it will be advantageous to redefine the path metric as twice the logarithm of the metric just defined:

$$M_k^{(i)} = 2\ln\left[P(\vec{S}_{k-1}^{(i)})P(u_k^{(i)})p(\vec{y}_k|s_k^{(i)}, s_{k-1}^{(i)})p(\vec{y}_{k-1}|s_{k-1}^{(i)}, s_{k-2}^{(i)}) \dots p(\vec{y}_0|s_0^{(i)})\right].$$

Because the logarithm is a monotonic increasing function, taking the logarithm on the path metric does not affect the maximization calculation, and similarly multiplying by two does not affect it. An advantage obtained from using logarithms is that multiplications in the metric are translated into additions:

$$M_k^{(i)} = 2 \ln P(u_k^{(i)}) + 2 \ln p(\vec{y}_k | s_k^{(i)}, s_{k-1}^{(i)}) + 2 \left[\ln P(\vec{S}_{k-1}^{(i)}) + \sum_{m=0}^{k-1} \ln p(\vec{y}_m | s_m^{(i)}, s_{m-1}^{(i)}) \right],$$

where, to keep the notation compact, we have implicitly written for the case of $m = 0$ that $p(\vec{y}_0 | s_0^{(i)}) = p(\vec{y}_0 | s_0^{(i)}, s_{-1}^{(i)})$. Inspecting the newly defined path metric reveals that the last two terms (appearing inside the brackets) equal $2 \ln P(\vec{S}_{k-1}^{(i)} | \vec{y})$, which is $M_{k-1}^{(i)}$, plus $2 \ln p(\vec{y})$ which acts as a constant for the maximization problem and can be ignored henceforth. Then, the path metric can be written in the form of a recursion apt for applying the Viterbi algorithm:

$$M_k^{(i)} = M_{k-1}^{(i)} + 2 \ln P(u_k^{(i)}) + 2 \ln p(\vec{y}_k | s_k^{(i)}, s_{k-1}^{(i)}).$$

This expression makes it clear that the terms $2 \ln P(u_k^{(i)})$ and $2 \ln p(\vec{y}_k | s_k^{(i)}, s_{k-1}^{(i)})$ are the update in the path metric resulting from the encoding and transmission of the most recent source-encoded bit. Moreover, since conditioning a probability on the channel encoder state transition from $s_{k-1}^{(i)}$ to $s_k^{(i)}$ over a trellis path i is equivalent to conditioning on the corresponding resulting channel encoding bits $\vec{x}_k^{(i)}$, we have that $p(\vec{y}_k | s_k^{(i)}, s_{k-1}^{(i)}) = p(\vec{y}_k | \vec{x}_k^{(i)})$. Also, since each received bit $y_{k,m}$ (for $m = 1, \ldots, N$) depends only on the corresponding transmitted bit $x_{k,m}^{(i)}$, $2 \ln p(\vec{y}_k | s_k^{(i)}, s_{k-1}^{(i)}) = 2 \ln p(\vec{y}_k | \vec{x}_k^{(i)}) = 2 \sum_{m=1}^{N} \ln p(y_{k,m} | x_{k,m}^{(i)})$. Each of the probabilities $p(y_{k,m} | x_{k,m}^{(i)})$ represents the probability that a transmitted bit $x_{k,m}^{(i)}$ is affected by the communication channel. Consequently, the path metric

$$M_k^{(i)} = M_{k-1}^{(i)} + 2 \ln P(u_k^{(i)}) + 2 \sum_{m=1}^{N} \ln p(y_{k,m} | x_{k,m}^{(i)}). \tag{10.7}$$

expresses the accumulation of a metric accounting for the probability of the most recent source-encoded bit binary values $2 \ln P(u_k^{(i)})$, a metric considering the probability distribution of the channel affecting the transmitted bits resulting from channel encoding the most recent source-encoded bit, $2 \sum_{m=1}^{N} \ln p(y_{k,m} | x_{k,m}^{(i)})$, and the path metric accumulated for all previous stages of the code trellis, $M_{k-1}^{(i)}$.

In order to apply the path metric (10.7) in joint source–channel decoding (JSCD) techniques, it is still necessary to express the metrics $2 \ln P(u_k^{(i)})$ and $2 \sum_{m=1}^{N} \ln p(y_{k,m} | x_{k,m}^{(i)})$ in terms of log-likelihood values. For convenience, we modify the metric (10.7) by adding a constant:

$$M_k^{(i)} = M_{k-1}^{(i)} + 2 \ln P(u_k^{(i)}) + \sum_{m=1}^{N} \left(2 \ln p(y_{k,m} | x_{k,m}^{(i)}) \right.$$

$$\left. - \ln p(y_{k,m} | x_{k,m} = +1) - \ln p(y_{k,m} | x_{k,m} = -1) \right). \tag{10.8}$$

Note that the new term is considered a "constant" for the purpose of maximizing the path metric, because the two probabilities $p(y_{k,m} | x_{k,m} = +1)$ and $p(y_{k,m} | x_{k,m} = -1)$ do not depend on the path choice. Because of this reason is that the notation does not include the trellis path index i for these terms.

Starting with the term $2 \ln P(u_k^{(i)})$, the source a priori log-likelihood ratio can be written as:

$$L(u_k) = \ln \frac{P(u_k = +1)}{P(u_k = -1)}. \tag{10.9}$$

Using this equation to solve for $P(u_k = +1)$ and $P(u_k = -1)$ results in

$$e^{L(u_k)} = \frac{P(u_k = +1)}{P(u_k = -1)} = \frac{P(u_k = +1)}{1 - P(u_k = +1)}.$$

With further algebraic operations

$$e^{L(u_k)} - e^{L(u_k)}P(u_k = +1) = P(u_k = +1),$$

which leads to the intermediate result

$$P(u_k = +1) = \frac{e^{L(u_k)}}{1 + e^{L(u_k)}}$$

We can also write

$$e^{L(u_k)} = \frac{1 - P(u_k = -1)}{P(u_k = -1)}$$

which means that

$$e^{-L(u_k)} = \frac{P(u_k = -1)}{1 - P(u_k = -1)}$$

and leads to

$$P(u_k = -1) = \frac{e^{-L(u_k)}}{1 + e^{-L(u_k)}}.$$

Combining both final expressions for $P(u_k = +1)$ and $P(u_k = -1)$ into one, we can compactly write the probability mass function of u_k as:

$$P(u_k) = \frac{e^{\pm L(u_k)}}{1 + e^{\pm L(u_k)}} = \frac{e^{u_k L(u_k)}}{1 + e^{u_k L(u_k)}}.$$

From this compact expression of $P(u_k)$, it is possible to write this probability in the case for the realization at time k over the trellis path i:

$$P(u_k^{(i)}) = \begin{cases} \frac{e^{L(u_k)}}{1+e^{L(u_k)}} = \frac{1}{1+e^{-L(u_k)}}, & \text{if } u_k^{(i)} = +1 \\ \frac{e^{-L(u_k)}}{1+e^{-L(u_k)}}, & \text{if } u_k^{(i)} = -1. \end{cases} \tag{10.10}$$

Finally, $P(u_k)$ can be written as:

$$P(u_k^{(i)}) = \underbrace{\left(\frac{e^{-L(u_k)/2}}{1 + e^{-L(u_k)}} \right)}_{A_k} e^{L(u_k)u_k^{(i)}/2} = A_k e^{L(u_k)u_k/2}; \quad u_k^{(i)} \in \{-1, 1\},$$

where A_k is a constant which depends only on the time instant k. Of course,

$$2 \ln P(u_k^{(i)}) = 2 \ln A_k + L(u_k)u_k^{(i)}, \tag{10.11}$$

which is the type of expression that was sought, expressing the metric $2 \ln P(u_k)$ in terms of a log-likelihood value.

Inspecting (10.8), it can be seen that it is necessary to derive an expression for the terms $2 \ln p(y_{k,m}|x_{k,m}^{(i)}) - \ln p(y_{k,m}|x_{k,m} = +1) - \ln p(y_{k,m}|x_{k,m} = -1)$ that can be used in the calculation of the trellis path metric. This derivation uses log-likelihood values to capture the effects of the channel on the transmitted bits in a simple way (and a simple calculation). Part of the derivation is specific to the statistics describing the channel characteristics. Let's assume a channel with additive white Gaussian noise (AWGN) and fading with a coefficient h, and a transmitted symbol energy equal to E_s. In this case, the relation between $y_{k,m}$ and $x_{k,m}^{(i)}$ is $y_{k,m} = hx_{k,m}^{(i)} + n_k$, with $n_k \sim \mathcal{N}(0, N_0/2)$. Considering this,

$$p(y_{k,m} \mid x_{k,m}^{(i)}) = \frac{1}{\sqrt{\frac{2\pi N_0}{2E_s}}} e^{-\frac{E_s}{N_0}(y_{k,m}-hx_{k,m}^{(i)})^2}$$

$$= \frac{1}{\sqrt{\frac{\pi N_0}{E_s}}} e^{-\frac{E_s}{N_0}(y_{k,m}^2+h^2)} e^{\frac{2E_s}{N_0} hy_{k,m}x_{k,m}^{(i)}}$$

$$= \Gamma_k(y_{k,m}) e^{\frac{2E_s}{N_0} hy_{k,m}x_{k,m}^{(i)}},$$

where we used the fact that $x_{k,m}^{(i)^2} = 1$ always (because $x_{k,m}^{(i)} = \pm 1$) and we introduced

$$\Gamma_k(y_{k,m}) = \frac{1}{\sqrt{\frac{\pi N_0}{E_s}}} e^{-\frac{E_s}{N_0}(y_{k,m}^2+h^2)}.$$

Applying these results to the likelihood ratio expression yields

$$L(y_{k,m} \mid x_{k,m}^{(i)}) = \ln \frac{p(y_{k,m} \mid x_{k,m}^{(i)} = +1)}{p(y_{k,m} \mid x_{k,m}^{(i)} = -1)} \tag{10.12}$$

$$= \ln \frac{\Gamma_k(y_{k,m}^2) e^{\frac{2E_s}{N_0} hy_{k,m}}}{\Gamma_k(y_{k,m}^2) e^{-\frac{2E_s}{N_0} hy_{k,m}}}.$$

Simplifying and with a bit more of algebraic operations, we obtain

$$L(y_{k,m} \mid x_{k,m}^{(i)}) = \ln e^{\frac{4E_s}{N_0} hy_{k,m}} = \frac{4hE_s}{N_0} y_{k,m} = L_c y_{k,m}, \tag{10.13}$$

where the *channel reliability*

$$L_c = \frac{4hE_s}{N_0} \tag{10.14}$$

is an indicator of channel quality. The channel reliability in (10.14) is specific to the case of a fading channel (or an AWGN channel when $h = 1$). For a binary symmetric channel (BSC), it can be shown that the channel reliability is the log-likelihood ratio of the crossover probabilities [2].

Next, consider the case when $x_{k,m}^{(i)} = 1$, where it follows from (10.12) and (10.13),

$$p(y_{k,m}|x_{k,m}^{(i)}) = e^{L_c y_{k,m}} p(y_{k,m}|x_{k,m}^{(i)} = -1).$$

Note that we do not write $x_{k,m}^{(i)} = +1$ on the left side of the equality. As it will become clear soon, the reason for this slight abuse of notation is to arrive at more compact expressions that are common for both cases of $x_{k,m}^{(i)} = +1$ and $x_{k,m}^{(i)} = -1$. Now, since $x_{k,m}^{(i)} = +1$, we can write

$$p(y_{k,m}|x_{k,m}^{(i)}) = e^{x_{k,m}^{(i)}L_c y_{k,m}} p(y_{k,m}|x_{k,m}^{(i)} = -1).$$ (10.15)

Similarly, when $x_{k,m}^{(i)} = -1$,

$$p(y_{k,m}|x_{k,m}^{(i)}) = e^{-L_c y_{k,m}} p(y_{k,m}|x_{k,m}^{(i)} = 1) = e^{x_{k,m}^{(i)}L_c y_{k,m}} p(y_{k,m}|x_{k,m}^{(i)} = 1).$$ (10.16)

Taking the terms from (10.8), $2\ln p(y_{k,m}|x_{k,m}^{(i)}) - \ln p(y_{k,m}|x_{k,m} = +1) - \ln p(y_{k,m}|x_{k,m} = -1)$, and using (10.15) and (10.16), we get that for $x_{k,m}^{(i)} = +1$,

$$2\ln p(y_{k,m}|x_{k,m}^{(i)}) - \ln p(y_{k,m}|x_{k,m} = +1) - \ln p(y_{k,m}|x_{k,m} = -1) =$$

$$2x_{k,m}^{(i)}L_c y_{k,m} + 2\ln p(y_{k,m}|x_{k,m}^{(i)} = -1) - \ln p(y_{k,m}|x_{k,m} = +1) - \ln p(y_{k,m}|x_{k,m} = -1) =$$

$$2x_{k,m}^{(i)}L_c y_{k,m} + \ln \frac{p(y_{k,m}|x_{k,m}^{(i)} = -1)}{p(y_{k,m}|x_{k,m}^{(i)} = +1)} =$$

$$2x_{k,m}^{(i)}L_c y_{k,m} - x_{k,m}^{(i)}L_c y_{k,m} =$$

$$x_{k,m}^{(i)}L_c y_{k,m}.$$

Similarly, when $x_{k,m}^{(i)} = -1$,

$$2\ln p(y_{k,m}|x_{k,m}^{(i)}) - \ln p(y_{k,m}|x_{k,m} = +1) - \ln p(y_{k,m}|x_{k,m} = -1) =$$

$$2x_{k,m}^{(i)}L_c y_{k,m} + 2\ln p(y_{k,m}|x_{k,m}^{(i)} = 1) - \ln p(y_{k,m}|x_{k,m} = +1) - \ln p(y_{k,m}|x_{k,m} = -1) =$$

$$2x_{k,m}^{(i)}L_c y_{k,m} + \ln \frac{p(y_{k,m}|x_{k,m}^{(i)} = +1)}{p(y_{k,m}|x_{k,m}^{(i)} = -1)} =$$

$$2x_{k,m}^{(i)}L_c y_{k,m} - x_{k,m}^{(i)}L_c y_{k,m} =$$

$$x_{k,m}^{(i)}L_c y_{k,m}.$$

Finally, using these last results and (10.11), the path metric (10.8) can be written as:

$$M_k^{(i)} = M_{k-1}^{(i)} + 2\ln A_k + L(u_k)u_k^{(i)} + \sum_{m=1}^{N} x_{k,m}^{(i)} L_c y_{k,m}.$$

Since the constant A_k does not depend on the chosen path, it will not play a role in the maximization and can be removed from the metric, leading to the final path metric expression:

$$M_k^{(i)} = M_{k-1}^{(i)} + L(u_k)u_k^{(i)} + \sum_{m=1}^{N} x_{k,m}^{(i)} L_c y_{k,m}.$$ (10.17)

It is interesting to compare the result of the path metric (10.17) with the one used in a "standard" Viterbi algorithm as reviewed in Section 3.2.3. While in the case of the Viterbi algorithm the metric appears different, the update to the path metric calculated for the latest trellis stage corresponds to the term $x_{k,m}^{(i)} L_c y_{k,m}$ in (10.17). The term $L(u_k)u_k^{(i)}$ does not appear in the Viterbi algorithm metric computation, because there is an implicit assumption that $p(u_k = +1) = p(u_k = -1)$, which leads to $L(u_k) = 0$. This assumption is frequently not true and ignores the a priori information about the source that could be used to improve the decoding. Since the path metric (10.17) includes the a priori information about the source, this implementation of the Viterbi algorithm was called APRI-VA in [1]. Furthermore, because of its simpler form in the "standard" Viterbi algorithm, the channel reliability L_c is not included in the path metric computation because it plays the role of a constant factor that does not affect the maximization. For the metric in (10.17), the log-likelihood values $L(u_k)$ and L_c work as weighting factors controlling the combination of a priori and channel effects metrics. This is, if the channel exhibits very good conditions, L_c will be much larger than $|L(u_k)|$ resulting in a path metric calculation that emphasizes the received bits and the inherent structure introduced by the channel encoding. If the channel is very poor, the decoding computations will make more use of the source a priori information.

The aforementioned intuition on how the addition of a priori information to the path metric computation improves performance can be put to the test by calculating the bit error rate (BER), approximated as the probability of a bit error, either when considering or when ignoring the a priori source likelihood. The performance of convolutional decoding when ignoring the a priori source information was reviewed in Section 3.2.4. The performance when considering the a priori source information can be derived following a similar procedure. This results in the BER being upper bounded as [3]:

$$P(\gamma) \le \sum_{d=d_f}^{\infty} c(d) P_e(d|\gamma), \tag{10.18}$$

where $c(d)$ indicates the total number of information errors at all paths with Hamming weight d, d_f is the code's free distance, and $P_e(d|\gamma)$ is the probability that two paths in the decoding trellis, a correct and an erroneous one, differ by d bits. Assuming binary phase shift keying (BPSK) modulation and following similar procedures as in Section 3.2.4, this probability can be shown to be

$$P_e(d|\gamma) = \frac{1}{2} \text{erfc}\left(\sqrt{d\gamma \left(1 + \frac{c(d)L(u)}{4d\gamma a(d)} \right)} \right), \tag{10.19}$$

where $\gamma = E_s/N_0$ is the received symbol signal-to-noise ratio, $a(d)$ is a function that returns the number of paths with d nonzero bits separating from the all-zero path and then rejoining again for the first time, and $\text{erfc}(\gamma)$ is the complementary error function: $\text{erfc}(\gamma) = 2/\pi \int_{\gamma}^{\infty} e^{-u^2} du$. Note that (10.19) reduces to the corresponding expression in Section 3.2.4 when $L(u) = 0$.

Figure 10.3 shows the result from combining (10.18) and (10.19) for $L(u)$ equal to 0, 1, and 2. For simplicty, the BER in (10.18) is calculated by considering only the term for $d = d_f$. The results are for a convolutional code with memory 4, channel coding rate $1/2$ and configured as shown in Figure 10.4. This convolutional code was studied in detail in [4]. The results in Figure 10.3 confirm that inclusion of a priori source information improves channel coding performance.

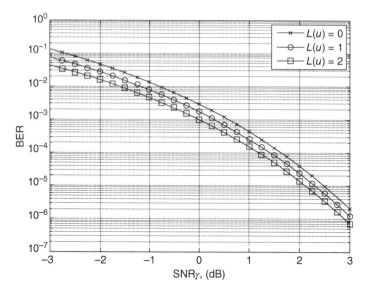

Figure 10.3 Approximate bit error rate performance for a memory 4, rate 1/2, convolutional code with different source a priori log-likelihood values $L(u)$.

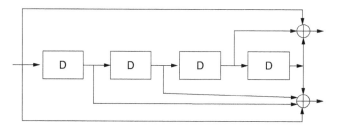

Figure 10.4 The convolutional encoder with memory 4 used in the results shown in Figure 10.3.

10.2 Exploiting Residual Redundancy at the Decoder

In the performance results shown in Figure 10.3, the a priori source information in the form of the a priori log-likelihood was assumed known beforehand. This is not always possible. In some cases when the a priori source information is not available, it is possible to still include a source log-likelihood metric in the decoding path computation by using a posteriori source information to compute $L(u)$. This has been done in a number of schemes that exploit the output from source encoders that do not eliminate all the source redundancy, as is the case in the works [1, 5–7]. This commonly arises from limitations in source encoding complexity, or from non-stationarity in the source statistics, for example. This redundancy that is left in the source-encoded stream, called a residual redundancy, can be observed as a correlation of the bits at the output of the source encoder in time (when the source is a time-dependent signal) or space (when the source is an image). For example, speech is encoded and transmitted by processing samples from a period of time ranging between 5 and 30 ms, depending on the coding algorithm. Because speech statistics change more slowly than this, bits in successive frames have some correlation (a property that can be used to conceal transmission errors). This situation can also be seen in image coding, where the first encoding step often consists of partitioning the image into square tiles. Usually, nearby tiles present similar visual characteristics and would result in coded bits that are correlated to some degree.

When the source information is formatted for transmission into blocks, frames, or packets, the correlation that may exist between bits in different blocks can be exploited to obtain the source a posteriori information. One approach to implement this is to include an indication of the likelihood, or reliability, for each of the estimated source-encoded bits (the output of the channel decoder) in the channel decoding. When the source-encoded bits are correlated in time across successive transmitted frames, the likelihood of bits that come from channel decoding one frame could be used to derive a log-likelihood metric for the bits in the following frame. To obtain a likelihood indicator from the channel decoding operation, the Viterbi algorithm needs to be modified to provide this information as part of the decoding result. The result is called in this case a *soft output*, because the decoding result in not just a *hard* ±1 result. With this change, the Viterbi algorithm is called a "soft output Viterbi algorithm" (SOVA).

10.2.1 The Soft Output Viterbi Algorithm (SOVA)

The output from the SOVA for any given bit is composed of two components:

- the usual estimation of source-encoded bits (the hard decision for a "+1" or a "−1" output), which will give the sign to the soft output value, and
- a metric that indicates the level of confidence for the estimation decision corresponding with each source-encoded bit. This metric will give the magnitude to the soft output value.

Figure 10.5 illustrates how each of the decisions in the algorithm implies the selection of the path with a larger metric. In the case illustrated in Figure 10.5, it is assumed that path i is chosen, instead of path i', because $M_k^{(i)} > M_k^{(i')}$. In choosing path i, there is always a

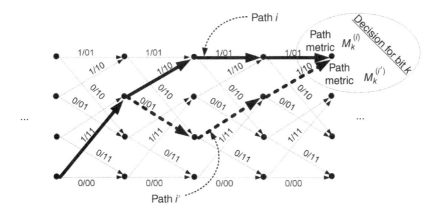

Figure 10.5 Trellis for the decoding of a rate 1/2 convolutional code, showing the selection between two competing paths.

probability that the decision is incorrect. The metric at the output of the channel decoder should be able to capture the probability that the chosen path i was, in fact, the correct choice.

The starting point to derive the SOVA is the observation that, while executing the Viterbi algorithm, at each decoding trellis node along the surviving path, there is another path that was discarded. This implies a binary decision at each node along the surviving path that might be right or wrong. In what follows, it will be useful to define the metric difference between the surviving path, with index i, and the non-surviving path, with index i'. For a decision made at the node corresponding to time k, we define the difference as:

$$\Delta_k^{(i)} = M_k^{(i)} - M_k^{(i')}. \tag{10.20}$$

This difference is always larger than 0, because the chosen path always has the larger metric. Next, in order to derive a metric that expresses the confidence that the decision at time k was the correct one, we calculate the probability that the path decision was correct:

$$P(\text{correct}) = \frac{P(\text{path } i)}{P(\text{path } i) + P(\text{path } i')}. \tag{10.21}$$

The probability of a path (e.g. $P(\text{path } i)$) is $P(\vec{S}_k^{(i)})$ as seen in (10.4). Tracing back the steps that led to deriving the path metric $M_k^{(i)}$ from $P(\vec{S}_k^{(i)})$, undoing the multiplication by 2, the introduction of the logarithm and the addition and subtraction of constants yields

$$P(\text{path } i) = P(\vec{S}_k^{(i)}) = \Upsilon_k e^{M_k^{(i)}/2},$$

where Υ_k is a constant originating in the constants that were added and subtracted from the path metric. Applying this result in (10.21) results in

$$\begin{aligned} P(\text{correct}) &= \frac{\Upsilon_k e^{M_k^{(i)}/2}}{\Upsilon_k e^{M_k^{(i)}/2} + \Upsilon_k e^{M_k^{(i')}/2}} \\ &= \frac{e^{\Delta_k^{(i)}}}{1 + e^{\Delta_k^{(i)}}}. \end{aligned} \tag{10.22}$$

This result can now be use to express the log-likelihood ratio for the binary decision between the two competing paths at a single node in the trellis, the correct one $\vec{S}_k^{(i)}$, with probability $P(\text{correct})$, and the incorrect one $\vec{S}_k^{(i')}$, with probability $1 - P(\text{correct})$, which is calculated as:

$$\ln \frac{P(\text{correct})}{1 - P(\text{correct})}.$$

After applying (10.22) and doing some algebraic operations, this log-likelihood ratio is shown to be equal to

$$\ln \frac{P(\text{correct})}{1 - P(\text{correct})} = \Delta_k^{(i)}.$$

Choosing a path i involves making a binary decision for that path at each of the nodes along it. When M such decisions have been made, there are M non-surviving paths (which we index with $m = 0, 1, \ldots, M - 1$) that were discarded. For the non-surviving paths, the difference of their path metrics $\Delta_k^{(m)}$ is positive. Picking some time instant (stage in the decoding trellis) $k - \delta$ along the path, the chosen path i will have associated with it a decoded bit $u_{k-\delta}$. Each discarded path will also have associated with it a decoded bit $u_{k-\delta}^{(m)}$, which by chance may equal $u_{k-\delta}$. If $u_{k-\delta} = u_{k-\delta}^{(m)}$, by lucky chance there is no error, and no difference when choosing path i or m. For these cases, the reliability (log-likelihood) of the bit decision is infinite. However, if $u_{k-\delta} \neq u_{k-\delta}^{(m)}$ the log-likelihood for the bit error is equal to $\Delta_k^{(m)}$.

The introduction of a bit error can be conveniently represented as an addition modulo 2 of the original (and correct) bit value with an error bit $e_{k-\delta}^{(m)}$. This representation allows writing the log-likelihood of a trellis node decision as the log-likelihood of the error bit:

$$
L(e_{k-\delta}^{(m)}) = \ln \frac{P(e_{k-\delta}^{(m)} = +1)}{P(e_{k-\delta}^{(m)} = -1)} =
\begin{cases}
\infty, & \text{if } u_{k-\delta}^{(m)} = u_{k-\delta} \\
\Delta_{k-\delta}^{(m)}, & \text{if } u_{k-\delta}^{(m)} \neq u_{k-\delta}.
\end{cases}
$$

Here, the modulo-2 addition is implemented as an addition in the Galois field GF(2) with the two elements $\{+1, -1\}$, where $+1$ is the null element under addition. Using the symbol $\oplus$ for the sum in this field (effectively, the modulo-2 addition), this means that for any value of a variable a in GF(2), $a \oplus +1 = a$. Note that when $u_{k-\delta}^{(m)} = u_{k-\delta}$, the log-likelihood needs to equal infinity because no decision error is made. In other words, the expression says when $u_{k-\delta}^{(m)} = u_{k-\delta}$, $P(e_{k-\delta}^{(m)} = +1) = 1$ and $P(e_{k-\delta}^{(m)} = -1) = 0$, implying that there are no chances of making a decision error. Still, the total error for bit $u_{k-\delta}$ resulting from all the discarded paths is

$$
e_{k-\delta} = \bigoplus_{m=0}^{M-1} e_{k-\delta}^{(m)}. \tag{10.23}
$$

The convenience in introducing the error representation as a modulo-2 addition will now become clear. It will become straightforward to introduce a special algebra that allows simple expressions and computation of log-likelihood results derived from expressions of the type $\oplus_{k=0}^{M-1}$. To see this, consider two bits, z_1 and z_2, defined in GF(2) with elements $\{+1, -1\}$. Assuming the bits are statistically independent of each other, the log-likelihood of their sum is

$$
L(z_1 \oplus z_2) = \ln \frac{P(z_1 \oplus z_2 = +1)}{P(z_1 \oplus z_2 = -1)}
$$

$$
= \ln \frac{P\left((z_1 = +1 \cap z_2 = +1) \cup (z_1 = -1 \cap z_2 = -1)\right)}{P\left((z_1 = +1 \cap z_2 = -1) \cup (z_1 = -1 \cap z_2 = +1)\right)}
$$

$$
= \ln \frac{P(z_1 = +1)P(z_2 = +1) + P(z_1 = -1)P(z_2 = -1)}{P(z_1 = +1)P(z_2 = -1) + P(z_1 = -1)P(z_2 = +1)}. \tag{10.24}
$$

Next, reusing the earlier intermediate result (10.10), we can write

$$
P(z_1 = +1) = \frac{e^{L(z_1)}}{1 + e^{L(z_1)}}, \quad P(z_2 = +1) = \frac{e^{L(z_2)}}{1 + e^{L(z_2)}},
$$

and

$$
P(z_1 = -1) = \frac{1}{1 + e^{L(z_1)}}, \quad P(z_2 = -1) = \frac{1}{1 + e^{L(z_2)}}.
$$

Replacing these expressions in (10.24), we can write

$$L(z_1 \oplus z_2) = \ln \frac{\dfrac{e^{L(z_1)+L(z_2)} + 1}{(1 + e^{L(z_1)})(1 + e^{L(z_2)})}}{\dfrac{e^{L(z_1)} + e^{L(z_2)}}{(1 + e^{L(z_1)})(1 + e^{L(z_2)})}}$$

$$= \ln \frac{e^{L(z_1)+L(z_2)} + 1}{e^{L(z_1)} + e^{L(z_2)}},$$

and by applying properties of logarithms, we can approximate

$$L(z_1 \oplus z_2) \approx \operatorname{sign}(L(z_1)) \operatorname{sign}(L(z_2)) \min (|L(z_1)|, |L(z_2)|).$$

To simplify the notation when using this result in the future, let us introduce the algebraic operator on log-likelihood values "⊞" defined as:

$$L(z_1) \boxplus L(z_2) \overset{\Delta}{=} L(z_1 \oplus z_2),$$

which can be generalized to its use on M independent random variables $z_1, z_2, \ldots, z_M$,

$$\boxplus_{k=1}^{M} L(z_k) = L \left(\bigoplus_{k=1}^{M} z_k \right) \approx \prod_{k=1}^{M} \operatorname{sign}(L(z_k)) \min_{k=1,2,\ldots,M} |L(z_k)|.$$

This new algebraic operator can be used to calculate the log-likelihood associated with a bit decision. If the random variables $\Delta_k^{(i)}$ and $e_k^{(i)}$ are statistically independent with respect to the indices i indicating different paths (which is usually found to be approximately true for well-conceived codes), it follows from (10.23) that the soft output from the Viterbi algorithm for the bit decoded at time $k - \delta$ is

$$L(u_{k-\delta}) = u_{k-\delta} \boxplus_{m=1}^{M} L(e_{k-\delta}^{(m)}),$$

$$L(u_{k-\delta}) \approx u_{k-\delta} \min_{m=1,2,\ldots,M} \Delta_{k-\delta}^{(m)}. \tag{10.25}$$

10.2.2 Exploiting Residual Redundancy to Estimate A Priori Information

A simple but instructive example on how to use the SOVA output as a posteriori information to improve decoding can be found in [1]. This example considers a scenario where information is organized in frames, and bits occupying the same position in frames transmitted at different times are correlated. The first-order Markov chain model is used to reflect the likely case where the correlation between bits decreases with time. To characterize the dynamics of this system, consider the bit $u_{k,q}$ which occupies position q in a frame at time k. Bit $u_{k,q}$ may have the same value as the preceding bit in that position, $u_{k-1,q}$, or its value may have changed. The probability of bits $u_{k-1,q}$ and $u_{k,q}$ being different is $P_{k,q}$. In order to apply a posteriori information using log-likelihood ratios, it will be convenient to model the change of a bit's value as an addition in GF(2) of $u_{k-1,q}$ with a "change" bit $c_{k,q}$: $u_{k,q} = u_{k-1,q} \oplus c_{k,q}$. The change bit $c_{k,q}$ is analogous to the error bit in the SOVA discussion earlier. Because of this, it is straightforward to see that the log-likelihood of the change bit is

$$L(c_{k,q}) = \ln \frac{P(c_{k,q} = +1)}{P(c_{k,q} = -1)} = \ln \frac{1 - P_{k,q}}{P_{k,q}}.$$

The difference with respect to the SOVA algorithm is that now the log-likelihood $L(c_{k,q})$ is not representing the possibility of an error but rather the likelihood of a change in the bit value between times $k-1$ and k. Furthermore, since $u_{k,q} = u_{k-1,q} \oplus c_{k,q}$, we can apply the algebraic operator $\boxplus$ on log-likelihood values and write

$$L(u_{k,q}) = L(u_{k-1,q}) \boxplus L(c_{k,q}).$$

This expression indicates how to calculate the a priori log-likelihood value for the bit $u_{k,q}$ using prior knowledge of the probability $P_{k,q}$ and the a posteriori log-likelihood value for the bit $u_{k-1,q}$, obtained as the SOVA output. The a priori log-likelihood value for the bit $u_{k,q}$ can be used to improve channel decoding using (10.17) and obtain performance improvements as illustrated in Figure 10.3.

This exploitation of correlation (in time) can be applied to improve channel decoding of images, by exploiting the correlation in space found in images. Figures 10.6 and 10.7 illustrate this correlation. The figures are created by partitioning three images, Lena, Barbara, and Sydney (same as Figure 5.5 in Chapter 5), into blocks of eight-by-eight pixels. Each block is transformed using the discrete cosine transform (DCT) transform. This is the first step in encoding images with JPEG. Figure 10.6 shows the most significant bit

| Barbara | Lena | Sydney |

Figure 10.6 Value of the most significant bit for the DCT DC coefficient in blocks of 8×8 pixels for three images: Barbara, Lena, and Sydney. Bits with logical values of "0" and "1" are represented with black and white pixels, respectively.

| Barbara | Lena | Sydney |

Figure 10.7 Value of the most significant bit for the DCT coefficient with coordinates (2,3) in blocks of 8×8 pixels for three images: Barbara, Lena, and Sydney. Bits with logical values of "0" and "1" are represented with black and white pixels, respectively.

for the DC coefficient from the resulting eight-by-eight DCT-transformed blocks. A bit with a logical value "0" is represented with a black pixel and a "1" bit is shown as white. Figure 10.7 shows the most significant bit of the DCT coefficient with coordinates (2,3) using the same representation.

The figures illustrate how the value of the most significant bits for the coefficients is the same over neighboring regions of the image, a sign of spatial correlation. This correlation can be exploited to compute a priori likelihood values using a posteriori likelihood values from neighboring blocks in an image. Specifically, it was proposed in [7] to improve the decoding performance of bits from an image block by computing the a priori log-likelihood using the a posteriori log-likelihood results from neighboring blocks and using the result as the factor $L(u_k)$ in (10.17). For convenience, the log-likelihood values are first written using a logarithmic base of 10:

$$L(u) = \log \frac{P(u = +1)}{P(u = -1)},$$

where u represents the value of any information bit, which has a priori probabilities $P(u = -1) = p$ and $P(u = +1) = 1 - p$. The change of base in the logarithms causes a few changes in the derivations done so far, for example, in the computations of channel reliabilities, (10.14) now becomes

$$L_c = \frac{4h}{\ln b} \frac{E_s}{N_0},$$

where b is the base of the logarithms used in the calculations (in this case $b = 10$).

Next, in order to have a simpler procedure to combine a posteriori information from more than one information bit, the log-likelihood function is approximated by a linear function as:

$$L(u) \approx \overline{L}(u) = K(1 - 2p),$$

This first-order polynomial approximation around the point $p = 0.5$ is based on the odd symmetry of the log-likelihood function about this point (see Figure 10.2). Because of this, the constant K is calculated so that the linear approximation has the same slope as the log-likelihood function at $p = 0.5$:

$$\frac{dL}{du}\bigg|_{p=0.5} = \frac{d\overline{L}}{du}\bigg|_{p=0.5} \implies K = 2/\ln b.$$

Figure 10.8 shows the result using the approximation for $K = 2/\ln 10$ on the left and $K = 1$ on the right. As expected, when $K = 2/\ln 10$, the relative error is 0 at $p = 0.5$ and grows toward the values of p equal to 0 and 1. Choosing $K = 1$ presents an advantage that will be discussed shortly, but at this point, it can be seen from Figure 10.8 that this choice yields similar ranges for relative error as with the choice $K = 2/\ln 10$. In analyzing the relative error seen with a linear approximation for the log-likelihood function, it is important to bear in mind that a deviation from the value of $L(u)$ as introduced by this approximation will not modify the behavior of the decoding algorithm, because the decoding decision is based on choosing the maximum $M_k^{(i)}$ in (10.17) and the error appears

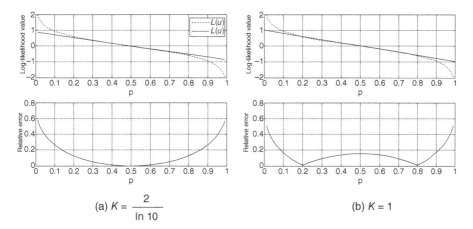

(a) $K = \dfrac{2}{\ln 10}$

(b) $K = 1$

Figure 10.8 Approximation of the log-likelihood function, using logarithms in base 10, with the linear function $K(1 - 2p)$. (a) Approximation with $K = 2/\ln 10$. (b) Approximation with $K = 1$.

as a constant term equally affecting all the metrics $M_k^{(i)}$. Also, when considering the relative error it is useful to recall that for an application such as images, the value of p usually is between 0.05 and 0.95 [7].

The advantage of choosing $K = 1$ is that in this case $\overline{L}(u)$ equals the mean value of the information bit u:

$$E[u] = (-1)p + (+1)(1 - p) = 1 - 2p = \overline{L}(u).$$

Because of this property, the value $\overline{L}(u)$ has been called the "m-value." The use of m-values simplifies the log-likelihood algebraic operations. For example, it can be shown that $\overline{L}(u \oplus v) = \overline{L}(u)\overline{L}(v)$.

The main idea to exploit the residual redundancy in coded images to improve channel decoding is based on combining a priori information that may be known for the source bits with a priori information derived from a posteriori information from spatially neighboring similar bits. Let the m-value corresponding to a priori information known about a bit be denoted $\overline{L}_0(u_{r,c})$, where the information bit $u_{r,c}$ is indexed through the row r and column c corresponding to the spatial coordinates of the block to which the bit belongs. Next, let $\overline{L}_1(u_{r,c})$ denote the a priori information that is estimated from a posteriori information from spatially neighboring similar bits. The value $\overline{L}_1(u_{r,c})$ can be computed as:

$$\overline{L}_1(u_{r,c}) = \rho_r \overline{L}_1(u_{r-1,c}) + \rho_c \overline{L}_1(u_{r,c-1}) - \rho_r \rho_c \overline{L}_1(u_{r-1,c-1}),$$

where ρ_r is the correlation coefficient in the vertical direction and ρ_c is the correlation coefficient in the horizontal direction. This formula indicates that $\overline{L}_1(u_{r,c})$ is derived as a weighted combination of the a posteriori log-likelihood values for the same bit in the blocks situated directly above, to the left, and diagonally above and to the left.

Both $\overline{L}_0(u_{r,c})$ and $\overline{L}_1(u_{r,c})$ indicate the reliability of a single bit using different sources of information. Since $\overline{L}_0(u_{r,c})$ and $\overline{L}_1(u_{r,c})$ are derived from sources of information that are independent from each other, $\overline{L}_0(u_{r,c})$ and $\overline{L}_1(u_{r,c})$ are themselves independent values and makes it reasonable to combine them using a weighted addition to obtain a final approximated a priori log-likelihood value:

$$\bar{L}(u_{r,c}) = \lambda_0 \bar{L}_0(u_{r,c}) + \lambda_1 \bar{L}_1(u_{r,c}). \tag{10.26}$$

The constants λ_0 and λ_1 allow for flexibility in combining $\bar{L}_0(u_{r,c})$ and $\bar{L}_1(u_{r,c})$, depending on their characteristics. For example, it was observed in [7] that the most significant bit of a DC coefficient has roughly the same probability of being equal to 0 or 1, making $\bar{L}_0(u_{r,c})$ close to 0 and $\bar{L}_1(u_{r,c})$ dominating in (10.26). In contrast, for other information bits from other coefficients, $\bar{L}_0(u_{r,c})$ and $\bar{L}_1(u_{r,c})$ play equally important roles. Consequently, in [7], it was proposed that a simple solution applicable across all cases was $\lambda_0 = \lambda_1 = 1$. The resulting $\bar{L}(u_{r,c})$ is used in the path metric calculation (10.17) as the a priori value $L(u_k)$.

Figure 10.9 shows, from [7], the reconstructed quality (measured in peak signal-to-noise ratio [PSNR] in dB) of the image "Lena" when it is transmitted using three different error protection schemes:

- *"No channel coding"* refers to the transmission of the image without any channel coding; this is included for comparison purposes.
- *"Viterbi algorithm"*: this scheme decodes using the Viterbi algorithm, with path metrics calculated using (10.17) but without considering the a priori information of the bits (so $L(u_k) = 0$).
- *"A priori information from SOVA"*: this scheme computes $\bar{L}(u_{r,c})$ and the result is used in the path metric calculation (10.17) as the a priori value $L(u_k)$.

In all cases, the image Lena is source-encoded using DCT coding at 1.5 bits per pixel. Where channel coding is used, it is a rate $1/2$ convolutional code. In addition to the numerical results depicted in Figure 10.9, [7] reported that the use of a priori/a posteriori information also resulted in a more graceful degradation of the subjective image quality.

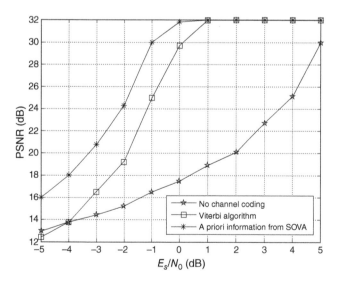

Figure 10.9 Peak signal-to-noise ratio (PSNR) for the image "Lena" when using different channel coding–decoding mechanisms.

10.3   Iterative Source–Channel Decoding

The techniques discussed in this chapter all use log-likelihood values as a tool to exchange information about the information bits between the decoders. Decoding techniques using log-likelihood values surged in research attention starting in 1993 when Turbo error correcting codes were introduced in [8]. Turbo codes have a bit error probability performance very close to Shannon's capacity limit (as close as 0.03 dB, as reported in [9]). Turbo codes derive their powerful error correcting capability from the "Turbo principle," which is an iterative decoding technique for concatenated channel codes. The original Turbo code implementation was based on a parallel concatenation of two convolutional encoders, as shown in Figure 10.10(a). The two convolutional encoders in the implementation may be identical. The third important component is the interleaver which reorganizes the bits at the input of the second encoder in a different order from those at the input of the first encoder (here, for simplicity, the encoder inputs are assumed to be bits, but other cases are possible). The function of the interleaver is to make the output of the second encoder uncorrelated with the output from the first encoder, key to achieving the strong error correction performance of Turbo codes. While the original implementation of turbo codes used convolutional codes concatenated in parallel, it is also possible to implement Turbo codes with serially concatenated error correcting codes, as for example, the convolutional codes shown in Figure 10.10(b). As will be seen, the serial configuration is of more interest for JSCD.

The Turbo principle was introduced as an iterative decoding technique for concatenated channel codes. The turbo decoder is formed by one-channel decoder for each channel encoder, connected in the same parallel or serial configuration used for encoding. Each of the decoders usually estimates the transmitted information bit stream as the sequence of bits that maximizes the a posteriori probability distribution. The use of log-likelihoods in the decoding process is key for the implementation of the Turbo principle, because the decoding results for each bit can be expressed as a soft value. That is, each bit is assigned a value that has both a sign (conveying the estimation of the source-encoded bit) and a magnitude equal to the bit log-likelihood value (indicating the level of confidence for the estimation decisions). Each bit's log-likelihood value can be expressed as the sum of three log-likelihood values:

$$L(\hat{u}_k) = L_{int}(u_k) + L(u_k) + L_e(u_k). \tag{10.27}$$

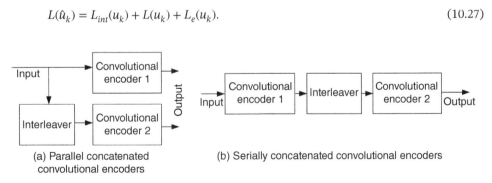

(a) Parallel concatenated
convolutional encoders

(b) Serially concatenated convolutional encoders

Figure 10.10   Concatenation of two convolutional encoders in (a) parallel and (b) series.

This expression shows how the log-likelihood value for a decoded bit, $L(\hat{u}_k)$, can be separated into the sum of three terms. The first term, $L_{int}(u_k)$, called the "intrinsic information," represents the part of $L(\hat{u}_k)$ that is calculated from the received coded symbol at time k. The second term, $L(u_k)$, is the a priori information on the bit u_k. The third term, called "extrinsic information," represents all the indirect information about u_k that the channel decoder can extract from the structure added at encoding time. It is made up of the information that can be derived from the likelihood information obtained from the received channel-encoded bits calculated from the source bits before and after u_k.

The definition of extrinsic information makes reference to the likelihood information that can be derived from the received channel-encoded bits calculated from the source bits *before* and *after* u_k. This information is implicitly conveyed by the decoding algorithm itself by finding the most likely path in the decoding trellis. Working with the entire trellis (that corresponds to the complete sequence of transmitted bits) would almost always imply a very large computational complexity and memory use at the decoder. In practice, and in order to balance computational complexity, memory consumption, and decoding performance, to estimate a bit at time k, the path is estimated from a few time stages before k until a few time stages after k. For example, the path might be traced (or estimated) from a time $k - 50$ to $k + 50$ (the actual limits depend on the code memory). This implementation issue was not considered in some of the previous expositions (for example in Figure 10.5). Deriving the expression (10.27) from the likelihood metric provided by the soft-output Viterbi algorithm also exposes how the likelihood information derived from coded bits before and after the estimated one relates to the extrinsic information. To see this, recall from (10.25) that the soft output from the SOVA for the bit decoded at time k is

$$L(\hat{u}_k) \approx \hat{u}_k \min_{m=1,2,\ldots,M} \Delta_k^{(m)}. \tag{10.28}$$

Using the definition of $\Delta_k^{(m)}$ in (10.20) and the definition of the path metric in (10.17), the likelihood value $\Delta_k^{(m)}$ can be written as the addition of three parts, the metrics prior to time k, the metrics after time k, and the transition contributions at time k. Doing this, we obtain

$$\Delta_k^{(m)} = \left(M_{j\leq k}^{(m)} - M_{j\leq k}^{(m')} \right) + \left(M_{k\leq j\leq k+l}^{(m)} - M_{k\leq j\leq k+l}^{(m')} \right)$$
$$+ \sum_{l=1}^{N} \left(x_{k,l}^{(m)} L_c y_{k,l} - x_{k,l}^{(m')} L_c y_{k,l} \right)$$
$$+ L(u_k)(u_k^{(m)} - (-u_k^{(m')})), \tag{10.29}$$

where the surviving path is m and the non-surviving path is m'. We have used the fact that $\hat{u}_k \equiv u_k^{(m)}$ and always $u_k^{(m')} = -u_k^{(m)}$ because $\Delta_k^{(m)}$ is calculated only for paths with differing estimated bits. More importantly, the aforementioned expression shows how the log-likelihood value calculated from the SOVA output follows (10.27), since it can be decomposed into three components. According to our earlier definitions, these are the a priori information (the term in the third line), the intrinsic information (the terms in the second line), and the extrinsic information (the terms in the first line).

Note that since the decision on u_k is based on the sign of $L(\hat{u}_k)$, the three components of the bit likelihood work in an additive way, working together to improve the decoding decision. This is not merely analogous to the use of a posteriori likelihood values as a priori likelihood values as seen earlier in the chapter, but it is indeed the same underlying principle. We now consider the use of the extrinsic information in the operation of both decoders acting together. At the beginning of the decoding algorithm, the likelihood output for the first decoder will be

$$L(\hat{u}_k) = L_{int}(u_k) + L(u_k) + L_e(u_k)_1.$$

Here the subscript "1" means that we are only considering the redundancy from encoder 1. If no a priori information is known for the bit u_k, we can assume that $P(u_k = +1) = P(u_k = +1)$ so $L(u_k) = 0$ and we will have

$$L(\hat{u}_k) = L_{int}(u_k) + L_e(u_k)_1.$$

Now, the extrinsic information $L_e(u_k)_1$ shall be seen as the indirect information that can be derived about encoder 1 from the encoding structure implied by the channel-encoded bits calculated from the source bits before and after u_k; it is a side information that the decoder 2 will not be able to calculate (since it will work with the coded bits generated by encoder 2, which were generated from a different sequence of input bits). This leads to the idea that the extrinsic information from decoder 1 can be provided to the second decoder as an additional input. Because of the independence of this log-likelihood value with the others, we can just add it to the input making it $L_c y_k + L_e(u_k)_1$. To implement this idea, it is necessary to pass $L_e(u_k)_1$ through the same interleaver used at the encoder, so the order of the information is the same as the bits at the input to the second encoder or decoder. After interleaving, the input to the second decoder will become $L_c y_l + L_e(u_l)_1$, with the use of a different time index meaning that these values are after interleaving. With this input, and continuing assuming that we still do not have any a priori information on u_l (i.e. $P(u_l = +1) = P(u_l = +1) = 1/2$, $L(u_l) = 0$) we have at the output of decoder 2:

$$L(\hat{u}_l) = L_{int}(u_l) + L_e(u_l)_1 + L_e(u_l)_2.$$

This equation, when compared with (10.27), shows us that the extrinsic information from the other decoder plays the role of the a priori likelihood ratio $L(u_l)$, and it is in this way that it is applied in the metric (10.17). From this point on, the use of the extrinsic information as side information obtained from one decoder and applied in the other can be continued, and we use $Le(u_k)_2$ (now de-interleaved) added to the input of decoder 1:

$$L(\hat{u}_k)_{1,(iteration\ 2)} = L_{int}(u_k) + L_e(u_k)_{1(iteration\ 2)} + L_e(u_k)_2$$

Because of the iterative process being built, at this point $Le(u_k)_2$ has some dependence with $L_{int}(u_k)$ and $L_e(u_k)_1$. Thus, combining all the likelihood values by simple additions is not strictly correct. Nevertheless, $Le(u_k)_2$ can still be added because it was de-interleaved, which helps in de-correlating the extrinsic information from one decoder to the likelihood values of the other decoder.

The aforementioned sequence of operations establishes an iterative procedure that gives the name "turbo" to these codes. As noted by many authors, they could more technically be called "two-dimensional iteratively decoded convolutional codes." Yet, the naming has helped to think that "turbo" decoding is not exclusive to parallel concatenated convolutional codes. As long as the decoding procedure can be expressed in a form such as (10.27), it will be possible to perform iterative ("turbo") decoding on any code and with any decoding method. This also shows that the basic idea behind "turbo" codes is to break up decoding of fairly complex codes into incremental steps. This is a consequence of the fact that a well-designed interleaver acts as a decorrelator between the feedback loop (carrying the extrinsic information) and the other inputs to each decoder.

In summary, the central idea in the Turbo decoding principle consists in separating the output of an individual decoding block into intrinsic, extrinsic, and a priori information parts, using the extrinsic information as additional side information that is fed into another decoding block and then repeating this operation between the decoders and over several iterations. Given the excellent performance that this idea yielded in the decoding of turbo error correcting codes, the central scheme was also considered for other decoding applications. First came the use of turbo decoding with error correcting codes other than convolutional ones, but soon after the turbo principle was being used to perform iterative equalization [10], and iterative demodulation and channel decoding [11].

The Turbo principle was also applied in JSCD. In this case, the turbo forward error correcting encoding–decoding scheme, with two (or more) concatenated channel codes, is replaced with the concatenation of the source and the channel codecs, with an interleaver between them. This idea is illustrated in Figure 10.11, where at the encoding side of the communication chain the source and the channel encoders are separated by an interleaver. The similarities in the structure of the encoding section in Figure 10.11 with that of the serially concatenated Turbo code shown in Figure 10.10b are clear. For iterative source–channel decoding, the source encoder plays the role of the first convolutional encoder in Figure 10.10b, and the channel encoder for iterative source–channel decoding plays the role of the second channel encoder in the serially concatenated Turbo encoder.

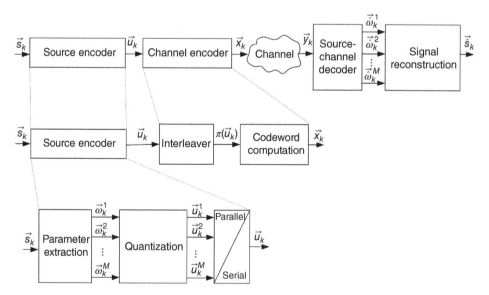

Figure 10.11 Block diagram for a system doing iterative JSCD. A comparison with Figure 10.10(b) shows how a configuration with the two encoders in series, separated by an interleaver, is common for a system for turbo channel decoding and is the enabler for the iterative decoding technique.

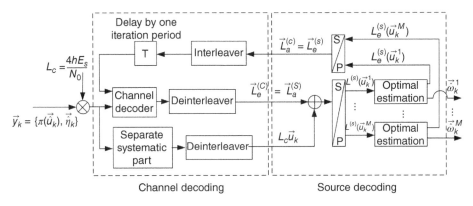

Figure 10.12 Block diagram for an iterative joint source–channel decoder.

This structural similarity points toward the mechanism to perform iterative JSCD. Channel decoding shall be done in such a way that extrinsic information can be extracted from the result. This extrinsic information is passed through a de-interleaver and be used as side information input to the source decoder. Also, it shall be possible to extract extrinsic information from the output of the source decoder and, after passing through an interleaver, to feed this as side information that the channel encoder can be used to improve the decoding result. The sequence of decoding operations interleaved with the exchange of extrinsic information constitute one decoding iteration cycle that can be repeated a number of times until the reliability of the decoding result cannot be further improved. The iterative JSCD operation is shown in Figure 10.12 and will be explained in the following pages.

The communication system formed by the transmitter and receiver shown in Figures 10.11 and 10.12 represents a typical model, where the source signal vector present at time k, $\vec{s}_k$, is source-encoded by first decomposing it into M parameter vectors $\vec{\omega}_k^1, \vec{\omega}_k^2, \ldots, \vec{\omega}_k^M$. For example, if the input source is a vector of speech samples, a parameter vector could be the linear predictive coding (LPC) coefficients or the vector could in fact be just a scalar value, such as the pitch estimation. After the decomposition, the parameters are quantized, resulting in M vectors of quantization indexes $\vec{u}_k^1, \vec{u}_k^2, \ldots, \vec{u}_k^M$. These indexes, output from the source encoder, are input to the channel encoder, where they first go through an interleaver. The interleaver applies a mapping $\pi(\cdot)$ that generates an output vector $\pi(\vec{u}_k)$ with the same elements as the input vector $\vec{u}_k$ but in a different order. The channel encoder computes from the vector $\pi(\vec{u}_k)$ a codeword $\vec{x}_k$ that is transmitted through an AWGN channel and received as the vector $\vec{y}_k$. We assume BPSK modulation is used and denote the lth bit in the N_x-bit codeword by $x_{k,l}$. The effect on $\vec{x}_k$ from the channel can be characterized by the joint conditional pdf of the received $\vec{y}_k$ given that $\vec{x}_k$ was transmitted and, because the channel noise is white, we can write

$$p(\vec{y}_k|\vec{x}_k) = \prod_{l=1}^{N_x} p(y_{k,l}|x_{k,l}).$$ (10.30)

The joint source channel-encoded decoding algorithm takes advantage of the residual redundancy that is left from the source coding stage due to practical implementation limitations. This redundancy is manifested by the existence of some level of dependency between a quantization index over consecutive time instances $\vec{u}_{k-1}^j$ and $\vec{u}_k^j$. In what follows, this dependency is modeled as a first-order stationary Markov process with index transition probabilities:

$$P(\vec{u}_k^j = v^j | \vec{u}_{k-1}^j = \psi^j),\tag{10.31}$$

where v^j and ψ^j take values $0, 1, \dots, 2^{N^j} - 1$ and $j = 1, \dots, M$.

On the receiver side, source and channel decoding should be seen as a joint operation. Mathematically, this JSCD operation is represented through the final decoding goal of finding the estimate of the parameters $\vec{\hat{\omega}}_k^1, \dots, \vec{\hat{\omega}}_k^M$ that minimize the expected end-to-end mean-squared parameter error:

$$\min_{\vec{\hat{\omega}}_k^1, \dots, \vec{\hat{\omega}}_k^M} E\left[\sum_{j=1}^M |\vec{\hat{\omega}}_k^j - \omega_q^j(\vec{u}_k^j)|^2 \Big| \vec{\mathbf{y}}_k \right],\tag{10.32}$$

where the expectation is conditioned on the set of words $\vec{\mathbf{y}}_k = \{\vec{y}_0, \dots, \vec{y}_{k-1}, \vec{y}_k\}$ received up to time k. In (10.32), $\omega_q^j(\vec{u}_k^j)$ is the entry in the quantization table corresponding to the index $\vec{u}_k^j$. Since the expected end-to-end mean-squared parameter error in (10.32) can be written as:

$$\sum_{\forall \vec{v} \in \mathcal{U}} \dots \sum \sum_{j=1}^M |\vec{\hat{\omega}}_k^j - \omega_q^j(\vec{u}_k^j)|^2 P(\vec{u}_k = \vec{v} | \vec{\mathbf{y}}_k),$$

where $\vec{v}$ represents one realization for the vector $\vec{u}_k$ and $P(\vec{u}_k = \vec{v} | \vec{\mathbf{y}}_k)$ is the a posteriori probability for all indexes, the solution to (10.32) is the mean-square estimator

$$\vec{\hat{\omega}}_k^j = \sum_{\forall \vec{v}^j \in \mathcal{U}^j} \omega_q^j(\vec{v}^j) P(\vec{u}_k^j = \vec{v}^j | \vec{\mathbf{y}}_k), \quad j = 1, 2, \dots, M.\tag{10.33}$$

Consequently, the solution to the JSCD in (10.32) hinges on computing the a posteriori probabilities of all indexes

$$P(\vec{u}_k = \vec{v} | \vec{\mathbf{y}}_k) = P(\vec{u}_k^1 = \vec{v}^1, \vec{u}_k^2 = \vec{v}^2, \dots, \vec{u}_k^M = \vec{v}^M | \vec{\mathbf{y}}_k),$$

or, in relation to (10.33), the a posteriori probabilities corresponding to each parameter, $P(\vec{u}_k^j = \vec{v}^j | \vec{\mathbf{y}}_k)$.

10.3.1 The Channel Coding Optimal Estimation Algorithm

The a posteriori probabilities corresponding to each parameter can be optimally computed through an algorithm called "channel coding optimal estimation" (CCOE). In a fashion very similar to the other techniques previously discussed in this chapter, the CCOE algorithm works recursively by separating the probability $P(\vec{u}_k = \vec{v} | \vec{\mathbf{y}}_k)$ into products of probabilities. In turn, the factorization of the probability $P(\vec{u}_k = \vec{v} | \vec{\mathbf{y}}_k)$ allows for the use of iterative source–channel decoding techniques in the calculation of some of the multiplying probabilities. In this subsection, we will describe this algorithm following the approach in [6], which showed the application to JSCD.

We start by considering the CCOE for the case when there is only one source parameter to estimate. In this case, $M = 1$, $\vec{\omega}_k^1 = \vec{\omega}_k$ and the probability $P(u_k = v|\vec{\mathbf{y}}_k)$ can be expressed in the recursive form [12]:

$$P(u_k = v|\vec{\mathbf{y}}_k) = B_k P\left(\vec{y}_k|\vec{x}_k^{(v)}\right) \sum_{\psi=0}^{2^N-1} P(u_k = v|u_{k-1} = \psi)P(u_{k-1} = \psi|\vec{\mathbf{y}}_{k-1}), \qquad (10.34)$$

for $v = 0, 1, \ldots, 2^N - 1$. In the expression on the right-hand side, the probability $P(u_k = v|\vec{\mathbf{y}}_k)$ is separated into three factors. The first factor, B_k, is a normalization constant that normalizes the computation of the probability and is analogous to other normalization factors seen earlier (e.g. A_k in the derivation leading to (10.11)). The second factor corresponds to the channel effects as described in (10.30), where $\vec{x}_k^{(v)}$ is the codeword resulting from applying channel coding to an index $u_k = v$. The third, and last, factor in (10.34) is a sum of the product of a factor that accounts for the leftover residual redundancy at the source encoder between two successive indexes, as given in (10.31), and a factor that is the a posteriori probability for the immediately preceding time $k - 1$.

Note that (10.34) shows how the a posteriori probability at time k can be calculated recursively from the a posteriori probability at time $k - 1$, the codeword received at the channel input and the known probabilities modeling the channel effects and the residual redundancy at the source encoder. The recursion can be initialized by using known statistics of the source: $P(u_{k=-1} = v|\vec{\mathbf{y}}_{k=-1}) = P(u = v)$, $v = 0, 1, 2, \ldots, 2^N - 1$.

The recursive calculation in (10.34) can be extended to the CCOE case when there are M source parameters to estimate. For the case of nonsystematic channel coders (the case of systematic codes will be considered shortly), the expression becomes

$$P(\vec{u}_k = \vec{v}|\vec{\mathbf{y}}_k) = B_k P\left(\vec{y}_k|\vec{x}_k^{(\vec{v})}\right) \sum_{\forall\vec{\psi}\in\mathcal{U}} \cdots \sum P(\vec{u}_k = \vec{v}|\vec{u}_{k-1} = \vec{\psi})$$

$$\cdot P(\vec{u}_{k-1} = \vec{\psi}|\vec{\mathbf{y}}_{k-1}). \qquad (10.35)$$

If indexes are mutually independent, which is the case when the source parameters are themselves independent (a condition commonly seen in many sources), it is possible to write

$$P(\vec{u}_k = \vec{v}|\vec{u}_{k-1} = \vec{\psi}) = \prod_{j=1}^{M} P(\vec{u}_k^j = \vec{v}^j|\vec{u}_{k-1}^j = \vec{\psi}^j),$$

$$P(\vec{u}_k = \vec{v}|\vec{\mathbf{y}}_k) = \prod_{j=1}^{M} P(\vec{u}_k^j = \vec{v}^j|\vec{\mathbf{y}}_k).$$

Using these expressions, (10.35) can now be written as:

$$\prod_{j=1}^{M} P(\vec{u}_k^j = \vec{v}^j|\vec{\mathbf{y}}_k) = B_k P\left(\vec{y}_k|\vec{x}_k^{(\vec{v})}\right) \prod_{j=1}^{M} \left(\sum_{\forall\vec{\psi}^j\in\mathcal{U}^j} P(\vec{u}_k^j = \vec{v}^j|\vec{u}_{k-1}^j = \vec{\psi}^j) \right.$$

$$\left. \cdot P(\vec{u}_{k-1} = \vec{\psi}|\vec{\mathbf{y}}_{k-1}) \right). \qquad (10.36)$$

This expression makes use of the a priori probability:

$$P(\vec{u}_k^j = \vec{v}^j|\vec{\mathbf{y}}_{k-1}) = \sum_{\forall\vec{\psi}^j\in\mathcal{U}^j} P(\vec{u}_k^j = \vec{v}^j|\vec{u}_{k-1}^j = \vec{\psi}^j)P(\vec{u}_{k-1} = \vec{\psi}|\vec{\mathbf{y}}_{k-1}). \qquad (10.37)$$

Then, (10.36) can be written in a more compact form as:

$$\prod_{j=1}^{M} P(\vec{u}_k^j = \vec{v}^j | \vec{\mathbf{y}}_k) = B_k P\left(\vec{\mathbf{y}}_k | \vec{x}_k^{(\vec{v})}\right) \prod_{j=1}^{M} P(\vec{u}_k^j = \vec{v}^j | \vec{\mathbf{y}}_{k-1}). \tag{10.38}$$

From (10.37) and (10.38), it is possible to compute the a posteriori probabilities correspond-ing to each parameter, $P(\vec{u}_k^j = \vec{v}^j | \vec{\mathbf{y}}_k)$, that are needed to compute (10.33). To accomplish this, (10.38) is summed over all the indexes that are not being considered, resulting in

$$P(\vec{u}_k^j = \vec{v}^j | \vec{\mathbf{y}}_k) = B_k \left[\sum_{\forall \vec{\psi} \in \mathcal{U} \,:\, \vec{\psi}^j = \vec{v}^j} \cdots \sum P\left(\vec{\mathbf{y}}_k | \vec{x}_k^{(\vec{\psi})}\right)\right.$$
$$\left. \cdot \prod_{\substack{m=1 \\ m \neq j}}^{M} P(\vec{u}_k = \vec{\psi}^m | \vec{\mathbf{y}}_{k-1}) \right] P(\vec{u}_k^j = \vec{v}^j | \vec{\mathbf{y}}_{k-1}). \tag{10.39}$$

If the channel code used is systematic, as is the case illustrated in Figure 10.12, the code-word bits can be separated into two parts: one part that is equal to the channel encoder input $\pi(\vec{u}_k)$ and a second part formed by the redundancy bits added by the channel encoder. Denoting the redundancy bits as $\vec{\eta}_k$, the codeword of the systematic code can be written as:

$$\vec{x}_k = \{\pi(\vec{u}_k), \vec{\eta}_k\},$$

and the corresponding received codeword as:

$$\vec{\mathbf{y}}_k = \{\pi(\vec{\tilde{u}}_k), \vec{\tilde{\eta}}_k\},$$

where $\pi(\vec{\tilde{u}}_k)$ denotes the received interleaved sequence of source indexes and $\vec{\tilde{\eta}}_k$ denotes the received redundancy bits. Using this notation, (10.39) can be written for the case of systematic codes as:

$$P(\vec{u}_k^j = \vec{v}^j | \vec{\mathbf{y}}_k) = B_k P_e(\vec{u}_k^j = \vec{v}^j) P(\vec{\tilde{u}}_k^j | \vec{u}_k^j = \vec{v}^j) P(\vec{u}_k^j = \vec{v}^j | \vec{\mathbf{y}}_{k-1}). \tag{10.40}$$

In this expression, the probability:

$$P_e(\vec{u}_k^j = \vec{v}^j) = B_k' \sum_{\forall \vec{\psi} \in \mathcal{U} \,:\, \vec{\psi}^j = \vec{v}^j} \cdots \sum P(\vec{\tilde{\eta}}_k = \vec{\eta}_k^{(\vec{v})})$$
$$\cdot \prod_{\substack{m=1 \\ m \neq j}}^{M} P(\vec{\tilde{u}}_k^m | \vec{u}_k^m = \vec{v}^m) P(\vec{u}_k = \vec{\psi}^m | \vec{\mathbf{y}}_{k-1}), \tag{10.41}$$

where $\vec{\eta}_k^{(\vec{v})}$ denotes the redundancy bits that result when the input to the systematic channel encoder is $\vec{v}$, is called the extrinsic index probability because it is calculated using redundancy bits and the correlation of other indexes, information derived from the coding procedure that is not directly related to the index $\vec{u}_k^j$.

10.3.2 Channel Coding Optimal Estimation Applied to JSCD

Practical implementation of the CCOE algorithm is limited by the significant computational complexity associated with calculating (10.41) when the number of channel encoded index bits is moderate or large and the codes are nonbinary. Therefore, it is reasonable to assume that the channel codes are binary, as is the case with the majority of codes used in modern wireless communications. Then, an index $\vec{u}_k^j$ is represented by the sequence of N bits

$$\vec{u}_k^j = \left\{ u_{l,k}^j \in \{0,1\} : l = 1, 2, \ldots, N^j \right\},$$

and the corresponding realizations of these random variables as:

$$\vec{v}_k^j = \left\{ v_l^j \in \{0,1\} : l = 1, 2, \ldots, N^j \right\}.$$

Since we are studying an iterative algorithm that implements JSCD, it will be convenient to differentiate probability values that are computed in the channel decoding stage and ones that are computed in the source decoding stage. To make this differentiation explicit in the notation, the probabilities will be annotated with a superscript "(C)" when computed in the channel decoding stage and with a superscript "(S)"when computed in the channel decoding stage.

Next, the goal is to find a mechanism that simplifies the calculation of the extrinsic information in (10.40) but that also maintains as much as possible the optimality of the CCOE algorithm. To do this, the extrinsic probability for indexes in (10.41) is simplified by using the extrinsic probabilities for bits:

$$P_e^{(C)}(\vec{u}_k^j = \vec{v}^j) \approx \prod_{l=1}^{N^j} P_e^{(C)}(u_{l,k}^j = v_l^j), \tag{10.42}$$

and where the extrinsic probabilities for bits is obtained by translating (10.41) from a computation based on indexes into one based on the corresponding bits:

$$P_e^{(C)}(u_{l,k}^j = v_l^j) = B_k' \sum_{\forall \psi_l^j = v_l^j} \cdots \sum P(\vec{\eta}_k = \vec{\eta}_k^{(\vec{v})}) \prod_{\substack{m=1 \\ m \neq j}}^{M} \prod_{\xi=1}^{N^m} P(\tilde{u}_{\xi,k}^m | u_{\xi,k}^m = \psi_\xi^m) P_a^{(C)}(u_{\xi,k}^m = \psi_\xi^m)$$

$$\cdot \prod_{\substack{v=1 \\ v \neq l}}^{M} P(\tilde{u}_{v,k}^j | u_{v,k}^j = \psi_v^j) P_a^{(C)}(u_{v,k}^j = \psi_v^j). \tag{10.43}$$

In this expression, $P_a^{(C)}(u_{v,k}^j = \psi_v^j)$ is the a priori probability, which can be computed as:

$$P_a^{(C)}(u_{l,k}^j = v_l^j) = \sum_{\substack{\forall \vec{v}^j \in \mathcal{U}^j \\ v_l^j = \psi_l^j}} P(\vec{u}_k = \vec{\psi}^j | \vec{y}_{k-1}), \tag{10.44}$$

$$j = 1, 2, \ldots, M, \qquad l = 1, 2, \ldots, N^j.$$

Using (10.42) and (10.30) for the case of systematic channel code bits, (10.40) can now be computed as:

$$P(\vec{u}_k^j = \vec{v}^j | \vec{y}_k) = B_k^{\natural} \prod_{l=1}^{N^j} \left[P_e^{(C)}(u_{l,k}^j = v_l^j) P(\tilde{u}_{l,k}^j | u_{l,k}^j = v_l^j) \right] P(\vec{u}_k^j = \vec{v}^j | \vec{y}_{k-1}). \tag{10.45}$$

Up to this point the discussion has focused on calculations at the channel decoding stage. To consider the source decoding stage and develop the joint source–channel decoder, we write the a posteriori probabilities as can be calculated at the source decoder. The a posteriori probabilities of the bits can be computed as:

$$P^{(S)}(u^j_{l,k} = v^j_l | \vec{y}_k) = \sum_{\substack{\forall \vec{v}^j \in \mathcal{U}^j \\ v^j_l = v^j_l}} P(\vec{u}_k = \vec{\psi}^j | \vec{y}_k). \tag{10.46}$$

Expanding this expression using (10.45), it becomes

$$P^{(S)}(u^j_{l,k} = v^j_l | \vec{y}_k) = B''_k P^{(S)}_e(u^j_{l,k} = v^j_l) P(\tilde{u}^j_{l,k} | u^j_{l,k} = v^j_l) P^{(S)}_a(u^j_{l,k} = v^j_l), \tag{10.47}$$

where the extrinsic probability $P^{(S)}_e(u^j_{l,k} = v^j_l)$ is calculated as:

$$P^{(S)}_e(u^j_{l,k} = v^j_l) = B'''_k \sum_{\substack{\forall \vec{\psi}^j \in \mathcal{U}^j \\ \psi^j_l = v^j_l}} P(\vec{u}_k = \vec{\psi}^j | \vec{y}_{k-1})$$

$$\cdot \prod_{\substack{v=1 \\ v \neq l}}^{N^j} P(\tilde{u}^j_{v,k} | u^j_{v,k} = \psi^j_v) P^{(S)}_a(u^j_{v,k} = \psi^j_v). \tag{10.48}$$

Here, $P^{(S)}_a(u^j_{v,k} = \psi^j_v)$ is the a priori probability that is applied in the source decoding stage. This probability is the link between channel and source decoding that enables the JSCD solution, because it is calculated as the extrinsic probability obtained from the channel decoding stage:

$$P^{(S)}_a(u^j_{v,k} = v^j_v) = P^{(C)}_e(u^j_{v,k} = v^j_v). \tag{10.49}$$

The JSCD cycle is completed by applying the same idea to the a priori probability used in the channel decoding stage. In this case,

$$P^{(C)}_a(u^j_{v,k} = v^j_v) = P^{(S)}_e(u^j_{v,k} = v^j_v). \tag{10.50}$$

In both cases where a priori probabilities are calculated from extrinsic probabilities, the operation is possible, because the calculations of extrinsic information in (10.43) and (10.48) do not include the channel transition probability $P(\tilde{u}^j_{l,k} | u^j_{l,k} = v^j_l)$. This not only gives its name to the extrinsic probability but also prevents the decoding information provided by the probabilities from being used more than once in the JSCD cycle.

Consequently, at each discrete time instant k, the JSCD algorithm will estimate the transmitted source parameter vectors from an iterative procedure. The kernel of the iterative algorithm is formed by two steps. In the first step, the channel extrinsic probability is computed from (10.43) using $P^{(C)}_a(u^j_{v,k} = v^j_v)$, the extrinsic probability from the source encoding stage, as a priori probability, as per (10.50). In the second step, the extrinsic probability from the source decoding stage is computed from (10.48) using $P^{(S)}_a(u^j_{v,k} = v^j_v)$, the extrinsic probability from the channel encoding stage, as a priori probability, as per (10.49). The two steps in the JSCD algorithm kernel are repeated iteratively multiple times. The number of iterations is typically a parameter that is chosen during implementation. The resulting channel decoding extrinsic probability is used to compute the a posteriori probabilities for the indexes using (10.45). The last step for the JSCD algorithm operation at time k is to use the a posteriori probabilities for the indexes to estimate the source parameter vectors based on (10.33). To initialize the iterations of the JSCD algorithm at time k, the index a priori probabilities are calculated using (10.37) with the a posteriori probabilities for the indexes calculated up to time $k-1$, and the bit a priori probabilities are calculated using (10.44).

While the derivation of the JSCD algorithm seen so far is based on the calculation of probability values, the implementation of the algorithm is based on the use of log-likelihood values for these probabilities. With this in mind, the algorithm operation at time k can be seen in Figure 10.12. In the figure, the notation used for the log-likelihood values is

$$L_a^{(C)}(u_{l,k}^j) = \ln \frac{P_a^{(C)}(u_{l,k}^j = 0)}{P_a^{(C)}(u_{l,k}^j = 1)}$$

$$L_e^{(C)}(u_{l,k}^j) = \ln \frac{P_e^{(C)}(u_{l,k}^j = 0)}{P_e^{(C)}(u_{l,k}^j = 1)}$$

$$L_a^{(S)}(u_{l,k}^j) = \ln \frac{P_a^{(S)}(u_{l,k}^j = 0)}{P_a^{(S)}(u_{l,k}^j = 1)}$$

$$L_e^{(S)}(u_{l,k}^j) = \ln \frac{P_e^{(S)}(u_{l,k}^j = 0)}{P_e^{(S)}(u_{l,k}^j = 1)},$$

and $\vec{L}_{a/e}^{(C/S)}$ is a condensed notation for the array of values $\{L_{a/e}^{(C/S)}(u_{l,k}^j), j = 1, \ldots, M, l = 1, \ldots, N^j\}$.

As expected, simulation results of this technique presented in [6] show that the use of JSCD improves source parameter estimation compared to using channel decoding and source decoding algorithms separately. More interestingly, these results show that the performance of the JSCD algorithm is close to the performance of the optimal CCOE estimation algorithm which has prohibitive computational complexity. Also, the results show little performance improvements when increasing the number of iterations. This indicates a quick convergence for the JSCD algorithm and a reduced computational complexity.

References

1 Hagenauer, J. (1995). Source-controlled channel decoding. *IEEE Transactions on Communications* 43 (9): 2449–2457.

2 Battail, G., Decouvelaere, M., and Godlewski, Ph. (1979). Replication decoding. *IEEE Transactions on Information Theory* 25 (3): 332–345.

3 Viterbi, A. (1971). Convolutional codes and their performance in communication systems. *IEEE Transactions on Communication Technology* 19 (5): 751–772.

4 Hagenauer, J. (1988). Rate compatible punctured convolutional (RCPC) codes and their applications. *IEEE Transactions on Communications* 36 (4): 389–399.

5 Kliewer, J. Thobaben, R. (2005). Iterative joint source-channel decoding of variable-length codes using residual source redundancy. *IEEE Transactions on Wireless Communications* 4 (3): 919–929.

6 Gortz, N. (2001). On the iterative approximation of optimal joint source-channel decoding. *IEEE Journal on Selected Areas in Communications* 19 (9): 1662–1670.

7 Xu, W., Hagenauer, J., and Hollmann, J. (1996). Joint source-channel decoding using the residual redundancy in compressed images. *1996 IEEE International Conference on Communications, 1996. ICC 96, Conference Record, Converging Technologies for Tomorrow's Applications*, Volume 1, pp. 142–148. IEEE.

8 Berrou, C., Glavieux, A., and Thitimajshima, P. (1993). Near Shannon limit error-correcting coding and decoding: turbo-codes. *IEEE International Conference on Communications (ICC)*, Volume 2, pp. 1064–1070. IEEE.

9 Boutros, J., Caire, G., Viterbo, E. et al. (2002). Turbo code at 0.03 dB from capacity limit. *2002 IEEE International Symposium on Information Theory, 2002. Proceedings*, p. 56. IEEE.

10 Douillard, C., Jézéquel, M., Berrou, C. et al. (1995). Iterative correction of intersymbol interference: turbo-equalization. *European Transactions on Telecommunications* 6 (5): 507–511.

11 Hoeher, P. and Lodge, J. (1999). "Turbo DPSK": iterative differential PSK demodulation and channel decoding. *IEEE Transactions on Communications* 47 (6): 837–843.

12 Gortz, N. (2000). Joint source-channel decoding by channel-coded optimal estimation. *3rd ITG Conference on Source and Channel Coding*, pp. nf267–272.

11

Recent Applications and Emerging Designs in Source–Channel Coding

To close this book, in this chapter, we will discuss some recent applications of joint source–channel coding (JSCC) coding, connecting the solutions with techniques seen in earlier chapters, and we explain an emerging design approach for JSCC. We will first discuss JSCC in wireless sensors networks (WSNs), and then discuss how JSCC's added flexibility allows for the control of a network's congestion state. At the current time, it is widely accepted that the next generation of wireless communication systems (dubbed 6G or Beyond-5G systems) will incorporate a good dose of machine learning and artificial intelligence technology into their design. Cognitive radios (CRs) are an important area of research and development in wireless communications not only because of introducing a new paradigm for the operation of radios, but also because of being the precursor technology in adopting the use of machine learning and artificial intelligence technology. As such, later in this chapter, we will discuss the use of JSCC in cross-layer CR technology. Finally, we conclude this chapter by discussing JSCC systems that are built using artificial neural networks. This is an emerging design technique that introduces a new paradigm where codes are not explicitly designed from models of the system but, rather, they are learned by an artificial neural network after following a training procedure.

11.1   Source–Channel Coding for Wireless Sensor Networks

While sensors have been used within wireless networks for many years, WSN research experienced a resurgence when it was recognized that WSNs present key architectural and conceptual differences from other common wireless networks. In common wireless networks such as mobile cellular networks, each node is typically driven by various degrees of its own selfish interest, caring only about achieving its best possible communication quality. In contrast, WSN nodes cooperate to achieve a common goal of sensing and communicating some variables over a geographical region and time span. This difference between cooperating nodes and selfish nodes leads to new technical possibilities and options to implement wireless networks.

In addition, WSNs stand out from other networks by the marked constraints in resources and operating conditions. WSN nodes are expected to operate autonomously over long periods of time, usually on battery power. The typical operating conditions in WSNs are those

Joint Source-Channel Coding, First Edition. Andres Kwasinski and Vinay Chande.
© 2023 John Wiley & Sons Ltd. Published 2023 by John Wiley & Sons Ltd.

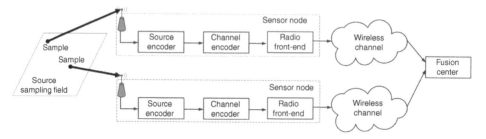

Figure 11.1 Overview of a wireless sensor network.

where JSCC techniques are an attractive alternative, so a number of JSCC techniques have been applied for the particular characteristics of WSNs.

Figure 11.1 shows a general architecture for WSNs, consisting of devices called *sensors* that collect physical phenomena measurements and transmit a processed version of them to a data collection station, called a *fusion center*. The processing at the sensors includes both sensing-related tasks, such as signal conditioning and noise removal, and communication-related operations, such as encoding the measurements. At the fusion center, the sensor measurements are combined to obtain some needed result. Frequently, the task at the fusion center to combine measurements involves hypothesis testing. At times, the hypothesis testing is done at each node and the binary result is transmitted to the fusion center to combine the hypothesis testing results from all the sensors into a single result. This approach can be considered a particular case of the general description of sensor processing, because the hypothesis testing computation is equivalent to encoding the measurement into a binary digit.

The problem of designing the encoder to process the measurements at the sensors is interesting from a JSCC viewpoint. Measurements from the sensors are usually correlated to some extent because they arise from physical phenomena at different but related locations or times (for example, sensing temperature over a geographical area). Consequently, the problem of designing sensor quantizers can be approached as designing a single encoder for samples distributed over sensors, instead of designing an encoder to operate at each sensor independently of the rest. Another perspective is to design sensor encoders that operate independently but consider that received measurements at the fusion center are affected by the channel, so sensor encoder design should take channel effects into account. This type of problem is studied in Chapter 7 in a more general context than WSNs. Among the studies of channel-matched encoders for WSNs, we will discuss [1] as a good representative example.

Consider a WSN with a configuration similar to Figure 11.1 (there is a difference from the figure which we will point out shortly). The goal of the sensor network is to sense some observable variable from the physical world and decide on the occurrence of two possible events H_0 and H_1, which are governed by an underlying probability distribution:

$$\pi_0 = P[H = H_0], \tag{11.1}$$

$$\pi_1 = P[H = H_1] = 1 - \pi_0. \tag{11.2}$$

The network has K sensor nodes. Sensor node k collects a sample s_k drawn from the random variable S_k at the source and encodes the sample for transmission. The samples are assumed to be conditionally independent given the hypothesis outcome, $P(S_1, S_2, \ldots, S_K | H_j) = \prod_{k=1}^{K} P(S_k | H_j)$, for $j = 0, 1$ (this is called a *naive Bayes assumption*).

Simplified from Figure 11.1, the sensor node has no channel encoder, and the source encoding operation consists of a scalar quantizer performing a mapping function $\gamma_k(\cdot)$ to generate an output $x_k = \gamma_k(s_k)$. The encoded sample x_k is transmitted to the fusion center through a wireless channel and received as the signal r_k. Channel effects are characterized by the probability distributions $P(R_k|X_k)$. All transmissions are done through mutually orthogonal channels, which means that

$$P(R_1, R_2, \ldots, R_K|X_1, X_2, \ldots, X_K) = \prod_{k=1}^{K} P(R_k|X_k).$$

To compensate for channel effects, the sensor encoder design will match channel characteristics. At the fusion center, the K received signals are combined into a single decision rule. For a Bayesian-type approach, the optimal fusion rule is the maximum *a posteriori* probability (MAP) decision rule [2]:

$$H = \begin{cases} H_0, & \text{if } L < \pi_0/\pi_1 \\ H_1, & \text{otherwise,} \end{cases}$$

where

$$L = \frac{f_{\vec{R}}(\vec{R}|H_1)}{f_{\vec{R}}(\vec{R}|H_0)}$$

is called the *likelihood ratio*, $\vec{R} = \{R_1, R_2, \ldots, R_K\}$ and $f_{\vec{R}}(\vec{R}|H_0)$ and $f_{\vec{R}}(\vec{R}|H_1)$ are the probability density functions for the samples being combined at the fusion center given the hypotheses H_0 and H_1, respectively. In this case, the quantizers at the sensor nodes are designed independently of each other. This is, one sensor node k is selected at a time and the quantizer is designed for this node given that the rest of the quantizers are already fixed in their design. Each individual quantizer is subsequently designed, now following the goal to minimize the probability of detection error at the fusion center, which includes the effects due to quantization and the channel. Assuming that the encoder has performed the mapping $x_k = \gamma_k(s_k) = i$, where $i \in \{0, 1, 2, \ldots, M - 1\}, M = 2^m$, the probability of detection error can be expressed as:

$$P_e = \pi_0 P\left(L > \pi_0/\pi_1|H_0\right) + \pi_1 P\left(L \le \pi_0/\pi_1|H_1\right)$$

$$= \pi_0 \int_{S_k} \left[\sum_{i=0}^{M-1} P(X_k = i|S_k) f_{S_k}(s_k|H_0) A_{0_k}\right] ds_k$$

$$+ \pi_1 \int_{S_k} \left[\sum_{i=0}^{M-1} P(X_k = i|S_k) f_{S_k}(s_k|H_1) A_{1_k}\right] ds_k, \tag{11.3}$$

where S_k is the support for the sampled signal s_k, $f_{S_k}(s_k|H_j)$ are the probability density functions for S_k conditioned on $H_j, j = 0, 1$, and

$$A_{0_k} = \int_{\vec{R}} P(\lambda = 1|\vec{r}) \sum_{\vec{X}^k} \left[f_{\vec{R}}(\vec{r}|\vec{X}^k, x_k = i) P(\vec{X}^k|H_0)\right] d\vec{r},$$

$$A_{1_k} = \int_{\vec{R}} P(\lambda = 0|\vec{r}) \sum_{\vec{X}^k} \left[f_{\vec{R}}(\vec{r}|\vec{X}^k, x_k = i) P(\vec{X}^k|H_1)\right] d\vec{r}.$$

Here, λ is the decision output from the fusion center ($\lambda = 0$ to signal $H = H_0$ and $\lambda = 1$ to signal $H = H_1$) and $\vec{X}^k$ is the vector $\vec{X} = \{X_1, X_2, \ldots, X_K\}$ without the component for X_k. Note that, in examining (11.3), the components A_{0_k} and A_{1_k} include the effects of the channel.

The first step in deriving the quantizer at node k consists of assigning the probabilities $P(X_k = i^* | S_k) = 1$, where i^* is the encoder mapping that minimizes the probability of error P_e, now becoming,

$$
P_e = \pi_0 \int_{S_k} \left[f_{S_k}(s_k|H_0) A_{0_k}^* \right] ds_k + \pi_1 \int_{S_k} \left[f_{S_k}(s_k|H_1) A_{1_k}^* \right] ds_k,
$$

$$
= \int_{S_k} \left[\pi_0 f_{S_k}(s_k|H_0) A_{0_k}^* + \pi_1 f_{S_k}(s_k|H_1) A_{1_k}^* \right] ds_k,
$$

where we have

$$
A_{0_k}^* = \int_{\vec{R}} P(\lambda = 1 | \vec{r}) \sum_{\vec{x}^k} \left[f_{\vec{R}}(\vec{r}|\vec{x}^k, x_k = i^*) P(\vec{x}^k|H_0) \right] d\vec{r},
$$

$$
A_{1_k}^* = \int_{\vec{R}} P(\lambda = 0 | \vec{r}) \sum_{\vec{x}^k} \left[f_{\vec{R}}(\vec{r}|\vec{x}^k, x_k = i^*) P(\vec{x}^k|H_1) \right] d\vec{r}.
$$

The next step is to minimize $\pi_0 f_{\vec{S}_k}(s_k|H_0) A_{0_k}^* + \pi_1 f_{\vec{S}_k}(s_k|H_1) A_{1_k}^*$ by converting the quantizing task at the sensor into a quantization of the local likelihood ratio:

$$
L_k(s_k) = \frac{f_{\vec{S}_k}(s_k|H_1)}{f_{\vec{S}_k}(s_k|H_0)}.
$$

This type of quantizer, called an *LR quantizer* (for likelihood ratio quantizer), has been shown to be optimal for the task at hand [1]. Designing the quantizer involves finding the optimal partitions Π_i, $\bigcup_{\forall i}\Pi_i = \Upsilon$ (where Υ is the support of $L_k(s_k)$), that minimize $\pi_0 f_{\vec{S}_k}(s_k|H_0) A_{0_k}^* + \pi_1 f_{\vec{S}_k}(s_k|H_1) A_{1_k}^*$. This entails finding the optimal LR quantization thresholds and index assignments, which first leads to the relation:

$$
\pi_0 f_{\vec{S}_k}(s_k|H_0)(A_{0_k}^* - A_{0_k}) + \pi_1 f_{\vec{S}_k}(s_k|H_1)(A_{1_k}^* - A_{1_k}) < 0
$$

$$
\pi_0(A_{0_k}^* - A_{0_k}) + \pi_1 \frac{f_{\vec{S}_k}(s_k|H_1)}{f_{\vec{S}_k}(s_k|H_0)}(A_{1_k}^* - A_{1_k}) < 0
$$

$$
\pi_0(A_{0_k}^* - A_{0_k}) + \pi_1 L_k(s_k)(A_{1_k}^* - A_{1_k}) < 0.
$$

From this, it is possible to conclude that the optimal local sensor scalar quantization must be a monotone LR partition, which means that for $i < j$, $L_k(s_k = i) < L_k(s_k = j)$. Furthermore, this result allows defining the set of LR partitions that results in minimum probability of detection error:

$$
\mathcal{A}_k = \left\{ L_k(s_k) : \pi_0 f_{\vec{S}_k}(s_k|H_0) A_{0_k}^* + \pi_1 f_{\vec{S}_k}(s_k|H_1) A_{1_k}^* < \pi_0 f_{\vec{S}_k}(s_k|H_0) A_{0_k} + \pi_1 f_{\vec{S}_k}(s_k|H_1) A_{1_k} \right\}.
$$

$$
(11.4)
$$

From here, the threshold that defines the partition is

$$
T_k = -\frac{\pi_0(A_{0_k}^* - A_{0_k})}{\pi_1(A_{1_k}^* - A_{1_k})},
$$

for $A_{1_k}^* - A_{1_k} \neq 0$. Based on the sign of $A_{1_k}^* - A_{1_k}$, the threshold T_k may be an upper or lower limit for the partition. Then, the partition is

$$
A_k = \begin{cases}
[0, T_k), & A_{1_k}^* - A_{1_k} > 0 \\
[T_k, \infty), & A_{1_k}^* - A_{1_k} < 0 \\
[0, \infty), & A_{1_k}^* - A_{1_k} = 0, A_{0_k}^* - A_{0_k} < 0 \\
\emptyset, & A_{1_k}^* - A_{1_k} = 0, A_{0_k}^* - A_{0_k} \geq 0,
\end{cases}
$$

where $\emptyset$ is the null interval or empty set. Combining the limits for all the partitions, the partitions become

$$
T_k^u = \min_{i:A_{1_k}^* - A_{1_k} > 0} \{T_k\},
$$

$$
T_k^l = \min_{i:A_{1_k}^* - A_{1_k} < 0} \{T_k\},
$$

and the optimal partitions to minimize the probability of error are defined as:

$$
A_k^* = \begin{cases}
\emptyset, & A_{1_k}^* - A_{1_k} = 0, A_{0_k}^* - A_{0_k} \geq 0, \text{ for some } i, \\
[0, \infty), & A_{1_k}^* - A_{1_k} = 0, A_{0_k}^* - A_{0_k} < 0, \text{ for all } i, \\
[T_k^l, T_k^u), & \text{otherwise.}
\end{cases}
$$

One can view the WSN's work as being a single distributed encoder, representing the sampling of a correlated variable from a physical phenomenon, and a single decoder, the fusion center, that is able to reconstruct the samples encoded in a distributed way. Figure 11.2 depicts this for the case of two encoders. The encoders work separately, with no physical communication between them. Because of the system configuration, this encoding technique is named **distributed source coding** (DSC). The study of two central problems characterize the performance of DSC: the Slepian–Wolf coding problem and the Wyner–Ziv coding problem.

The main result in Slepian–Wolf coding is a theorem that states the achievable performance. Let $\{X_{1_k}\}_{k=1}^{\infty}$ and $\{X_{2_k}\}_{k=1}^{\infty}$ be two sequences of correlated random variables. If the two sequences are jointly encoded for lossless compression (i.e. the reconstruction values

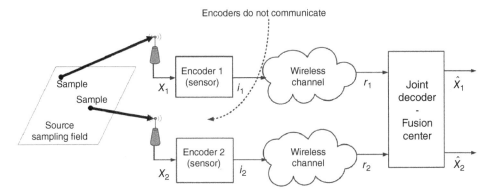

Figure 11.2 Overview of a distributed source coding (DSC) scheme.

$\hat{X}_1$ and $\hat{X}_2$, are such that $\hat{X}_1 = X_1$ and $\hat{X}_2 = X_2$), using DSC, the required coding rate regions are $R_1 \geq H(X_1|X_2)$, $R_2 \geq H(X_2|X_1)$ and $R_1 + R_2 \geq H(X_1, X_2)$ where $H(\cdot)$ is the entropy function. Comparing the result with Shannon's coding theorem, the Slepian–Wolf theorem says that there is no penalty in performance when encoding the correlated sources using DSC, compared to a configuration where the two encoders communicate with each other.

Consider the following example. The two sources X and Y are discrete, each generating one of four possible equiprobable values for each sample. Each source can be encoded independently using $H(X) = H(Y) = 2$ bits, and the two sources can be encoded using four bits. Now, assume that the two sources are not independent; the Hamming distance (see Chapter 3) between a sample from X and the sample from Y is always $d_H = 1$. With this relation, for any given sample Y, there are two possible values for the X sample. For example, if the sample outcome for Y is $y = 00$, the sample X can only be $x = 01$ or 10. Given Y, encoding X requires only one bit to differentiate between the two possible cases. By the theorem, using lossless DSC does not require four bits, but a total rate of only $H(X, Y) = H(Y) + H(X|Y) = 2 + 1 = 3$. The question is how to design the lossless DSC encoder for X. An interesting insight about DSC now appears in the intimate relation between channel coding and DSC. We build the encoder for X using the simplest parity-check matrix $\mathbf{H} = [1\ 0]$. Recall from Chapter 3 that the parity-check matrix is used to decode a block code and correct errors (when possible) by checking whether the syndrome $\vec{s} = \vec{r}\mathbf{H}^T$ equals zero. In the case of DSC, the calculation of the syndrome is the actual encoding operation. The syndromes for the four possible values of X are as follows:

$$\vec{s}_{00} = [0\ 0]\begin{bmatrix}1\\0\end{bmatrix} = 0,$$

$$\vec{s}_{11} = [1\ 1]\begin{bmatrix}1\\0\end{bmatrix} = 1,$$

$$\vec{s}_{01} = [0\ 1]\begin{bmatrix}1\\0\end{bmatrix} = 0,$$

$$\vec{s}_{10} = [1\ 0]\begin{bmatrix}1\\0\end{bmatrix} = 1.$$

The single-bit encoding of X resulting from computing the syndrome is enough to resolve the ambiguity in the value of X that exists when knowing the sample from Y. Figure 11.3 summarizes the encoding of X and Y to illustrate these points. The close relationship

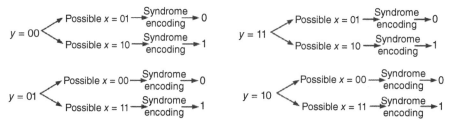

Figure 11.3 DSC using syndrome calculation.

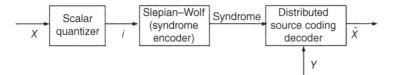

Figure 11.4 A Wyner–Ziv codec.

between DSC and channel coding makes the DSC naturally robust to channel errors and a case of JSCC, because the decoding of X relies on the extra information that is available from Y.

While the Slepian–Wolf theorem considers lossless DSC, the Wyner–Ziv theorem considers lossy encoding. Specifically, Wyner–Ziv coding studies the rate-distortion performance (recall that this is how many bits are needed to achieve some distortion performance) achievable in encoding a source X when there is some side information from the source available at the decoder. This side information takes the form of another random variable Y that is correlated with X in a known way. As with traditional rate-distortion theory, the results from Wyner–Ziv coding depend on the system setup, and the two most studied cases are binary symmetric sources [3] and Gaussian memoryless sources.

Consider two zero-mean and stationary Gaussian memoryless sources X and Y with variances σ_X^2 and σ_Y^2, respectively. The two sources are dependent with an interrelation expressed through the covariance matrix:

$$\Sigma_{XY} = \begin{bmatrix} \sigma_X^2 & \rho\sigma_X\sigma_Y \\ \rho\sigma_X\sigma_Y & \sigma_Y^2 \end{bmatrix},$$

where ρ is the correlation coefficient. If the distortion metric is mean squared error (MSE), for these settings it was shown in [4] that there is no extra performance penalty in the rate-distortion function when using Wyner–Ziv coding and that this function equals the one calculated from the distribution of X conditioned on Y:

$$R(D) = \frac{1}{2} \max \left\{ 0, \log_2 \left(\frac{\sigma_X^2(1 - \rho^2)}{D} \right) \right\}.$$

A Wyner–Ziv codec can be built using a JSCC approach as shown in Figure 11.4, where on the encoder side, a standard quantizer is connected in tandem to a Slepian–Wolf encoder that uses syndrome encoding [5]. At the decoder, the correlated random variable is used as side information to resolve the ambiguity in the original encoded variable.

With the growth of interest in WSNs, many studies have examined DSC techniques. We conclude this section by describing one of these [6] using channel-matched source encoder design. Consider a two-sensor network as in Figure 11.2. Both samples at the sensors measure the same underlying random variable, Y, but they are affected by independent and identically distributed random noise:

$$X_1 = Y + N_1,$$
$$X_2 = Y + N_2.$$

Here, Y, N_1 and N_2 are independent zero-mean Gaussian distributed random variables with variances σ_Y^2 and $\sigma_{N_1}^2 = \sigma_{N_2}^2 = \sigma_N^2$. At the sensors, the encoders input samples of X_1 or X_2 and output an m-bit codeword $i_1 = q_1(X_1)$ or $i_2 = q_2(X_2)$, with $i_1, i_2 \in \{0, 1, 2, \ldots, M = 2^m - 1\}$. The encoded codewords i_1 and i_2 are transmitted through a binary symmetric channel with crossover probability p and received as codewords r_1 and r_2, respectively. The decoder combines r_1 and r_2 and through decoding functions g_1 and g_2 obtains reconstructed estimates $\hat{X}_1$ and $\hat{X}_2$ of X_1 and X_2:

$$\hat{X}_1 = g_1(r_1, r_2),$$
$$\hat{X}_2 = g_2(r_1, r_2).$$

The design goal is to find the optimal distributed source encoders and joint decoders that minimize the MSE distortion between the sensed samples and their reconstructed estimates. Because of the symmetry between the processing chains for sensors 1 and 2, for design purposes, it is only necessary to work on one of the two (we choose sensor 1), and the distortion can be written as:

$$D = D_1 + D_2 = E[(X_1 - \hat{X}_1)^2] + E[(X_2 - \hat{X}_2)^2], \tag{11.5}$$

where

$$D_1 = \int_{x_1} f_{X_1}(x_1) \sum_{r_1=0}^{M} P(r_1|q_1(x_1)) \sum_{i_2=0}^{M} P(i_2|x_1) \sum_{r_2=0}^{M} (x_1 - g_1(r_1, r_2))^2 P(r_2|i_2) dx_1,$$

$$D_2 = \int_{x_1} f_{X_1}(x_1) \sum_{r_1=0}^{M} P(r_1|q_1(x_1)) \int_{x_2} f_{X_2}(x_2|x_1) \sum_{r_2=0}^{M} (x_2 - g_2(r_1, r_2))^2 P(r_2|q_2(x_2)) dx_2 dx_1,$$

where $f_{X_1}(x_1)$ is the probability density function for source X_1, $f_{X_2}(x_2|x_1)$ is the probability density function that incorporates the correlation between the two sources, and $P(i_2|x_1)$ also incorporates the encoder interrelation in DSC, as it is the probability that source sample x_2 is encoded to index i_2 given the sensor sample x_1. Note that the presence of the probability $P(r_1|q_1(x_1))$ in the calculation of distortion incorporates the channel effects into the design. As such, it places the result not only within the framework of DSC, but also of channel-matched coder design. In the design of the encoder q_1, given encoder q_2 and decoders g_1 and g_2, it is sufficient to minimize the MSE for each possible input sample value x_1, because the probability density function $f_{X_1}(x_1)$ is always positive. Therefore, the range of values of x_1 that are assigned the optimal index i_1^* is

$$A_{i_1}^* = \{x_1 : D'(x_1, i_1^*) \leq D'(x_1, i_1), \forall i_1 \neq i_1^*\}, \tag{11.6}$$

where

$$D'(x_1, i_1) = D_1(x_1, i_1) + D_2(x_1, i_1), \tag{11.7}$$

$$D_1(x_1, i_1) = E[(X_1 - \hat{X}_1)^2 | x_i, i_1]$$
$$= \sum_{r_1=0}^{M} P(r_1|q_1(x_1)) \sum_{i_2=0}^{M} P(i_2|x_1) \sum_{r_2=0}^{M} (x_1 - g_1(r_1, r_2))^2 P(r_2|i_2),$$

$$D_2(x_1, i_1) = E[(X_2 - \hat{X}_2)^2 | x_i, i_1]$$
$$= \sum_{r_1=0}^{M} P(r_1|q_1(x_1)) \int_{x_2} f_{X_2}(x_2|x_1) \sum_{r_2=0}^{M} (x_2 - g_2(r_1, r_2))^2 P(r_2|q_2(x_2)) dx_2.$$

The symmetry in the system setup makes the design of encoder q_2 the same as that for q_1. As for the optimal decoder g_1 (or, equivalently, g_2 due to symmetry), the reconstruction values $\hat{x}_1$ assigned to each received index r_1 (and, for a joint decoder, r_2) are given by:

$$\hat{x}_1(r_1, r_2) = E[x_1 | r_1, r_2]. \tag{11.8}$$

To find the encoders and decoders, in [6] an iterative algorithm was presented that considers the interdependencies of the encoders and decoders. Each iteration first computes the decoders g_1 and g_2 using (11.8), given the encoders q_1 and q_2. Then, it computes the optimal encoder q_1, given by (11.6), followed by computing the optimal decoders g_1 and g_2 and then computing the optimal encoder q_2, also given by (11.6). The iterations stop once the change in distortion between two successive iterations is small enough. As pointed out in [6], the solution to this iterative procedure is locally optimal and not necessarily a global optimum. The algorithm could be enhanced through different well-known methods, such as simulating annealing, for the initial choices of encoders q_1 and q_2.

In summary, sensor networks because of being formed by nodes intended to accomplished a common goal present a number of interesting variations to concepts seen earlier in this book. Such is the case with the design of channel-optimized quantizers, which now add the variation that at the receiver the messages from multiple transmitters that are received after being affected by the transmission through a channel will be combined to make a single detection decision. Multiple encoding nodes working for a common goal also lead to the concept of DSC with an implementation that is based on a duality between the source encoding of correlated sources (or, equivalently, of sources that act as side information for each other) and channel encoding. Here again, we see elements of channel-optimized quantizer design.

11.2 Extending Network Capacity Through JSCC

Wireless networks are normally seen as having some limit on the maximum number of calls that can be supported. This may be due to the number of available radio channels or time slots. In a multiuser network, the limit appears because each new active call adds to the level of interference seen by all other users; the limit is reached when this interference grows to a maximum acceptable level. Since in multiuser networks the interference plays a key role in the number of active calls, there is always a process to control the admission of new calls so as to limit the interference. However, it becomes valuable for network resiliency to be able to dynamically extend network operation beyond the strict limits set by narrow call admission control criteria based on interference thresholds. Such a feature could allow for flexible network adaptation to situations where it is more important to service a call than to guarantee its best quality. Examples of such situations are the servicing of a cellular user entering the coverage area of a congested cell, military communications, servicing of emergency teams in a disaster area, and increasing network resiliency by assigning to in-service base stations the coverage area and load of out-of-service stations. Extending the number of calls accepted in a network will reduce the quality of service (QoS), so this degradation should be smooth and controlled. This would allow network operators to dynamically reconfigure the network, based on different external conditions, in a controlled way without the need for costly new hardware.

In this section, we will study how this idea can be realized by incorporating JSCC adaptation into a multiuser network carrying real-time traffic. For clarity of the presentation, in what follows we will focus the presentation on the specific case of a code-division multiple access (CDMA) multiuser network. The scheme to be studied goes beyond JSCC; it is a cross-layer design that also adapts the physical layer to dynamically extend operation beyond the congestion point. This problem naturally leads to the general problem of optimal adaptation to resolve interference-generated congestion for an arbitrary set of real-time source encoders with arbitrary signal-to-interference-plus-noise ratio (SINR) goals and variable transmit bit rates. The discussion will also address the simplified problem of transmit rate allocation for an arbitrary set of variable-rate real-time source coders. We will see that the problem of jointly adapting source and channel coding in a multiuser environment subject to a power feasibility constraint is equivalent to the problem of efficient bit budget allocation to an arbitrary set of quantizers [7] and can be further considered as the optimal source-controlled statistical multiplexing solution in a CDMA system. The interested reader can find in [8] and [9] works that apply the concepts explained in this section.

In the rest of this section, we will consider the uplink of a single cell, and a chip-sampled direct-sequence code-division multiple access (DS-CDMA) system with bandwidth W. This will be the setting to explain how JSCC can be combined with adapting other network resources to trade network capacity for end-to-end quality. Assume there are N users in the system, each carrying on an independent conversational call for which the end-to-end delay is less than 200 ms, allowing participant interaction similar to that in telephony applications. As there are many delays that contribute to the end-to-end value, the requirement of conversational delay means that for speech or video sources, the JSCC scheme cannot add much more than a source coding period (e.g. the duration of a video frame) to the overall delay.

Figure 11.5 shows the main components of the system, consisting of a tandem JSCC component, connected to a spreader (for DS-CDMA operation), followed by the RF front-end.

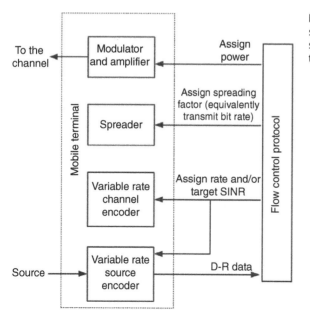

Figure 11.5 Block diagram of the system that relies on a concatenated source–channel scheme to manage the congestion state of the network.

The setup allows for different types of sources; we assume that the source is conversational video. The configuration for the concatenated JSCC is the same as that discussed in Chapter 4. A block of samples from a real-time source is encoded into a *source frame* using an encoder for which a flow control protocol or unit externally controls the encoding rate. We assume there is a finite set of possible source encoding rates. Each user could operate with a different source codec, and although a user will not change encoders during a call, its distortion-rate (D-R) performance may change from frame to frame based on changing source statistics (as was explained in Chapter 2). A variable-rate channel encoder provides channel error protection for the source frame. The system setup does not impose a restriction of equally protecting all source bits; one could apply an unequal error protection (UEP) scheme. The channel encoder output, with transmit bit rate r_i, is fed into the spreader for transmission. This is a variable spreading factor (VSF) spreader, which means that while keeping the transmit bandwidth (or system bandwidth) fixed, the transmit bit rate can vary. This results in a DS-CDMA system that can adapt the call's processing gain. This adaptation of a physical layer parameter will translate into the ability to control, to some extent, the level of interference contributed by a transmit node to the overall system.

This system design allows each user to dynamically switch among different combinations of source and channel coding rates. Each such combination forms an *operating mode*. A flow control protocol at the base station allocates and communicates an operating mode and power to each mobile in the coverage area. The flow control's function is analogous to that of a statistical multiplexer. Most of the discussion in this section will focus on the design and analysis of this protocol. The main design problem is to find the adaptation rules and the resource allocation by the flow control protocol that minimize mean end-to-end distortion subject to traffic demands. This allocation needs to also satisfy a *channel impairment limit*, a condition that ensures that the communication will not be noticeably impaired by channel-introduced errors. This limit could be specified using a target maximum frame error rate (FER) or bit error rate (BER).

Because the D-R performance for each call may change from one transmission period to the next, each call needs to send information about this performance to the flow control protocol. Using the estimates of traffic demands from each call's D-R performance information, the flow control protocol performs resource allocation analogous to optimal statistical multiplexing. Therefore, during each transmission period, each mobile sends not only the encoded source data sampled during the previous period but also information about the source encoder D-R performance corresponding to the source data sampled during the current period. In effect, transmission of a source frame is delayed by exactly one frame duration with respect to the time when data was sampled. We will explain later how the transmission overhead associated with the D-R information can be kept small.

11.2.1 Video Telephony Calls as Application Example

In the system of Figure 11.5, assume that all mobile terminals use an MPEG-4 fine granularity scalable (FGS) coder [10] that is error-protected with rate-compatible punctured convolutional (RCPC) codes [11]. Transmission parameters are chosen so that the end-to-end quality is good for conversational communication, which means that source encoding distortion should be kept small and channel-induced errors should not introduce annoying effects. The MPEG-4 FGS encoder generates a two-layer (base and enhancement)

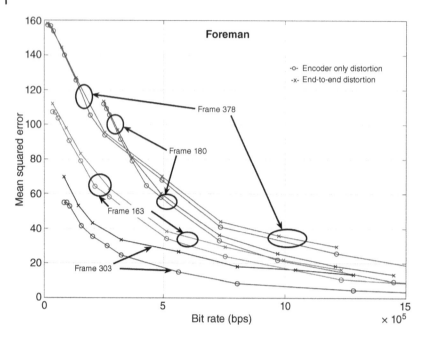

Figure 11.6 Distortion-rate performance for different video frames of the video sequence "Foreman."

source-coded bit stream. The enhancement layer bit stream is embedded, allowing easy control of encoding bit rate.

The D-R performance may change for each call and frame because it depends on various characteristics such as frame texture and type of temporal prediction (I or P) used and the amount of motion in the video sequence. This is illustrated in Figures 11.6 and 11.7, which shows the D-R performance of several representatives frames from two video sequences: *Foreman* and *Akiyo*. The sequences are QCIF at 30 frames per second (fps). *Foreman* has relatively high motion and *Akiyo* is low motion. Both sequences were encoded with 29 P-frames between each I-frame. The figure shows results with and without channel errors. Channel errors were introduced at a BER equal to 5×10^{-6} and 10^{-5} for the base and enhancement layers, respectively. These values were chosen following the channel impairment limit criterion: it was found after exhaustive simulations using the MPEG-4 FGS encoder with error resilience and concealment that these BER values correspond to the limit where the end-to-end subjective quality was good (comparable to "toll" quality in telephone communications) and channel errors did not introduce annoying artifacts or impaired understandability of the source. In this case, the channel impairment limit is specified in terms of BER. The differentiated target BER specification for the base and enhancement layers constitutes a simple but effective UEP scheme.

At 30 fps, the single-frame delay of approximately 33 ms that is being introduced is acceptable for conversational video. We note in Figures 11.6 and 11.7, the D-R performance changes from frame to frame but is more stable for sequences with low motion (*Akiyo* in Figure 11.7). While the contribution of channel errors to the end-to-end distortion

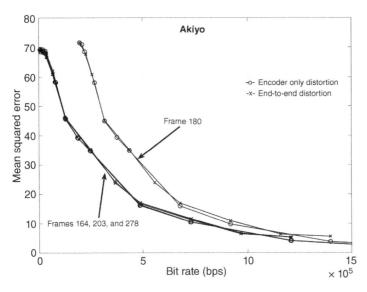

Figure 11.7 Distortion-rate performance for different video frames of the video sequence "Akiyo."

is usually negligible, there are cases, as in Foreman's frame 303 (part of a camera panning section), where this is not true. In these cases, channel-introduced distortion is approximately the same for all encoding rates.

11.2.2 CDMA Statistical Multiplexing Resource Allocation and Flow Control

This section discusses allocating operating modes and power to users based on traffic load, aiming to minimize the average source distortion per call while ensuring a maximum distortion is not exceeded and BER requirements are satisfied. We will use a technique that allows the extension of the network operation beyond the nominal congestion point by controllably trading off distortion. This trade-off can be more desirable than the other option of just blocking calls. Of course, while the minimization of average source distortion and BER requirements will provide for a graceful degradation of quality to accommodate traffic that otherwise would not be possible to service, the condition of a maximum distortion still makes it possible that an allocation solution will not be possible, in which case, as will be seen within the details of the described technique, there still will be a network outage. However, this network outage will occur under conditions far beyond the nominal congestion point.

For the system setup, assumed with ideal power control, an additive white Gaussian noise (AWGN) channel and a matched filter at the receiver, the power assignment and interference from other users are related to the target SINR required by each call as [12]:

$$\beta_i \geq \frac{(W/r_i)\, P_i}{\sigma^2 + \sum_{j \neq i} P_j}, \qquad i = 1, 2, \ldots, N, \tag{11.9}$$

where W/r_i is the processing gain, P_i is the power assigned to user i, as measured at the receiver, necessary to obtain the target SINR β_i, and σ^2 is the background noise variance, which accounts for intercell interference [12]. The target SINRs are set based on the channel impairment limit so as to not exceed the BER threshold that limits channel errors to an acceptable magnitude. If it is possible to find feasible power assignments that satisfy the N inequalities (11.9) with equality, then these assignments minimize the sum of the transmitted powers, [13]. Taking (11.9) as an equality and solving the ensuing linear system, we obtain the power assignment:

$$P_i = \frac{\Psi_i \sigma^2}{1 - \sum_{j=1}^{N} \Psi_j}, \quad i = 1, 2, \ldots, N. \tag{11.10}$$

where

$$\Psi_i = \left(1 + \frac{W}{r_i \beta_i}\right)^{-1}. \tag{11.11}$$

The solution in (11.10) is the power at the receiver necessary to obtain an SINR such that the expected channel induced distortion remains below a preset limit. It is clear that we need $\sum_{j=1}^{N} \Psi_j \leq 1$ for the power P_i to be positive. This condition determines the maximum number of users that can be accepted into the system. Furthermore, when $\sum_{j=1}^{N} \Psi_j \approx 1$ the power assignments in (11.10) are likely too large to be practically feasible. As more users are admitted into the system, the sum $\sum_{j=1}^{N} \Psi_j$ grows until it exceeds a threshold $1 - \epsilon$, where ϵ represents the congestion point and is a small positive number set during design. Therefore, a practical limit on the user capacity will be determined by the condition:

$$\sum_{i=1}^{N} \Psi_i \leq 1 - \epsilon. \tag{11.12}$$

We want the flow control adaptation rule to minimize the average distortion per call. Let $f_i(x_i)$ be the D-R function of the ith user source encoder at rate x_i. Then, the optimization goal can be written as:

$$\min_{x_1, x_2, \ldots, x_N} \sum_{i=1}^{N} f_i(x_i), \tag{11.13}$$

Typically, $f_i(x_i)$ would be a decreasing function. Since the goal (11.13) does not prevent each user from experiencing excessive distortion, we will ensure each user's source rate exceeds some minimum value. At the other extreme, $f_i(x_i)$ is minimum when the rate is maximum; $x_i = x_{max}$ would be the source rate assignment for all users in the absence of congestion. We take $f_i(x_i) = \alpha_i 2^{-k_i x_i}$, a general form of D-R function that, as we have seen, applies to Gaussian sources with squared error distortion and also when the high-rate approximation holds. In practice, real encoders are complex and their D-R functions do not strictly follow this rule for all rates. Nevertheless, by carefully choosing α_i and k_i, this function approximates well a close upper bound for real D-R characteristics such as those in Figures 11.6 and 11.7. The optimization problem can be written as:

$$\min_{x_1, r_1, x_2, r_2, \ldots, x_N, r_N} \sum_{i=1}^{N} f_i(x_i) \text{ subject to } \sum_{i=1}^{N} \Psi_i(\beta_i, r_i) \leq 1 - \epsilon. \tag{11.14}$$

The formulation also needs to consider channel-introduced distortion. Similar to the analysis discussed in Chapter 4, when the design follows the goal of preventing annoying channel

impairments, in most cases channel distortion can be neglected. However, in some scenarios, the channel-introduced distortion does not necessarily is small enough that can
be directly ignored. In some video sources, specially those with relatively large levels of
motion, this distortion could be allowed to be numerically significant because the motion in
the video perceptibly reduces to some extent the effect of the impairment. Because in these
cases the channel-introduced distortion is independent of the source encoding rate (i.e. it
is approximately constant for all encoding rates) with a magnitude that is made acceptable
by system design, we can still ignore it in our formulation without loss of optimality. Note
that meeting the channel impairment limit is implied in the constraint of (11.14) since Ψ_i
depends on the target SINR.

Let $b_i = (r_i, \beta_i)$ be user i's transmit rate and target SINR allocation pair. This will turn out
to be all the parameters needed to specify the operating mode. Then, the problem (11.14) of
optimal operating mode and power allocation to minimize average end-to-end distortion in
the uplink of a multiuser, single-cell CDMA system with an arbitrary set of source encoders
can be stated as:

$$\min_{b_1, b_2, \ldots, b_N} \sum_{i=1}^{N} D_i(b_i), \quad \text{s.t.} \quad \sum_{i=1}^{N} \Psi_i(b_i) \leq 1 - \epsilon, \tag{11.15}$$

where D_i is a distortion function. This problem formulation implicitly states that it is equivalent to specify the operating mode as the combination of a source and channel rate or as
the pair $b_i = (r_i, \beta_i)$.

The problem formulation also implicitly states that there is a functional relation between
b_i and D_i. From its definition, Ψ_i is a function of only r_i and β_i. Let D_i be user i's distortion,
which depends only on user i's source rate x_i (when the target SINR is set to prevent annoy-
ing channel impairment, so channel distortion is negligible). We assume a user follows the
natural guideline of maximizing the transmit bit rate utilization, i.e. sending as many bits
as allowed by the transmit bit rate. Also, if β_i is given, user i's channel coding rate is automatically determined as the one that provides enough error protection to meet the channel
impairment limit. Since each call's transmit bit rate is divided between source coding and
channel error protection, given any two of user i's source encoding, channel coding and
transmit bit rate, the third is automatically determined. Therefore, if user's i target SINR is
changed when keeping the transmit bit rate r_i fixed, the channel rate will need to change
to meet the channel impairment limit, and so the source coding rate x_i will also change.
Likewise, if r_i is changed, x_i will need to change for a fixed β_i. In summary, the source coding rate x_i implicitly depends on r_i and β_i through the requirements of meeting a channel
impairment limit and maximizing transmit bit rate utilization. Also, each pair b_i has only
one associated value $D_i(b_i)$. Therefore, D_i is a function of b_i.

The problem stated in (11.15) is analogous to the problem studied in [7], where the goal
is to allocate a bit quota R_b to an arbitrary set of quantizers. The problem is also analogous
to the one studied in [14] of allocating a fixed bandwidth among a number of users in a
TDMA network. These analogies allow us to see that Ψ_i can be considered as the *equivalent
bandwidth* assigned to user i out of the total $1 - \epsilon$. The concept of an equivalent bandwidth
was studied for CDMA as a tool for designing algorithms for resource allocation and call
admission control. The definition just derived by analogy to the problems in [7] and [14]
is consistent with the definition of equivalent bandwidth in the earlier CDMA work [15]

and with the definition of effective interference in [16]. Equivalent bandwidth represents the portion of resources being used, and we can think of the solution to (11.15) discussed next as the solution to effective bandwidth assignment among real-time calls in a multiuser CDMA system. Clearly, there is a direct analogy between problem (11.15) and statistical multiplexing. In essence, (11.15) states that the problem of modes and power allocation in a CDMA uplink can be considered as performing statistical multiplexing in a multiuser CDMA setup. The formulation is general but powerful enough that as a conceptual charac-terization of system resources, it allows for the inclusion of other related resource allocation problems in CDMA. Importantly, the problem (11.15) differs from the one in [7] in that dis-tortion now is a function of two variables, namely transmit bit rate and target SINR (as opposed to source bit rate only) and that the constraint function is the sum of functions of transmit bit rate and target SINR instead of just sum of allocated bits. An optimal solution to (11.15) can be found by extending and adapting the results in [7], as explained next.

Let $S_r^{(i)}$ and $S_\beta^{(i)}$ be the finite sets of all possible transmit rates and target SINRs, respec-tively, for user i. Let $S^{(i)}$ be the set of all user i's possible allocation vectors $b_i = (r_i, \beta_i)$ and S be the set of all possible allocations $B = \{b_1, b_2, \dots, b_N\}$. Let $H(B)$ be some real-valued func-tion, called the objective function of B, defined for all $B \in S$. Let $R(B)$ be some real-valued function, called the constraint function of B, defined for all $B \in S$. From this setting, we can study the following theorem:

Theorem 11.1 There exists a $\lambda \geq 0$ such that the optimal solution, $B^*(\lambda)$, to the constrained problem:

$$\min_{B \in S} H(B), \quad \text{subject to } R(B) \leq R_c,$$

with $R(B^*(\lambda)) = R_c$ is also the solution to the unconstrained problem $\min_{B \in S} \{H(B) + \lambda R(B)\}$

Proof: The proof is in [7]. We summarize the proof to emphasize that it still holds for the extended problem considered here.

$$H(B^*) + \lambda R(B^*) \leq H(B) + \lambda R(B)$$

for all B in S. Then we have

$$H(B^*) - H(B) \leq \lambda R(B) - \lambda R(B^*),$$

which is true for all B in S. Thus, (11.16) is true for all B in the subset of S, $S^* = \{B : R(B^*) \leq R(B^*)\}$. Since $\lambda \geq 0$

$$H(B^*) - H(B) \leq 0.$$

This means that B^* is the solution to the constrained problem with $R_c = R(B^*)$. □

For the particular case of problem (11.15), we can say the following: Let $H(B)$ and $R(B)$ be of the forms $H(B) = \sum_{i=1}^{N} D_i(b_i)$ and $R(B) = \sum_{i=1}^{N} \Psi_i(b_i)$. Then the unconstrained problem $\min_{B \in S} \{H(B) + \lambda R(B)\}$, $\lambda \geq 0$, can be written as:

$$\min_{B \in S} \left\{ \sum_{i=1}^{N} D_i(b_i) + \lambda \sum_{i=1}^{N} \Psi_i(b_i) \right\}. \tag{11.16}$$

Note that the solution $B^*(\lambda) = \{b_1^*(\lambda), \dots, b_N^*(\lambda)\}$ can be obtained by minimizing each term of the sum in the unconstrained problem separately, i.e. $b_k^*(\lambda)$ solves

$$\min_{b_i \in S^{(i)}} \{D_i(b_i) + \lambda \Psi_i(b_i)\}. \tag{11.17}$$

From this result, we can take one step further and consider the following theorem and corollary that will help derive the optimum solution algorithm.

Theorem 11.2 Let $D_i(b_i)$ and $\Psi_i(b_i)$ be real-valued functions over some closed domain on the real line. Let $b_i^*(\lambda_1)$ be a solution to $\min_{b_i \in S} \{D_i(b_i) + \lambda_1 \Psi_i(b_i)\}$ and let $b_i^*(\lambda_2)$ be a solution to $\min_{b_i \in S} \{D_i(b_i) + \lambda_2 \Psi_i(b_i)\}$. Then for any function $D_i(b_i)$,

$$0 \leq (\lambda_2 - \lambda_1)\left(\Psi_i\left(b_i^*(\lambda_1)\right) - \Psi_i\left(b_i^*(\lambda_2)\right)\right).$$

Proof: Following [7], by definition of $b_i^*(\lambda_1)$ and $b_i^*(\lambda_2)$, we have

$$D_i(b_i^*(\lambda_2)) + \lambda_2 \Psi_i(b_i^*(\lambda_2)) \leq D_i(b_i^*(\lambda_1)) + \lambda_2 \Psi_i(b_i^*(\lambda_1))$$
$$D_i(b_i^*(\lambda_1)) + \lambda_1 \Psi_i(b_i^*(\lambda_1)) \leq D_i(b_i^*(\lambda_2)) + \lambda_1 \Psi_i(b_i^*(\lambda_2)).$$

From this, we get

$$D_i(b_i^*(\lambda_1)) - D_i(b_i^*(\lambda_2)) \leq \lambda_1 \left[\Psi_i(b_i^*(\lambda_2)) - \Psi_i(b_i^*(\lambda_1))\right]$$
$$D_i(b_i^*(\lambda_2)) - D_i(b_i^*(\lambda_1)) \leq \lambda_2 \left[\Psi_i(b_i^*(\lambda_1)) - \Psi_i(b_i^*(\lambda_2))\right].$$

Adding both sides of these inequalities proves the theorem. $\square$

Corollary 11.3 The solutions $\Psi_i(b_i^*(\lambda))$, for all i, and the corresponding constraint function $R^*(\lambda) = \sum_{i=1}^{N} \Psi_i(b_i^*(\lambda))$ are monotonically non-increasing with λ, i.e. if $\lambda_2 \geq \lambda_1 > 0$, then $\Psi_i(b_i^*(\lambda_2)) \leq \Psi_i(b_i^*(\lambda_1))$, and $R^*(\lambda_2) \leq R^*(\lambda_1)$.

Proof: Theorem 11.2 says that as λ increases, the minimizing value for $b(\lambda)$ makes the resulting $\Psi_i(b_i^*(\lambda))$ either increase or stay the same. This proves the corollary for $\Psi_i(b_i^*(\lambda))$ and hence for the sum $R^*(\lambda) = \sum_{i=1}^{N} \Psi_i(b_i^*(\lambda))$. $\square$

Figure 11.8 shows a typical behavior of $\sum_{i=1}^{N} \Psi_i(b_i)$ as a function of λ. The curve in this figure was obtained from simulations which are detailed later in this section.

We next apply this theory to develop two algorithms that solve problem (11.15) by optimally allocating resources among calls. We first describe the case where the transmit bit rate is adapted, while the target SINR is kept unchanged. This is a simplified problem that does not perform JSCC but is an important first step because it focuses on controlling the interference generated (recall that processing gain is inversely proportional to transmit bit rate and that it is a factor directly scaling the achieved SINR). After this initial case, we will discuss the JSCC joint adaptation of both transmit bit rate and target SINR.

When adapting the transmit bit rate only, the channel coding rate and, consequently, target SINR are assumed fixed, thus $b_i = r_i$. We assume that each call's D-R performance is known at the base station. Based on this information, the flow control protocol will

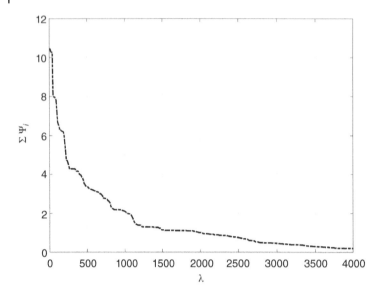

Figure 11.8 Total optimal equivalent bandwidth $\sum_{i=1}^{N} \Psi_i(b_i)$ as a function of λ.

choose, for each call, a transmit bit rate, $r_i = \hat{r}_i$, large enough to reach a target small distortion. If the network is lightly loaded and the total requests from all calls meet (11.12), all users are allocated resources so that they operate with best quality. If the network is congested ((11.12) fails), then a congestion resolution algorithm needs to solve (11.15). A low-complexity greedy but optimal solution exists for the problem (11.15) as a derivation from the analogous problems [14] and [7].

Given a finite set of available transmit rates $\mathbf{r} = \{r_{t_1}, r_{t_2}, \ldots, r_{t_M}\}$, we define $\Delta_i^{(j)}$, the ith incremental distortion associated with call j, as the reduction in distortion caused by increasing the transmit rate one discrete step $\Delta q = r_{t_{i+1}} - r_{t_i}$, i.e.,

$$\Delta_i^{(j)} = D_j\left(\Psi_j(r_{t_{i+1}})\right) - D_j\left(\Psi_j(r_{t_i})\right).$$

The rate allocation algorithm uses a table associated with the incremental distortions. The table stores all pairs of indices (j, i) in increasing order of their associated incremental distortions, while also respecting each user's rate reduction order. The pair (j_1, i_1) precedes in the table the pair (j_2, i_2) if $i_1 > i_2$ and $j_1 = j_2$, or if $\Delta_{i_1}^{(j_1)} < \Delta_{i_2}^{(j_2)}$ and $j_1 \neq j_2$. The first location of the table has a "0," and the second location has a pair (j, i) corresponding to the smallest possible incremental distortion. A pointer p addresses a location in the table.

The network active calls overflow resolution algorithm proceeds through iterations indexed by m. For the mth iteration, we let $r_i^{(m)}$ and $\Psi_i^{(m)}$ denote the transmit rate and effective bandwidth associated with the ith call, respectively. Also, let $\mathbb{S}^{(m)} = \sum_{i=1}^{N} \Psi_i^{(m)}$ denote the total effective bandwidth assigned at the mth iteration. This overflow resolution algorithm is described as Algorithm 11.1.

Although Algorithm 11.1 is greedy, it finds an optimal allocation of transmit bit rate. The following theorem proves this point.

Algorithm 11.1 Overflow resolution algorithm

Initialization: $m = 0$, $r_i^{(0)} = \hat{r}_i$, $\Psi_i^{(0)}(\hat{r}_i)$ for $i = 1, 2, \ldots, N$, and $\mathbb{S}^{(0)} = \sum_{i=1}^{N} \Psi_i^{(0)}$.

        Set the pointer $p = 1$ to the first entry in the table.

1: **if** $\mathbb{S}^{(m)} \leq 1 - \epsilon$ **then**

2:     There is no overflow, **go to line** 17

3: **else**

4:     An overflow has occurred, **go to line** 6.

5: **end if**

6: $p \leftarrow p + 1$

7: **if** p exceeds the table length **then**

8:     It is not possible to perform allocation subject to the minimum per-user distortion constraint,

9:     report **outage**,

10:     **go to line** 17

11: **else**

12:     The pth entry of the table is a pair (j, i), indicating that to optimally resolve the overflow, the transmit rate of the jth user needs to be updated,

13:     $r_j^{(m+1)} \leftarrow r_{l_i}$ for $r_j^{(m)} = r_{l_{i+1}}$, and $r_l^{(m+1)} \leftarrow r_l^{(m)}$ for $l \neq j$.

14:     **update** $\Psi_i^{(m+1)}$ and $\mathbb{S}^{(m+1)}$.

15:     Proceed with the next iteration: **go to line** 1.

16: **end if**

17: **Exit step**: if $\mathbb{S}^{(m)} < 1 - \epsilon$, it means that the network is not fully loaded. Also, if $p \neq 1$, it means that overflow has occurred and has been resolved.

Theorem 11.4 If the D-R functions $D_i(r_i)$ are convex and decreasing (as is commonly the case), then the proposed greedy algorithm for overflow resolution provides the optimal rate assignment minimizing the average distortion per call.

Proof: We establish this assertion by mathematical induction. Since initialization assures the minimum absolute average distortion, the claim holds true for $m = 1$. Assuming optimal assignment at the mth iteration, we prove that at iteration $m + 1$ the algorithm is optimal too. To do this, it suffices to show that the rate reduction in the optimal assignment in the mth step will not be required for the optimal assignment in the $m + 1$st step. In other words, the optimal rate assigned to each call at the mth step is no less than the optimal rate assigned to each call at the $m + 1$st step.

If β_i is fixed, so is the amount of error protection (channel coding rate) necessary to meet the channel impairment limit. Also, in the most general setting, it is possible to use a UEP scheme where different source-encoded bits receive different error protection. In this case, the source encoding rate is the result of an affine mapping $x_i = \sum_k x_{ik} = \sum_k R_{ik} r_{ik}$, where R_{ik} is the fixed channel coding rate for each group of source bits and r_{ik} is the corresponding transmit bit rate with $r_i = \sum_k r_{ik}$. Also, it is assumed that the functions $h_k(r_i)$ that specify how a change in r_i affects each r_{ik} are of the form $r_{ik} = h_k r_i + r_{ik}^o$, where $r_{ik}^o \geq 0$, $h_k \geq 0$ and $\sum_k h_k = 1$. Then, $D_i(r_i) = \alpha_i 2^{-k_i(r_i \sum_k h_k R_{ik} + \sum_k r_{ik}^o R_{ik})}$ is a convex function, decreasing in r_i. We have asserted that the solution to the constrained problem (11.15) can be obtained by minimizing each term of the sum separately, i.e.

$$\min_{r_i} \left\{ D_i(r_i) + \lambda \Psi_i(r_i) \right\}, \quad i = 1, 2, \dots, N. \tag{11.18}$$

The same λ appears for all the terms independently of i, and $\Psi_i(r_i)$ is increasing (recall (11.11)). At the mth step of the algorithm, the network obtains the optimal solution for a total assigned equivalent bandwidth $\mathbb{S}^{(m)}$. Equivalently, from Theorem 11.1, at the mth step of the algorithm, there exists a positive $\lambda^{(m)}$ corresponding to $\min \left\{ \sum_{i=1}^{N} D_i(r_i^{(m)}) + \lambda^{(m)} \sum_{i=1}^{N} \Psi_i(r_i^{(m)}) \right\}$. At the next step, if overflow persists, the rate assignment to at least one of the calls has to decrease so as to lower its equivalent bandwidth. Because, from Corollary 11.3, the optimal equivalent bandwidth for this rate assignment is a nonincreasing function of $\lambda^{(m+1)}$, this suggests that $\lambda^{(m+1)} > \lambda^{(m)}$. However, $\lambda^{(m+1)}$ is the same for all calls, independent of i. Therefore, as the algorithm proceeds, the Lagrange multiplier coefficient increases or remains unchanged, and the optimal rates (and equivalent bandwidths) for all calls decrease or remain unchanged. As a result, to achieve the optimal solution in a given step, the algorithm never needs to increase back the rate assignment to a call whose rate was reduced in previous steps; therefore, the proposed iterative greedy algorithm provides the optimal solution. □

We now focus on the more general case where JSCC is used for transmit bit rate and target SINR adaptation. In this case, the adaptation of transmit bit rate and target SINR requires a different algorithm than the one just described because it cannot be asserted that $D(b_i)$ is convex and decreasing on the pair b_i. Nevertheless, based on the theory discussed earlier, Algorithm 11.2 describes the procedure to optimally allocate the pairs $b_i = (r_i, \beta_i)$ to all calls, where we again use m as the iteration index.

In a practical system, the numbers of both available transmit bit rates and target SINRs are finite and typically small. For example, a typical transmitting node may have to choose among half a dozen or a dozen possible channel coding rates (equivalently, target SINRs), and among a similar number of transmit bit rates. Therefore, each minimization in step 1 of Algorithm 11.2 is easily solved by exhaustive search, where each possible $\Psi_i(b_i)$ could be calculated offline and each $D_i(b_i)$ is essentially the D-R performance information communicated to the base station. For example, if there are eight possible channel coding rates and target SINRs each, the problem reduces to choosing the smallest element in a matrix resulting from adding two eight-by-eight matrices. The updates of λ in steps 3 and 6 of Algorithm 11.2 can be done following any of the methods suggested in the literature [7]. One simple option, that will be assumed next, is the use of a simple bisection.

Algorithm 11.2 finds an optimal solution, which can be proved by noticing that the algorithm performs an iterative search for the allocation that meets the optimality criteria in Theorem 11.1.

11.2.3 Overhead from Communicating Rate-Distortion Data

Algorithms 11.1 and 11.2 involve some communication overhead to send each call's D-R performance information to the centralized element where the algorithms are run. This overhead is larger for Algorithm 11.2 because it needs the distortion values corresponding

---

Algorithm 11.2 Optimal allocation of the pairs $b_i = (r_i, \beta_i)$.

Initialize $\lambda^{(0)}$ with some small positive number.

1: Solve each of the N unconstrained problems

$$\min_{b_i(\lambda^{(m)})} \left\{ D_i(b_i) + \lambda^{(m)} \Psi_i(b_i) \right\}, \tag{11.19}$$

and update $\mathbb{S}^{(m)} = \sum_{i=1}^{N} \Psi_i \left(b_i(\lambda^{(m)}) \right)$.

2: **if** $\mathbb{S}^{(m)} > 1 - \epsilon$ and the stopping criterion is false **then**

3:     Update $\lambda^{(m+1)}$ such that $\lambda^{(m+1)} > \lambda^{(m)}$, so that the effective interferences, $\Psi_i(b_i)$, will be reduced.

4:     **go to line** 1.

5: **else if** $\mathbb{S}^{(m)} < 1 - \epsilon$ and the stopping criterion is false **then**

6:     Update $\lambda^{(m+1)}$ such that $\lambda^{(m+1)} < \lambda^{(m)}$, so that the effective interferences, $\Psi_i(b_i)$, will be increased.

7:     **go to line** 1.

8: **end if**

9: **Stopping criteria**: Iterations stop whenever one of the following occurs:

    a $\mathbb{S}^{(m)}$ is sufficiently close to, but less than, $1 - \epsilon$; this is optimal allocation in the presence of congestion;

    b $\mathbb{S}^{(m)} < 1 - \epsilon$ and all allocations b_i correspond to the threshold minimum target distortion; this is a lightly loaded network;

    c $\mathbb{S}^{(m)} > 1 - \epsilon$ and all allocations b_i correspond to the maximum allowable distortion; this is an outage condition and could be avoided with high probability by proper admission control.

---

to each possible pair b_i of transmit bit rate and target SINR. This also adds to the complexity of each mobile terminal's encoder, since it needs to compute each of these distortion values (this is a problem in a distributed algorithm also). Both problems can be addressed by summarizing the D-R performance information that is being calculated and transmitted. This is, instead of sending the distortion and rate data for each operating mode, each transmitting node sends the following: one reference encoding bit rate and three distortion values at predefined bit rate points. These three predefined bit rate points are separated from the reference encoding bit rate (the one that is sent) by fixed bit rate values which are suitably picked to represent high, medium, and low distortion values. The base station uses the transmitted data to approximate two curves of the form $f(x) = \alpha 2^{-kx}$, one for the high distortion section of the D-R performance using the high and medium bit rate-distortion points and another for the low distortion section using the low and medium bit rate distortion points. The rest of the distortion-rate points are calculated by interpolation using the approximate D-R curves. This scheme allows a representation for the D-R performance that has low overhead, involves computing only three distortion points at the mobile station, and still allows good performance of the overall allocation algorithm.

11.2.4 Analysis for Dynamic Call Traffic and Admission Control

So far we have considered a static network model where the number of calls N is fixed. In reality, this number is a random variable that depends on the traffic in the cell. We want next to address the problem of admission control when the number of calls dynamically changes over time. This subsection shows a study case of how JSCC and a cross-layer mechanism can be used to shape or control network traffic.

We assume that calls enter the cell at a rate v following a Poisson arrival process and that the random call duration follows an exponential distribution with mean $1/\mu$. Then, it is possible to model the DS-CDMA network as an $M/M/\infty$ queue. Using Kendall's notation for describing a queuing system, an $M/M/\infty$ queue is one with a Poisson random arrival process, random service times with an exponential distribution, and a potentially infinite number of servers.

The admission control policy will be based on an outage probability, where outage occurs when the system exceeds some operational parameter. The failure of (11.12) has been typically considered as an outage condition in DS-CDMA systems [17, 18]. One key concept driving the JSCC application we are discussing is that it prevents condition (11.12) from failing at the cost of a smooth increase in end-to-end distortion. Therefore, the relevant operation for call admission control is to limit the number of calls to a maximum N_L, where N_L is set so that $\overline{D}_{N_L} = D_{max}$, D_{max} being the maximum tolerable expected distortion and $\overline{D}_N$ the expected distortion per call when there are N calls. For the purpose of call admission control, it is more pertinent to model the network as an $M/M/N_L/N_L$ blocking system. This is, the system is limited to N_L ongoing calls and the number of servers is not infinite any more, But it is limited to N_L. This model better represents the problem of call admission control from the network operator's viewpoint, because it rejects new calls once a maximum number has been reached so as to maintain quality for the existing calls.

It is common practice to represent most known queue models, including $M/M/N_L/N_L$, in the form of a state transition diagram as in Figure 11.9 [19, 20]. From queuing theory, the steady-state probability that there are N calls in the network is [21]:

$$q_N \overset{\Delta}{=} P[n = N] = \frac{\phi_N}{\sum_{i=0}^{N_L} \phi_i},$$

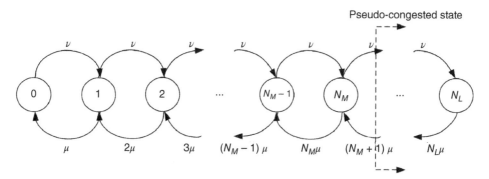

Figure 11.9 Markov chain representation of the $M/M/N_L/N_L$ traffic model.

where $\phi_i = \rho^i/i!$ and $\rho = v/\mu$ is the called the *offered load*. Therefore, assuming ergodic processes, over a sufficiently large period of time the average distortion per call as a function of the offered load will be

$$E\left[\overline{D}_n\right] = \frac{\sum_{n=0}^{N_L} \overline{D}_n \phi_n}{\sum_{i=0}^{N_L} \phi_i}. \tag{11.20}$$

The fact that we are considering arbitrary source encoders, coupled with the use of the aforementioned algorithms to allocate resources, makes it difficult to obtain a closed-form solution for $\overline{D}_N$. This challenge is common when analyzing queuing systems and is usually addressed, as in this section, by estimating $\overline{D}_N$ from Monte Carlo simulations.

It is possible to recognize three different operating conditions in the network:

- *Congestion-free state*: The network is lightly loaded, and it is possible to allocate resources to all users such that they all meet their target distortion.
- *Congestion state*: There are N_L calls in the network and new calls arriving are denied service and dropped from the system.
- *Pseudo-congestion state*: New calls are accepted but at least one call cannot be granted resources to operate at the target distortion. Because we are dealing with an arbitrary set of source coders, in the pseudo-congestion state some calls may be operating at target distortion (while the rest not). As such, the name for this state conveys the notion that in the cases corresponding to this state it is not possible to allocate resources to all users such that they all meet their target distortion and the network is not congested.

The *pseudo-congestion state* is defined as the operational state when one or more users are operating at a source rate lower than that corresponding to the target distortion goal. Defining the probability of (11.12) failing as:

$$P_{out} \stackrel{\Delta}{=} P\left[\sum_{i=1}^{N} \Psi_i(\hat{b}_i) > 1 - \epsilon\right], \tag{11.21}$$

the steady-state probability of this state is given by:

$$P_{sc} = \sum_{N=0}^{N_L} P_{out} \, q_N = \frac{\sum_{N=0}^{N_L} P_{out} \phi_N}{\sum_{i=0}^{N_L} \phi_i}, \tag{11.22}$$

where $\hat{b}_i$ is the allocation to user i such that it meets its target distortion.

The congestion state corresponds to the event when new incoming calls need to be blocked, because $N = N_L$ and a new call arrives. The probability of this event is queuing theory's *blocking probability*. From the PASTA property (Poisson Arrivals See Time Averages) [21], this probability is given by $P_b = q_{N_L}$.

From these definitions, we see that the maximum number of users and the blocking probability can be determined from the maximum tolerable expected distortion. Also the probability of operating in the pseudo-congestion state depends on the traffic load and each user's target distortion and source encoder R-D performance. Note that in (11.22), P_{out} corresponds to the outage probability in $M/M/\infty$ models, [17, 18]. The outage states in [17, 18] have now been divided into a congestion state and a pseudo-congestion state, where, for the latter, new calls are still admitted, with the probability of at least one call operating with a distortion larger than the minimum target given by P_{sc}. The system we are studying here performs statistical multiplexing by shaping the traffic in such a way that it avoids congestion (or outage) by smoothly increasing source distortion up to the point where a maximum expected distortion is reached. Further discussion of this issue, for the case of real-time MPEG-4 FGS video, follows in the next subsection.

11.2.5 Performance Results

Performance for this application can be obtained from Monte Carlo simulations based on a CDMA system carrying video calls. The simulations are set so that the system can support a reasonable number of video calls at a good quality (details will be discussed shortly but this setup design aims at supporting roughly a dozen video calls with average peak signal-to-noise ratio (PSNR) of at least 36 dB). Roughly half of the calls use the "Foreman" sequence and the rest use "Akiyo" (QCIF resolution, 30 fps, with 29 P-frames between each I-frame). To ensure that all user's sequences were desynchronized with respect to each other, every sequence starts at a random frame and follows a circular loop, i.e. the first frame follows the last frame once the sequence end is reached. The discontinuity at the point of looping back is a welcome artifact as it simulates an abrupt scene change. The sequences are encoded using an MPEG-4 FGS coder (see Chapter 2 and [10]). To transmit the source-encoded sequence over a noisy channel, the bit stream is partitioned into packets as an error resiliency feature. Those packets for which errors are detected at the receiver after the error control coding block are discarded and concealed. The concealment replaces lost packets by using the corresponding correctly received previous packet and then applying motion compensation as necessary. In effect, this error concealment approach implements what is often called *zero-motion error concealment* (ZMEC), which conceals the macroblocks in a lost packet by holding over the macroblocks in the colocated positions in the previous frame. The variable-rate channel coder is an RCPC code with mother code rate 1/4, $K = 9$, and puncturing period 8. The code family and main parameters are defined in [11]. The system bandwidth is 40 MHz with available transmit rates of 5000, 2500, 1250, 625, 312.5, 156.25, 78.125, and 39.0625 kbps. This corresponds to a choice for possible VSFs similar to the orthogonal variable spreading factor (OVSF) chosen for the UMTS standard [22]. Each user requests resources so as to achieve a target PSNR of at least 36 dB, corresponding to reasonably good quality for both high and low motion sequences. Recall that the target BER is set to 5×10^{-6} for the base layer and 10^{-5} for the enhancement layer, which corresponds to a simple form of UEP. For the solutions that could adapt the target SINR, the possible values are 1.93, 1.76, 1.63, 1.47, 1.35, 1.16, 1, and 0.81 dBs. For these target SINRs and in order to guarantee the target channel BER, the corresponding available channel coding rates are 8/16, 8/17, 8/18, 8/20, 8/21, 8/24, 8/27, and 8/32 for the base layer and 8/16, 8/17, 8/18, 8/19, 8/20, 8/23, 8/26, and 8/31 for the enhancement layer. Other simulation parameters are $\sigma^2 = 10^{-6}$ and $\epsilon = .1$, unless otherwise noted.

We consider results for three different systems. With "No Adaptation," the calls request resources so as to achieve some quality level but cannot perform any adaptation. In "Transmit Rate Adaptation," calls can change transmit rate by changing source encoding rate with a fixed target SINR using Algorithm 11.1. Lastly, with "Full Adaptation" both transmit rate and target SINR are adapted and are allocated using Algorithm 11.2. The systems with no

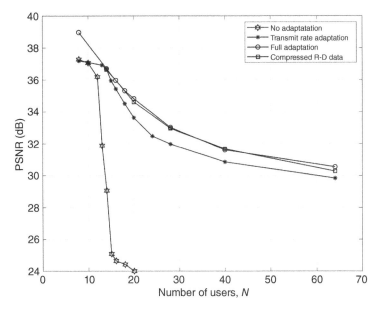

Figure 11.10 Comparison of three CDMA systems supporting video calls.

adaptation and with only transmit rate adaptation have the channel coding rate fixed with the highest possible values (which will lead to a requirement for higher operating target SINRs). Figure 11.10 shows the simulation result, where the number of users in the system is gradually increased and performance is measured through the PSNR averaged over all calls and frames. The performance of the system with no adaptation rapidly degrades as the number of calls increases. This behavior was reported in [15]; it contradicts the accepted notion that CDMA systems exhibit graceful performance degradation with an increasing number of active calls, an idea called "soft capacity." While the notion of soft capacity for CDMA systems is true from the perspective of transmission performance measures, such as BER, when this degradation is applied to a real-time source and its source encoder operation, the effect is the observed rapid decline in performance. As more users are admitted into the system, it eventually becomes impossible to set the powers at the levels needed to meet all the end-to-end performance goals given by the combined target SINRs and condition (11.12). When this happens, each mobile user becomes constrained by the power amplifier dynamic range limit, which leads to a decrease in their SINRs and consequent quality degradation due to the BER increase. The simulations for this system stopped at 19 calls; at that point, the degradation was so severe that the error concealment scheme essentially kept reproducing a frozen frame with no change in the sequence. Also, Figure 11.10 shows that both adapted systems are able to achieve both the target SINR and condition (11.12) for a larger number of users and that as more users are admitted into the system, the distortion increases smoothly allowing an increase of roughly three times in the number of calls. We see that "Full Adaptation" outperforms "Transmit Rate Adaptation" by roughly 0.8 dB in most of the operating points.

In Section 11.2.3, we discussed how to reduce the communication overhead necessary to send each call's source encoder R-D data. Figure 11.10 also includes, labeled as "Compressed R-D data," the simulation results for the same system with full adaptation but with the R-D performance information sent using the low overhead scheme described in Section 11.2.3. In this case, the required overhead is equal to only five bytes per frame, one byte for each of the three distortion values and two to represent the number of bits in the base layer (rate data). In contrast, full overhead would mean sending 64 R-D values per frame because in the system setup there are eight possible channel coding rates and eight possible target SINRs. What Figure 11.10 cannot show is that some simulation outcomes exhibit errors in estimating distortion when using the reduced R-D representation as high as 10%; however, the algorithm is robust enough that there is negligible performance loss.

Both adaptive systems allow an increase in distortion that is smooth and controllable. This is because channel-induced errors are kept at a small and perceptually acceptable value while distortion mostly follows the predictable rate-distortion function. This manifests mostly as a gradual blurring of the video frames. This is not the case for the system with no adaptation. In this case, the increase in distortion is a consequence of the uncontrolled increase in the BER and the associated random effects from increased channel-induced errors which are subjectively more annoying (mostly appearing as noticeable blocking artifacts and freezing of frame sections). This observation is illustrated in Figures 11.11–11.16 showing representative frames. Figures 11.11–11.13 show results for the system with no adaptation. The increase in BER and channel errors creates artifacts that are clearly noticeable and, in many cases, affect understandability of the frame content. As expected, these artifacts are more frequent as the number of ongoing calls increases. Figures 11.14 and 11.16 show results when using Algorithm 11.2. In Figure 11.14, the blurring associated with the smooth increase in distortion starts to become noticeable, especially in the region

Figure 11.11 A frame from the sequence "Foreman" when the network operates with no adaptation and there are 13 ongoing calls in the network.

Figure 11.12 A frame from the sequence "Foreman" when the network operates with no adaptation and there are 14 ongoing calls in the network.

Figure 11.13 A frame from the sequence "Foreman" when the network operates with no adaptation and there are 18 ongoing calls in the network.

Figure 11.14 A frame from the sequence "'Foreman" when the network operates with Algorithm 11.2 and there are 30 ongoing calls in the network.

Figure 11.15 A frame from the sequence "Foreman" when the network operates with Algorithm 11.1 and there are 30 ongoing calls in the network.

Figure 11.16 A frame from the sequence "Foreman" when the network operates with using Algorithm 11.2 and there are 60 ongoing calls in the network.

of the eyes and eyebrows. Nevertheless, there are no annoying artifacts or important loss of understandability. In Figure 11.16, we can see how the blurring increases with the number of calls. Finally, in Figure 11.15, we see the result of Algorithm 11.1 with the setup as in Figure 11.14. As expected from the objective measurements, there is some degradation in the performance of Algorithm 11.1 compared to Algorithm 11.2, but many of the behavioral properties hold. Lastly, on the subject of smooth increase of distortion, note that this

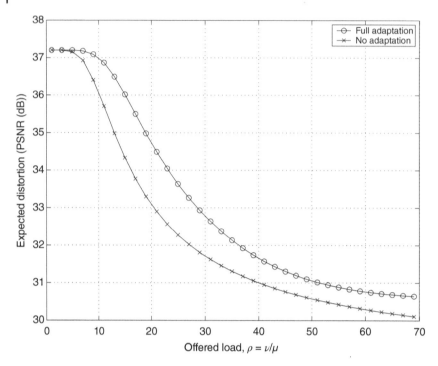

Figure 11.17   Expected distortion (PSNR in dB) as a function of the offered load.

graceful degradation of quality being observed here for the JSCC schemes is a theme that has been noted numerous times throughout this book and one of the most useful properties present in JSCC schemes.

Since the problem at hand focuses on real-time communications, it is not enough to evaluate results by fixing the number of calls. It is also important to consider a dynamic system and evaluate performance as a function of the traffic load. The results in Figures 11.17–11.20 focus on this evaluation approach, following the system setup described in Section 11.2.4 with $\mu = 3$ and $D_{max} = 30$ dB. Figure 11.17 shows the expected distortion per call as a function of the offered load. We see that the "Full Adaptation" system can support an offered load roughly 50% larger than the "No Adaptation" system. Figure 11.18 shows the probability of pseudo-congestion as a function of the offered load for different values of ϵ. We see here that there is a range of offered loads where the probability of pseudo-congestion transitions from 0 to 1. This is the typical region of focus when studying the Erlang capacity of $M/M/\infty$ CDMA networks [17, 18]. For larger offered loads, although the probability of pseudo-congestion is 1 (equivalently the outage probability in [17, 18]) the system is still able to accept more calls. Showing expected distortion as a function of the probability of pseudo-congestion, Figure 11.19 highlights the fact that extending operation into the pseudo-congestion state occurs with a smooth and controlled degradation of quality. Finally, Figure 11.20 shows blocking probability as a function of the offered load when call admission control for both the "Full Adaptation" and the "No Adaptation" systems is performed so that the average distortion when the number of calls is maximum does not exceed 30 dB. For this performance measure, the difference

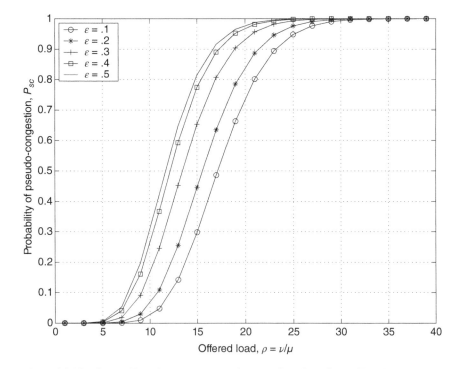

Figure 11.18 Probability of pseudo-congestion as a function of the offered load.

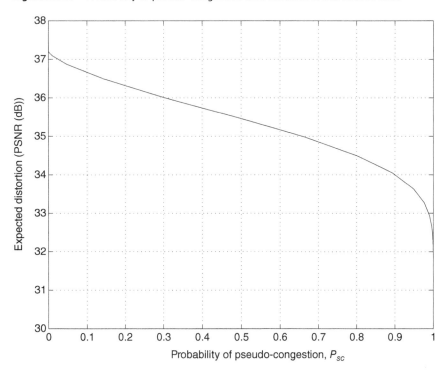

Figure 11.19 Expected distortion (PSNR in dB) as a function of probability of pseudo-congestion.

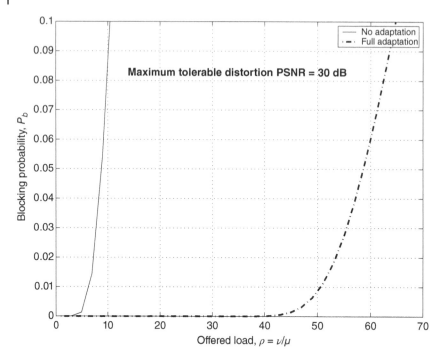

Figure 11.20 Blocking probability as a function of the offered load.

between a smooth increase in distortion, as with the "Full Adaptation" system, and the steep increase, as for the "No Adaptation" system, translates into the "Full Adaptation" system supporting, for the same limiting blocking probability more than five times the offered load than the "No Adaptation" system.

11.3 Source–Channel Coding and Cognitive Radios

The CR paradigm evolved from developments in software-defined radio. A software-defined radio is a wireless device where most of the processing functions, save for a limited number of components such as the RF power amplifier, are implemented as software algorithms running on digital signal processing devices. The CR paradigm extends the adaptation of the wireless device by executing a cognitive cycle of operations known as the "observe–decide–act" (ODA) loop. The loop's first function is to observe the wireless environment to gain awareness of its dynamic status. At the core of a CR module, an algorithm takes as input the awareness of the wireless environment and decides on the adaptation of the device's operating parameters. Examples for these algorithms are machine learning algorithms, artificial intelligence algorithms, game theory algorithms, and statistical signal processing algorithms.

Since the inception of CRs, considerable work has centered on techniques to sense the RF medium, interpret its level of usage, and decide on how to share the wireless medium with other wireless devices. The development of CR technology has been limited mostly to the Physical and Data Link layers. However, awareness of the wireless environment can be shared with all layers, and adaptation of layer parameters can be integrated and dictated from the ODA loop. In this way, the CR framework can dynamically adapt different components of a software-defined radio in an integrated way. In this section, we will study how using cross-layer CR functionalities improves performance while retaining a useful and flexible modular configuration. At the same time, the main challenge with cross-layer CR designs is the increased complexity in the action and observation spaces used in the ODA cycle. We will discuss approaches to reduce the complexity of cross-layer CR.

The most studied CR application scenario has two coexisting wireless networks. The *secondary network* (SN) seeks to operate on the same radio spectrum owned by the *primary network* (PN). SN operation is constrained to follow coexistence rules so that PN communications are essentially unaffected by the SN. We will consider the scenario of an SN carrying real-time multimedia traffic. We consider the case where the secondary users (SUs) in the SN communicate using an *underlay* spectrum access approach, in which the PN and SN use the same spectrum band simultaneously, but SN transmissions are unobtrusive to the PN by ensuring that they do not exceed an interference limit set by the PN [23]. The spectrum band under consideration (assumed to be a channel used by a single transmitter–receiver pair in the PN) has a bandwidth W. The SN nodes communicate with each other using DS-CDMA. The SN nodes are CRs, of which there will N active ones. The discussion will concentrate on the uplink of the PN and SN. The transmission over the SN will follow a cross-layer CR scheme that jointly adapts transmit bit rate (spreading factor) and source and channel coding rate at the nodes (concatenated JSCC) in the SN for transmission of a real-time multimedia source. The overall goal for the SN nodes is to minimize the average end-to-end distortion.

Figure 11.21 illustrates the block diagram for a typical modern wireless device. It is a complex hardware–software system formed by several processors and software subsystems. The networking software likely runs the processes for different protocol layers on different processors (e.g. a main CPU and a baseband radio processor). This results in software modules lacking a framework to allow different layers to exchange the data and commands needed to implement cross-layer solutions. Figure 11.21 also shows how CR provides a framework that enables these cross-layer techniques. The ODA cycle is applied with a scope encompassing all layers of the networking stack. The wireless device continuously executes an ODA cycle of observations to gain awareness of the wireless environment and adapt. In this cross-layer framework, the environment includes all layers of the networking stack, from the application layer to the physical layer.

Typical CR applications operate only on the physical and medium access control (MAC) layers and so the ODA cycle is most frequently found in the baseband modem. In the case at hand now, the ODA cycle is promoted from the baseband radio processor to a "cognitive cycle process" (CCP) running in the kernel of the main operating system (OS). The ODA operations across layers are integrated into the CCP and enabled through established methods of interaction between the OS kernel and other software–hardware system layers. For the lowest networking layers, typically implemented in a co-processor, interaction with the CCP is done by modifying the device drivers to include new application programming interface (API) function calls. The CCP interacts with other OS kernel processes (e.g. those in the networking stack) by modifying the library of functions that provide the abstraction to control them (the sockets) to exchange the needed variables. For processes associated with higher layers, such as the application layer, interaction with the CCP is through the

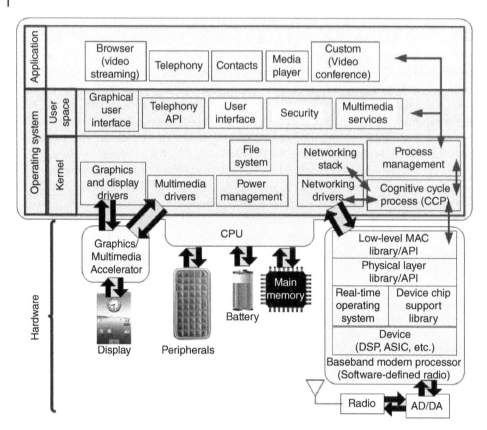

Figure 11.21 A typical system architecture for a wireless device, including the CCP to enable cross-layer technology.

OS Process Management. While this architecture seamlessly integrates cross-layering into the wireless device, it has the added advantage of maintaining a modular organization and of being agnostic to the specific algorithms implementing the cognitive cycle, to the extent that these algorithms could be modified or updated as done with any OS process. Also, the architecture is scalable, because the CCP is only involved with parameter adaptation but not with the actual processing done in each layer.

All transmissions in the application scenario are assumed to be over a flat quasi-static fading channel, constant during a sensing and transmission period. Since the SUs transmit using DS-CDMA, each transmission will add some interference to the PN. Therefore, coexistence between the two networks is determined by the condition that the SUs' interference is such that the PN SINR does not go below a set limit β_0:

$$\beta_0 \leq \frac{G_0^{(p)} P_0}{\sigma^2 + \sum_{j=1}^{N} G_j^{(p)} P_j}, \tag{11.23}$$

where P_0 is the primary user (PU) transmit power, $G_0^{(p)}$ is the PU transmission channel gain, $G_j^{(p)}$ is the channel gain from the jth SU to the primary base station (PBS), P_j is the jth SU's transmit power, and σ^2 is the background noise power. If we assume that PUs transmit using adaptive modulation and coding (a widespread technique in modern wireless communication systems, also known as link adaptation, where transmitters adapt based on the link SINR the channel coding rate and the modulation order to maximize throughput), the SUs will be able to estimate the interference they create on the PN, and from here derive the channel gains, through the use of the procedure explained in [24]. Similarly, SU resource allocation needs to ensure QoS requirements, leading to the power assignment condition (analogous to one already seen):

$$\beta_i \le \frac{(W/r_i^{(s)})G_i^{(s)}P_i}{\sigma^2 + G_0^{(s)}P_0 + \sum_{j \ne i}G_j^{(s)}P_j}, \quad i = 1, 2, ..., N, \tag{11.24}$$

where $r_i^{(s)}$ is the transmit rate, $G_i^{(s)}$ and P_i are the SU's channel gain and transmit power, $G_0^{(s)}$ is the PU's channel gain to the secondary base station (SBS) and β_i is the required SINR for the ith SU. As discussed earlier in this chapter, the optimal power assignment for the SU is the one such that all the inequalities (11.24) are met with equality:

$$P_i = \frac{\Psi_i(\sigma^2 + G_0^{(s)}P_0)}{G_i^{(s)}(1 - \sum_{j=1}^{N}\Psi_j)}, \quad i = 1, 2, ..., N, \tag{11.25}$$

where, as earlier,

$$\Psi_i = \left(1 + \frac{W}{r_i^{(s)}\beta_i}\right)^{-1}, \tag{11.26}$$

From here, it can be seen that a condition for the power assignment is $\sum_{i=1}^{N}\Psi_i < 1$. This condition is considered in practical terms as:

$$\sum_{i=1}^{N}\Psi_i = 1 - \epsilon, \tag{11.27}$$

where ϵ is a small number chosen at design time to control the maximum power assignment. Following this, and with the power assignment (11.26), the SINR constraint from the PN becomes

$$\sum_{j=1}^{N}\alpha_j\Psi_j \le 1, \text{where } \alpha_j = \frac{G_j^{(p)}(\sigma^2 + G_0^{(s)}P_0)}{G_j^{(s)}(G_0^{(p)}P_0/\beta_0 - \sigma^2)} + 1. \tag{11.28}$$

The application case in the previous section serves as a good example of the need to optimize the transmission of real-time multimedia using end-to-end performance measures. Then, the transmission of SUs will adapt parameters seeking to minimize the average end-to-end distortion, measured as the expected source coding distortion plus channel-induced distortion. It is assumed that distortion is measured using MSE. By assuming a synthetic Gaussian source with zero mean and unit variance as a generic representation of a variety of practical sources, as seen in Chapter 1 the R-D function is $D_s(x_i) = 2^{-2x_i}$, where x_i is the source coding rate to be determined for the ith SU. To simplify the analysis, we assume that the channel-introduced distortion D_c can be approximated as a constant that does not depend on the source encoding rate. Then, the end-to-end distortion can be expressed as:

$$D_i(x_i, \beta_i) = D_cP_e(\beta_i) + 2^{-2x_i}(1 - P_e(\beta_i)), \tag{11.29}$$

where $P_e(\beta_i)$ is the probability of channel errors, hence a function of the SINR at the receiver. Note again that when any two of the source encoding rate, x_i, the channel coding rate, R_i, and the transmit rate, $r_i^{(s)}$, are given, the third parameter is directly determined because the three are linked through the relation $x_i = r_i^{(s)} R_i$. As a consequence of this and (11.24), D_i becomes a function of the parameter pair $(r_i^{(s)}, \beta_i)$. Ultimately, finding the optimal JSCC rate allocation, transmit bit rate, and power assignment becomes the problem of finding the optimal parameter pair $(\hat{r}_i^{(s)}, \hat{\beta}_i)$ for each SU to minimize their own distortion, while the SINR constraints (11.27) and (11.28) are satisfied.

A CR framework is introduced to solve this problem by considering the joint decision process of finding the pair $(\hat{r}_i^{(s)}, \hat{\beta}_i)$ for each SU as an N-player general-sum stochastic game. Such a game is defined by the set:

$$\Gamma = \{S, A_1, \dots, A_N, c_1, \dots, c_N, p\},$$

in which S is the state space, $A_i = \{(r_i, \beta_i)\}$ is the independent action that a SU takes, c_i is the immediate cost that a SU pays as a function of D_i for its decision, and p represents the conditional transition probability between a state s to a state s', $P(s'|s, a_1, \dots, a_N)$ (the states will be defined soon). We presume that SUs treat other SUs as part of the environment and do not know anothers' actions or the effect of joint actions on states. Let the policy π_i denote the probability for the ith SU to take action a_i. The optimal solution is approached with the SUs repeatedly adjusting their actions to minimize the discounted expected cost:

$$V_i(s, \vec{\pi}) = \sum_{t=0}^{\infty} \gamma^t E(c_t^{(i)} | \vec{\pi}, s_0 = s), \qquad i = 1, 2, \dots, N, \tag{11.30}$$

where $\vec{\pi} = (\pi_1, \pi_2 \dots, \pi_N)$ is the strategy vector and $s_0 = s$ is the initial state. The optimal solution can be obtained through iterative search using Bellman's equation by all the SUs:

$$V_i(s, \vec{\pi}^*) = \min_{\vec{a}} [c(s, \vec{a}) + \gamma \sum_{s'} P(s'|s, \vec{a}) V_i(s', \vec{\pi}^*)]. \tag{11.31}$$

Further, $V_i(s, \vec{\pi}^*)$ in (11.31) can be approached by the Nash Q-function following Nash Q-learning as in [25]. The Q-value is formed as:

$$Q_{t+1}^i(s, \vec{a}) = (1 - \alpha_t) Q_t^i(s, \vec{a}) + \alpha_t [c_t^i(s, \vec{a}) + \gamma Q_t^{i*}(s')],$$

where α_t is the learning rate, $0 < \alpha_t(s, \vec{a}) < 1$, and $Q_t^{i*}(s')$ is the ith SU's Q-value corresponding to the stage Nash equilibrium (NE) $(\pi_1^*(s'), \dots, \pi_N^*(s'))$ in the new state s' after $\vec{a}$ is taken.

This framework has led to implementing the cross-layer joint adaptation of parameters using a reinforcement learning algorithm known as *Q-learning*. Q-learning is a learning algorithm that belongs to the same family as dynamic programming algorithms used at other times in this book, the difference between them being that in dynamic programming the environment state transition probabilities are known, while in the use case of Q-learning these probabilities are not known at the time of running the algorithm and, instead, are implicitly learned in the process of finding the solution to the optimization problem. To reduce the computational complexity, the original problem with constraints (11.27) and (11.28) can be decomposed by introducing Lagrangian multipliers λ_1 and λ_2 and forming a group of independent optimization problems for each SU:

$$\{(\hat{r}_i^{(s)}, \hat{\beta}_i)\} = \arg\min_{r_i^{(s)}, \beta_i} \sum_{i=1}^{N} [D_i(r_i^{(s)}, \beta_i) + (\lambda_1 + \lambda_2 \alpha_i) \Psi(r_i^{(s)}, \beta_i)]. \tag{11.32}$$

With this, the joint action stochastic process becomes a group of independent Markov decision processes, if λ_1 and λ_2 are determined. Consequently, this means that searching for
$Q_t^{i*}(s')$ will only be dependent on each SU's own action:

$$Q_{t+1}^i(s, a_t^i) = (1 - \alpha_t)Q_t^i(s, a_t^i) + \alpha_t[c_t^i(s, a_t^i) + \gamma \min_b Q_t^i(s', b)], \tag{11.33}$$

In the learning game, the system state is defined as $s = (I, L)$, where from (11.27) and
(11.28),

$$I_t = \begin{cases} 0, & \text{if } \sum_{i=1}^{N} \Psi_i(r_t^{(i)}, \beta_t^{(i)}) < 1 - \epsilon \\ 1, & \text{otherwise,} \end{cases} \tag{11.34}$$

represents the interference state between the cognitive secondary nodes and

$$L_t = \begin{cases} 0, & \text{if } \sum_{i=1}^{N} \alpha_i \Psi_i(r_t^{(i)}, \beta_t^{(i)}) \leq 1 \\ 1, & \text{otherwise.} \end{cases} \tag{11.35}$$

represents the interference state on the PN. Next, the cost function is defined as:

$$c_t^{(i)}(a_t) = \begin{cases} M, & \text{if } I_{t+1} + L_{t+1} > 0 \\ D(r_t^{(i)}, \beta_t^{(i)}), & \text{otherwise,} \end{cases} \tag{11.36}$$

where M is a number larger than the maximum possible distortion (used to introduce maximum cost when either of the interference goals is not satisfied).

Having defined the general-sum stochastic game and the learning algorithm to find the
solution, the learning algorithm followed by each SU is summarized in Algorithm 11.3.

Algorithm 11.3 Learning algorithm for cross-layer CR resource allocation.

Initialization: $Q_0^i = 0$ for all the SUs

    for $t < t_{\max}$ **do**

      for SU_i, $i = 1, \ldots, N$ **do**

        Select the action $a_t^i = \arg\min_{a_t^i} Q^i(s_t^{(i)}, a_t^i)$

        Update the state s_{t+1}^i (11.34), (11.35) and the cost c_t^i (11.36)

        Update Q-value $Q_{t+1}^i(s_t, a_t)$ with (11.33)

      end for

    end for

Here, the cross-layer scheme differs from a physical-layer-only scheme in that the former operates an ODA cycle with action space from $S_r^{(i)} \times S_\beta^{(i)}$, while the latter operates an
ODA cycle with action space from $S_r^{(i)}$. The cross-layer CR scheme results in much better
performance at the cost of increased complexity.

This can be addressed by taking advantage of the properties of the data used by each SU
to internally represent the learned long-term cost/reward function that constitutes wireless
environment awareness. In the algorithm described here, these are the Q-tables, with

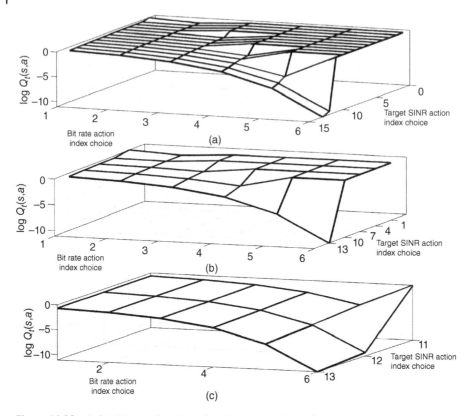

Figure 11.22 A Q-table as a function of action space indexes for a given state.

their constituent elements $Q_t(s, a)$. By calculating Q-values as in (11.33), the Q-learning algorithm (a type of reinforcement learning) described earlier explores the long-term discounted rewards/costs resulting from a sequence of action choices. The results from the learned experience in terms of learned Q-values are stored in a Q-table, whose size grows linearly with the set size of any of the action variables ($S_r^{(i)}$ or $S_\beta^{(i)}$ in our case). Figure 11.22a shows a typical Q-table, in this case for fixed state $s = (0, 0)$, as a function of the two action variables. Here the entries of the Q-table, the Q-values in the table, are shown as the height of a surface. Figure 11.22a illustrates that a typical Q-table presents smooth variations with the choice of different actions. These observations lead to a technique to reduce the increased complexity in cross-layer CRs.

To reduce implementation complexity, a modified algorithm performs the Q-learning Algorithm 11.3 in two distinct stages. Let $A = S_r^{(i)} \times S_\beta^{(i)}$ denote the set of all possible actions. During stage 1 (illustrated in Figure 11.22b), the learning algorithm operates with a subset of all possible action choices, denoted $A^{(1)}$. This subset represents with a coarser resolution the complete set of possible actions. As a result of stage 1, the learning algorithm at each SU derives an approximate Q-table, which is nonetheless still useful to identify an approximate minimum Q-value. Figure 11.22b illustrates the subset of actions used during stage 1 as the subset formed from some of the original actions. The subset of actions chosen for stage 1 is formed by eliminating other actions corresponding to parameter choices similar to the selected actions; the subset involves a coarser representation of the surface made from the Q-table values. For this example, the actions $\beta_i \in S_\beta^{(i)}$ were assumed to be indexed according to increasing values of β_i, with indexes from 1 to 13 (seen in Figure 11.22a), $A = \{\beta_{i_1}, \ldots, \beta_{i_{13}}\}$. The actions chosen to be used in stage 1 are those with indexes 1, 4, 7, 10, and 13 (seen in Figure 11.22b), $A^{(1)} = \{\beta_{i_1}, \beta_{i_4}, \beta_{i_7}, \beta_{i_{10}}, \beta_{i_{13}}\}$. In this way, the selected actions will approximately represent the results (Q-values) for the nonselected similar action values.

The reduced complexity algorithm performs a second stage of Q-learning after completing the learning process in stage 1. During stage 2, the same learning algorithm is executed, this time operating in another subset of action choices, $A^{(2)}$, which depends on the approximate minimum Q-value calculated in stage 1. $A^{(2)}$ is formed by the action corresponding to the approximate minimum Q-value identified in stage 1 and the actions similar in value to it that were not in $A^{(1)}$. Therefore, while stage 1 obtains an approximation for the Q-tables good enough to identify an approximate minimum, stage 2 searches for a more accurate minimum in the neighborhood of the approximate minimum. This stage is illustrated in Figure 11.22c, where the minimum from stage 1 was assumed to correspond to action β_i with index 13 and, consequently, $A^{(2)} = \{\beta_{i_{11}}, \beta_{i_{12}}, \beta_{i_{13}}\}$.

In practice, the design task of selecting actions from A for $A^{(1)}$ and $A^{(2)}$ can be approached using any of the methods for the design of quantizers for signals of two variables, such as images, which were reviewed in Chapter 2. We will next use a simple scalar quantizer design that is based on the frequency for each action in A to be chosen as the solution, as measured through Monte Carlo simulations. From this information, those actions with highest frequency were chosen for $A^{(1)}$. The design of the possible subsets $A^{(2)}$ followed the criteria described earlier. Finally, Algorithm 11.4 summarizes the reduced complexity learning algorithm.

Algorithm 11.4 Two-stage learning algorithm for reduced complexity cross-layer CR

Initialization:

- $Q_0^i = 0$ for all the SUs.
- Assign to all SUs the subset $A^{(1)}$ of all possible actions $S_r \times S_\beta$.

Stage 1: Run Algorithm 11.3 at all SUs.
Stage 2:

- Assign to each SU a subset of all possible actions given by the solution action (the action corresponding to the minimum Q-value) in stage 1 and the actions, not used in stage 1, with parameters similar to the solution action.
- Run Algorithm 11.3 at all SUs.

The performance of the different algorithms can be compared through Monte Carlo simulations. In the simulation scenario, a single channel had a bandwidth of 10 MHz. The system noise power was set to −60 dBm. The primary user limited SINR to 10 dB and had a transmit power of 10 dBm. The primary user and the SUs were placed randomly around their respective base stations within a circle of radius 250 m. Channel gains followed a log-distance path loss model with path loss exponent equal to 2.8. The SN was based on a VSF DS-CDMA system using binary phase shift keying (BPSK) modulation. The SUs used a RCPC channel code with mother code rate 1/4, constraint length $K = 9$, and puncturing period 8. For the learning algorithm, all the SUs set a learning rate $\alpha = 0.1$ and a discounting factor $\gamma = 0.4$.

We can compare the next three different systems: "Single Stage," "Two Stage," and "Physical layer only." "Single Stage" implements the cross-layer CR Algorithm 11.3 with possible action choices {9.77, 19.53, 39.06, 78.13, 156.25, 312.5}Kbps for transmit bit rate and {8/9, 4/5, 2/3, 4/7, 1/2, 4/9, 4/10, 4/11, 1/3, 4/13, 2/7, 4/15, 1/4} for channel coding rates (corresponding each to a different target SINR choice). The "two Stage" system implements the cross-layer CR Algorithm 11.4. For stage 1, it was found that the best action choices were {9.77, 39.06, 78.13, 312.5}kbps for transmit bit rate and {8/9, 4/9, 4/15, 1/4} for channel coding rates. During stage 2, the transmit bit rate action choices were {9.77, 19.53}kbps, {19.53, 39.06}kbps, {78.13, 156.25}kbps and {156.25, 312.5}kbps, when the resulting action from stage 1 was 9.77 kbps, 39.06 kbps, 78.12 kbps, or 312.5 kbps, respectively. During stage 2, the possible choices of channel coding rate actions were {8/9, 4/5}, {2/3, 4/7, 1/2, 4/9, 4/10, 4/11, 1/3, 4/13}, {2/7, 4/15}, and 1/4 when the resulting action from stage 1 was 8/9, 4/9, 4/15 or 1/4, respectively. The "Physical layer only" system provides a layered scheme benchmark by executing Algorithm 11.3 but only adapting the transmit bit rate.

The performance can be evaluated based on three metrics: average SUs end-to-end distortion, congestion rate in the SN, and average number of iterations needed by Algorithms 11.3 and 11.4 to reach a solution. In the evaluation, congestion rate is the percentage of cases for which one or both of the SINR constraints would be violated when all SUs remain within acceptable levels of distortion. For the physical layer only scheme, the channel coding rate was fixed at 1/2. All three performance measurements were evaluated as a function of the number of SUs in the network. The maximum number of SUs was set to 20 because this corresponded in the best case to a congestion rate of 15%, a value that can be considered borderline large for networks.

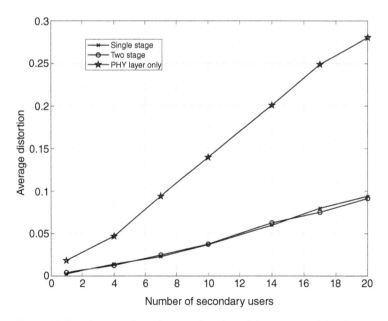

Figure 11.23 Average distortion as a function of the number of SUs in the network.

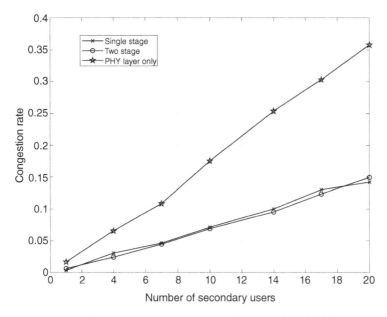

Figure 11.24 Congestion rate as a function of the number of SUs in the network.

The results are shown in Figures 11.23–11.25 and show that while the cross-layer schemes notably outperform the physical layer only scheme (two and a half to three times distortion reduction and support for more than twice the number of SUs for a given target congestion rate), the advantage in performance comes at the expense of an increase in the number of iterations, as seen in Figure 11.25 (e.g. six times increase in the number of iterations for 10

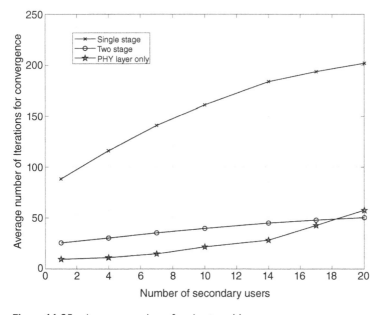

Figure 11.25 Average number of cycles to achieve convergence.

SUs). This issue is effectively addressed with the reduced complexity two-stage cross-layer CR solution. Indeed, the low-complexity solution in Algorithm 11.4 achieves an important reduction in the number of iterations. Specifically, when comparing the cross-layer schemes, the two-stage scheme reduces the number of iterations by factors ranging between three and four times, approximately. The advantage of the two-stage approach, where each stage operates on a reduced action space, is highlighted by the fact that at 20 SUs, the two-stage solution requires fewer iterations than even the physical layer only scheme. More importantly, the two-stage solution achieves the same average end-to-end distortion and congestion rate performance as the single-stage schemes, thus achieving a notable reduction in complexity without sacrificing performance.

11.4 Design of JSCC Schemes Based on Artificial Neural Networks

An exciting recent development in JSCC is the use of artificial neural networks for the design of JSCC communication systems. We explained two instances of this design approach in Chapter 9. We now revisit this subject to go beyond the specifics of analog or hybrid digital–analog JSCC and provide a wider view of the connection between JSCC and neural networks. General concepts of neural networks were presented in Chapter 9. The key paradigm change introduced with neural networks is that JSCC design is no longer following an analytical procedure based on existing models (e.g. of channel impairment) but, instead, is learned by the neural network based on examples of input values and corresponding outputs. In this emergent learning-based approach to JSCC design, the same general neural network structure would result in different codes when presented during learning with input–output examples with different underlying statistical behavior. Nevertheless, a benefit associated with this learning-based approach is that appropriately designed and trained neural networks have a "generalization" capability, which as opposed to a "memorization" capability, means that the neural network can accomplish the task for which it was trained when presented with never-before-seen input data (of course, this new data still needs to adhere to the statistical characteristics of the examples presented during training). In the context of JSCC design, the generalization capability means that the learned code would be effective even for combinations of source data and channel scenarios that were not present in the training data.

We will consider in this section the work of [26], a fully digital JSCC system that corresponds to concatenated JSCC but is implemented through a neural network. The neural network has the goal of allocating a fixed bit budget between source compression and error protection. In this JSCC system (Figure 11.26), a source message V^n formed as a random

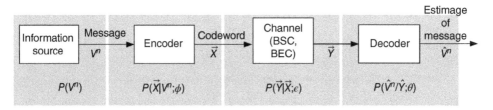

Figure 11.26 Block diagram abstracting the JSCC system with the model for the key variables.

sequence of symbols from a discrete source associated with a finite alphabet $\mathcal{V}$ is encoded into a codeword $\vec{X}$. The codeword $\vec{X}$ is sent through a discrete channel and received as the codeword $\vec{Y}$. Codewords in this JSCC system are sequences of m bits. After decoding $\vec{Y}$, the receiver output is the estimate $\hat{V}^n$ of the message. The relation of causality between the different variables in the JSCC system can be mathematically described through a Markov chain. Relative to the underlying phenomenon that introduces the stochastic behavior in the system (in this case, the generation of messages at the source), the Markov chain is $V^n \rightarrow \vec{X} \rightarrow \vec{Y} \rightarrow \hat{V}^n$. This Markov chain describes how the variable V^n is generated at the source, then mapped through the encoding process into $\vec{X}$ which is in turn transformed into $\vec{Y}$ after transmission through the channel. Finally, at the receiver, the variable $\vec{Y}$ is mapped into $\hat{V}^n$ through decoding.

At a root level, the design of the JSCC system entails managing the statistical behavior of the system variables to achieve a reproduction of the source message as faithful as possible given the channel effects and communication constraints. The statistics of all variables are fully described by the joint probability distribution $P(V^n, \vec{X}, \vec{Y}, \hat{V}^n)$. For the convenience of the design, we expand this probability as:

$$P(V^n, \vec{X}, \vec{Y}, \hat{V}^n) = P(V^n)P(\vec{X}|V^n)P(\vec{Y}|\vec{X}, V^n)P(\hat{V}^n|\vec{Y}, \vec{X}, V^n).$$

Because of the Markov chain relation between the variables, $\hat{V}^n$ is conditionally independent of V^n and $\vec{X}$ given $\vec{Y}$, and $\vec{Y}$ itself is conditionally independent of V^n given $\vec{X}$. Consequently, the joint probability distribution $P(V^n, \vec{X}, \vec{Y}, \hat{V}^n)$ can be written as:

$$P(V^n, \vec{X}, \vec{Y}, \hat{V}^n) = P(V^n)P(\vec{X}|V^n)P(\vec{Y}|\vec{X})P(\hat{V}^n|\vec{Y}). \tag{11.37}$$

In this expression, $P(V^n)$ describes the statistics for the source message generation. When following a learning-based design approach, it is not necessary to explicitly know the source statistics. During the design process (the learning process), the statistics of the source are implicitly presented to the JSCC system in the form of training examples, which are drawn from sampling the source. Here, it is assumed the sampling is done correctly, without introducing biases, drawing an ensemble of samples from the source that represent the source statistics. Similarly, the statistical behavior of the channel, $P(\vec{Y}|\vec{X})$, does not need to be known explicitly; instead, the neural network implicitly learns a representation of the channel behavior from examples during training. In this case, this is done by injecting noise (characterized according to the channel statistics) into the representation of the latent variable $\vec{Y}$ within the neural network model (a latent variable is not directly observable, but it is inferred from other variables that are observable). Since the JSCC system in [26] is entirely digital, the channel is also considered as discrete. Examples of discrete channel models are the binary erasure channel and binary symmetric channel (the ones considered in [26]) that were described in Section 1.4. In what follows, we denote the parameter for the channel model (the probability of error for the binary symmetric channel or the erasure probability for the binary erasure channel) as ϵ and include it in the channel model notation by writing $P(\vec{Y}|\vec{X}; \epsilon)$.

We can now see the reason of the formulation leading to (11.37). As seen in (11.37) and illustrated in Figure 11.26, in the spirit of JSCC, the joint statistical characterization of the variables in the system $(P(V^n, \vec{X}, \vec{Y}, \hat{V}^n))$ can be mathematically expressed using their Markov chain relation as a "train" of probabilities, each characterizing one of the components of the JSCC system: $P(V^n)$ for the statistics for the source message generation, $P(\vec{X}|V^n)$ for the mapping operation in the encoder, $P(\vec{Y}|\vec{X}; \epsilon)$ for the channel model, and

$P(\hat{V}^n|\vec{Y})$ for the mapping operation at the decoder. This characterization leads to encoding and decoding mappings that are stochastic (as they are defined through a probability distribution). As such, this JSCC system is built with a *stochastic* encoder and decoder, both of which, following the learning-based approach, are developed during training. The encoder in [26] is built as a neural network, similar in general structure to those in Chapter 9, with layers of convolutional and fully connected layers (the specific structure is tailored to the characteristics of the source messages). The neural network's property of being a universal function approximator is exploited. The function approximation done at the encoder, which is the generation of a codeword $\vec{X}$ given the message V^n according to the probability distribution $P(\vec{X}|V^n)$, is learned during training by setting the values for the weights and biases of the neural network. Both the weights and biases are called the *parameters* of the neural network. We will denote all the encoder parameters as ϕ and write $P(\vec{X}|V^n; \phi)$ to describe both the encoder mapping and its implementation through a neural network with parameters ϕ. For the specific implementation of the encoder, recall that its output is a codeword of length m bits. The codeword is characterized as a sequence of m independent Bernoulli random variables, and the probability distribution for the resulting random sequence is modeled with a neural network with parameters ϕ that acts as a parameterized approximator of the function $f_\phi(\cdot)$. Denoting as x_i the ith bit in the codeword $\vec{x}$, the model for the encoder can be expressed as:

$$P(\vec{X} = \vec{x}|V^n = v^n; \phi) = \prod_{i=1}^{m} \sigma(f_\phi(v^n))^{x_i}(1 - \sigma(f_\phi(v^n)))^{(1-x_i)}, \qquad (11.38)$$

where $\sigma(\cdot)$ is the sigmoid function of the output layer activation function:

$$\sigma(z) = \frac{1}{1 + e^{-x}}.$$

a common choice for neural networks acting as approximators of probability distributions. In the operation of the encoder, the codeword itself is generated by taking a sample from the random sequence with the probability distribution given by (11.38).

The learning-based approach to JSCC design entails finding the values of the parameters ϕ that configure the encoder's neural network. These parameter are obtained following a process of iterative training where, broadly speaking, each iteration consists of the following steps: *(1)* samples from a data set formed by many examples of input data and the corresponding target output are presented to the neural network, *(2)* the output of the neural network is observed, *(3)* a "cost" associated with the difference between the observed output and the target output is calculated, and *(4)* the neural network parameters (weights and biases) are updated with the goal of minimizing the cost associated with the difference between target and actual output. The training process for the neural networks in [26] follows this general approach but is written in terms of a maximization. Specifically, the training procedure in [26] aims at maximizing the mutual information between the source output V^n and the received noisy codeword $\vec{Y}$. To formulate the mutual information between V^n and $\vec{Y}$, $I(V^n, \vec{Y})$, the encoder and channel blocks in the JSCC system were logically combined into a single block by deriving the probability distribution of $\vec{Y}$ given V^n:

$$P(\vec{Y}|V^n; \phi, \epsilon) = \sum_{\vec{X} \in \mathcal{X}} P(\vec{X}|V^n; \phi)P(\vec{Y}|\vec{X}; \epsilon), \qquad (11.39)$$

where $\mathcal{X}$ is the codebook (the set of all length-m binary sequences in this case). From here, the training procedure aims at solving the maximization problem:

$$\max_{\phi} I(V^n, \vec{Y}; \epsilon, \phi) = \max_{\phi} H(V^n) - H(V^n|\vec{Y}; \phi, \epsilon)$$

$$= \max_{\phi} H(V^n) + \sum_{V^n}\sum_{\vec{Y}} P(V^n, \vec{Y}) \log\left(P(V^n|\vec{Y}; \phi, \epsilon)\right)$$

$$= \max_{\phi} H(V^n) + \sum_{V^n} P(V^n) \sum_{\vec{Y}} P(\vec{Y}|V^n; \phi, \epsilon) \log\left(P(V^n|\vec{Y}; \phi, \epsilon)\right),$$

$$= \max_{\phi} E_{(V^n)}\left[E_{(\vec{Y}|V^n;\phi,\epsilon)}\left[\log\left(P(V^n|\vec{Y}; \phi, \epsilon)\right)\right]\right], \qquad (11.40)$$

where $P(V^n|\vec{Y}; \phi, \epsilon)$ is parameterized by ϕ and ϵ, the same as $P(\vec{Y}|V^n; \phi, \epsilon)$. Because of the relation between these two conditional probabilities based on Bayes' rule, in the maximization expression, we have eliminated terms that do not depend on the maximization variables (the parameters ϕ) and have made explicit in the last equation that the first expectation is on the probability distribution $P(V^n)$ and the second expectation is on the probability distribution $P(\vec{Y}|V^n; \phi, \epsilon)$ that is calculated in (11.39).

Solving the maximization in (11.40) faces the challenge that $P(V^n|\vec{Y}; \phi, \epsilon)$ is usually not amenable to be used in the training process as outlined earlier. To address this, $P(V^n|\vec{Y}; \phi, \epsilon)$ is approximated with the distribution $P(\hat{V}^n|\vec{Y})$ that models the stochastic decoder. The decoder too is implemented through a neural network with parameters denoted θ. As with the encoder, we make explicit the parameterization of $P(\hat{V}^n|\vec{Y})$ on the decoder's model by writing $P(\hat{V}^n|\vec{Y}; \theta)$. By applying the approximation of $P(V^n|\vec{Y}; \phi, \epsilon)$ in (11.40), the maximization problem to be solved through learning becomes

$$\max_{\phi,\theta} I(V^n, \vec{Y}; \epsilon, \phi, \theta) = \max_{\phi,\theta} E_{(V^n)}\left[E_{(\vec{Y}|V^n;\phi,\epsilon)}\left[\log\left(P(\hat{V}^n|\vec{Y}; \theta)\right)\right]\right], \qquad (11.41)$$

which now incorporates both the encoder and decoder design. This implies that the training procedure will find the parameters for both the encoder and decoder.

It is important to consider the conceptual implication behind (11.39) and (11.40) as used for encoder design. Reflecting on Shannon's channel capacity theorem (reviewed in Chapter 1) and its relation to the present encoder design, recall that channel capacity is defined as the maximum of the mutual information between the channel input and output. In this encoder design, the encoder and the channel are combined into one single element. So the encoder is being designed to maximize the mutual information between the input to, and output from, the combined encoder-and-channel; the encoder is designed to maximize a "channel capacity" for a "logical" channel that is set between the source output and the actual channel output. This approach achieves an encoder design that is matched to the channel characteristics so that as much information of the message V^n is preserved as possible in the noisy received codeword $\vec{Y}$. At the same time, source compression is built into the encoder design, because codewords are designed with a fixed length of m bits. In this way, the resulting code compresses the source while simultaneously producing a codeword that has some robustness to channel impairments.

This section highlighted a new paradigm in the design of JSCC systems. While we focused on a representative example, with a more general view, we see that the universal function approximator characteristic of neural networks allow them to be applied in other capacities in JSCC. For example, the reinforcement learning approach discussed in Section 11.3 can be implemented using neural networks as function approximators of the Q-values (interested readers can find more details about this approach in [27] and [28]). Besides the universal function approximator characteristics of neural networks, their learning-based paradigm comprises valuable properties including robustness and adaptability of the resulting codes to unforeseen operating conditions (as enabled by the generalization capability of machine learning techniques) and the possibility of designing codes without needing an explicit model of the source or the channel. Moreover, the natural introduction of stochastic encoders and decoders as seen in this section points toward the possibilities that this learning-based paradigm presents new opportunities to code design and implementation.

References

1 Liu, B. and Chen, B. (2006). Channel-optimized quantizers for decentralized detection in sensor networks. *IEEE Transactions on Information Theory* 52 (7): 3349–3358.
2 Nguyen, K.C., Alpcan, T., and Basar, T. (2008). Distributed hypothesis testing with a fusion center: the conditionally dependent case. *47th IEEE Conference on Decision and Control, 2008. CDC 2008*, pp. 4164–4169.
3 Cover, T. and Thomas, J. (1991). *Elements of Information Theory*. Wiley.
4 Wyner, A.D. (1978). The rate-distortion function for source coding with side information at the decoder-II: general sources. *Information and Control* 38 (1): 60–80.
5 Pradhan, S.S. and Ramchandran, K. (2003). Distributed source coding using syndromes (DISCUS): design and construction. *IEEE Transactions on Information Theory* 49 (3): 626–643.
6 Karlsson, J., Wernersson, N., and Skoglund, M. (2007). Distributed scalar quantizers for noisy channels. *IEEE International Conference on Acoustics, Speech and Signal Processing, 2007. ICASSP 2007*, Volume 3, pp. III-633–III-636.
7 Shoham, Y. and Gersho, A. (1988). Efficient bit allocation for an arbitrary set of quantizers. *IEEE Transactions on Acoustics, Speech, and Signal Processing* 36 (9): 1445–1453.
8 Han, Z., Kwasinski, A., and Liu, K.J.R. (2006). A near-optimal multiuser joint speech source-channel resource-allocation scheme over downlink CDMA networks. *IEEE Transactions on Communications* 54 (9): 1682–1692.
9 Han, Z., Su, G.-m., Kwasinski, A. et al. (2006). Multiuser distortion management of layered video over resource limited downlink multicode-CDMA. *IEEE Transactions on Wireless Communications* 5 (11): 3056–3067.
10 Li, W. (2001). Overview of fine granularity scalability in MPEG-4 video standard. *IEEE Transactions on Circuits and Systems for Video Technology* 11 (3): 301–317.
11 Frenger, P., Orten, P., Ottosson, T., and Svensson, A.B. (1999). Rate-compatible convolutional codes for multirate DS-CDMA systems. *IEEE Transactions on Communications* 47: 828–836.
12 Kim, J.B., Honig, M.L., and Jordan, S. (2001). Dynamic resource allocation for integrated voice and data traffic in DS-CDMA. *IEEE 54th Vehicular Technology Conference*, Volume 1, pp. 42–46.

13 Sampath, A., Mandayam, N.B., and Holtzman, J.M. (1995). Power control and resource management for a multimedia CDMA wireless system. *PIMRC'95*, Toronto, Canada.

14 Alasti, M. and Farvardin, N. (2000). SEAMA: a source encoding assisted multiple access protocol for wireless communications. *IEEE Journal on Selected Areas in Communications* 18 (9): 1682–1700.

15 Chan, Y.S. and Modestino, J.W. (2001). Transport of scalable video over CDMA wireless networks: a joint source coding and power control approach. *IEEE International Conference on Image Processing (ICIP)*, Volume 2, pp. 973–976.

16 Tse, D.N.C. and Hanly, S.V. (1999). Linear multiuser receiver: effective interference, effective bandwidth and user capacity. *IEEE Transactions on Information Theory* 45 (2): 641–657.

17 Sampath, A., Mandayam, N.B., and Holtzman, J.M. (1997). Erlang capacity of a power controlled integrated voice and data CDMA system. *IEEE 47th Vehicular Technology Conference*, Volume 3, pp. 1557–1561.

18 Viterbi, A.J. (1995). *CDMA, Principles of Spread Spectrum Communications, Addison-Wesley Wireless Communications Series*. Addison-Wesley Wireless Communications Series.

19 Kleinrock, L. (1975). *Queueing Systems: Theory*, vol. 1. Wiley-Interscience.

20 Kleinrock, L. (1976). *Queueing Systems: Computer Applications*, vol. 2. Wiley-Interscience.

21 Akimaru, H. and Kawashima, K. (1999). *Teletraffic*, 2e. Springer-Verlag.

22 Walke, B., Seidenberg, P., and Althoff, M.P. (2003). *UMTS, The Fundamentals*. Wiley.

23 Goldsmith, A., Jafar, S.A., Maric, I., and Srinivasa, S. (2009). Breaking spectrum gridlock with cognitive radios: an information theoretic perspective. *Proceedings of the IEEE* 97 (5): 894–914.

24 Shah-Mohammadi, F., Enaami, H.H., and Kwasinski, A. (2021). Neural network cognitive engine for autonomous and distributed underlay dynamic spectrum access. *IEEE Open Journal of the Communications Society* 2: 719–737.

25 Hu, J. and Wellman, M.P. (2003). Nash Q-learning for general-sum stochastic games. *Journal of Machine Learning Research* 4: 1039–1069.

26 Choi, K., Tatwawadi, K., Grover, A. et al. (2019). Neural joint source-channel coding. *International Conference on Machine Learning*, pp. 1182–1192. PMLR.

27 Kwasinski, A., Wang, W., and Mohammadi, F.S. (2020). Reinforcement learning for resource allocation in cognitive radio networks. In: *Machine Learning for Future Wireless Communications*, Chapter 2 (ed. F.-L. Luo), 27–44. Wiley.

28 Shah-Mohammadi, F. and Kwasinski, A. (2018). Deep reinforcement learning approach to QoE-driven resource allocation for spectrum underlay in cognitive radio networks. *2018 IEEE International Conference on Communications Workshops (ICC Workshops)*, pp. 1–6.

Index

Joint Source-Channel Coding, First Edition. Andres Kwasinski and Vinay Chande.
© 2023 John Wiley & Sons Ltd. Published 2023 by John Wiley & Sons Ltd.